Detlef A.W. Wendisch

Acronyms and Abbreviations in Molecular Spectroscopy

An Enzyclopedic Dictionary

Springer-Verlag
Berlin Heidelberg New York
London Paris Tokyo Hong Kong

Dr. Detlef A. W. Wendisch
Central Research Services
Bayer AG
D-5090 Leverkusen-Bayerwerk
West Germany

ISBN 3-540-51348-5 Springer-Verlag Berlin Heidelberg New York
ISBN 0-387-51348-5 Springer-Verlag New York Berlin Heidelberg

Library of Congress Cataloging in Publication Data.
Wendisch, Detlef A.W. (Detlef Arthur Wilhelm),
Acronyms and abbreviations in molecular spectroscopy: an
enzyclopedic dictionary/Detlef A.W. Wendisch.
Includes bibliographical references.
ISBN 0-387-51348-5 (alk. paper: U.S.)
1. Molecular spectroscopy—Acronyms.   2. Molecular spectroscopy—
Abbreviations.   3. Molecular spectroscopy—Dictionaries.
I. Title.
QC454.M6W46   1990

Typesetting: Macmillan India Ltd., Bangalore-25; Printing: Colordruck Dorfi GmbH, Berlin;
Bookbinding: Lüderitz & Bauer, Berlin
2152/3020-543210—Printed on acid-free paper

Preface

The author's intention is to give correct definitions of acronyms, abbreviations and symbols commonly used in the field of molecular spectroscopy.

The reader will find acronyms used especially in nuclear magnetic resonance spectroscopy, nuclear magnetic resonance imaging combined with the so-called *in vivo* spectroscopy, infrared spectroscopy, Raman spectroscopy, electron spin resonance, and other molecular spectroscopic methods.

Structuring of each item—acronym, abbreviation, and symbol—according to the following scheme: i. definition, ii. description of the effect, iii. application, and iv. literature may be helpful for people starting in the field of molecular spectroscopy.

Linguistic interpretation or merely translation of these items may be found elsewhere.

This dictionary can be used alphabetically. There is, however, an index corresponding to the different kinds of molecular spectroscopy, too.

Leverkusen/Germany
December 1989

D. Wendisch

1D-NMR

Abbreviation of **O**ne-**D**imensional **N**uclear **M**agnetic **R**esonance Spectroscopy

This abbreviation became first necessary after the introduction of the new two-dimensional methods in nuclear magnetic resonance spectroscopy (see 2D-NMR). However, this abbreviation gained in relevance because of the recently published possibility of transforming homonuclear 2D-NMR techniques into one-dimensional methods. Kessler and co-workers [1] have described how two-dimensional NMR experiments can be converted into analogous one-dimensional experiments with the aid of GAUSSian pulses (see p 119), first used in NMR in 1984 [2]. These new 1D-NMR techniques are particularly advantageous in cases where only a limited amount of information is required for solving a chemical or structural problem, e.g. a few connectivities, coupling constants, or NOE values (see NOE and GOE). This is often the case for medium-size molecules.

The requirements in storage capacity and acquisition as well as data manipulation time for these 1D-NMR experiments are drastically reduced compared to corresponding two-dimensional procedures. In addition, some of these new 1D techniques proposed by Kessler [1] have unique properties for which no practical equivalent in 2D-NMR exists.

Applications of 1D-COSY (see COSY), 1D-NOESY (see NOESY and 1D-NOE), one-dimensional homonuclear relayed techniques [3,4] (see RELAY), and 1D-TOCSY (see TOCSY) have been demonstrated [1]. Furthermore, a variant of the one-dimensional COSY, the so-called refocused 1D-COSY with z filter has been suggested, which in combination with 1D-COSY allows the application of the DISCO technique (see DISCO) for determination of coupling constants from multiplets in very crowded spectral regions. The suggested one-dimensional NOESY sequence allows the measurement of NOE's for very short mixing times and thus provides reliable values for buildup rates of transient NOE's.

Selective excitation of a certain resonance may be impossible by conventional techniques, because it lies in a crowded spectral region. In such difficult cases by exploiting the J dependence of transfer functions magnetizations can be transferred selectively from a coupled nucleus to the one of interest. Its NMR parameters (J coupling, NOE data) are then accessible via relayed COSY or relayed NOESY, respectively. Quantitation of peak areas in 1D-NOE measurements and NOE factors has been studied in presence of noise and digitization errors when the lineshape was LORENTZian [5].

The use of apodization functions in the numerical estimation of peak areas has also been investigated [6]. The formalism in this study has been introduced and the effects of exponential and GAUSSian apodization have been analyzed. Two effects have been found to be very important: the bias due to the inability of executing an exact integration over the peak, and the effects arising from instrumental noise. A generalized variance that combines bias and noise has been applied to show advantage of GAUSSian over exponential apodization [6].

1D

References

1. Kessler H, Oschkinat H, Griesinger C, Bermel W (1986) J. Magn. Reson. 70: 106
2. Bauer CJ, Freeman R, Frenkiel T, Keeler J, Shaka AJ (1984) J. Magn. Reson. 58: 442
3. Eich G, Bodenhausen G, Ernst RR (1982) J. Am. Chem. Soc. 104: 3731
4. Wagner G (1984) J. Magn. Reson. 57: 497
5. Weiss GH, Ferretti JA (1983) J. Magn. Reson. 55: 397
6. Weiss GH, Ferretti JA, Byrd RA (1987) J. Magn. Reson. 71: 97

1D-NOE

Abbreviation of **One-D**imensional **N**uclear **OVERHAUSER E**nhancement or **E**ffect

This abbreviation became necessary after Kessler and co-workers [1] described how two-dimensional NMR experiments (see 2D-NMR) can be converted into analogous one-dimensional methods. This new 1D-NOE technique is particularly advantageous in cases where only a low number of NOE values is required for solving a chemical or structural problem, as for medium-size molecules.

The requirements in storage capacity and acquisition as well as data handling times for these 1D-NOE experiments are essentially reduced compared to the corresponding two-dimensional methods (see 2D-NOE and NOESY) [1].

References

1. Kessler H, Oschkinat H, Griesinger C, Bermel W (1986) J. Magn. Reson. 70: 106

2D-CIDNP
Acronym for **Two-D**imensional Chemically Induced Dynamic Nuclear Polarization

2D

In chemically induced dynamic nuclear polarization (see CIDNP) experiments, nuclear spin states of the products of radical reactions are normally created with non-equilibrium populations [1,2].

This can be observed as selectively changed intensities in the nuclear magnetic resonance (see NMR) spectrum. Two types of polarization are usually distinguished: net polarization (enhanced absorption or emission) and multiplet effects (both absorption and emission in a multiplet).

These two effects depend, in a critical manner, on three important factors

- the type of the radical reaction,
- the route of product formation, and
- the magnetic parameters of the radicals involved.

Qualitatively net polarization and multiplet effect can be estimated by two of KAPTEIN's sign rules [3,4].

Polarization created in the product of a given radical reaction can be transferred to other nuclei of the molecule either by dipolar cross-relaxation [5,6] or by J coupling [6,7]. Sheck et al. [7,8] have demonstrated the application of two-dimensional NMR (see 2D-NMR) for the observation of such polarization transfer processes.

Net polarization and multiplet effect cannot be easily separated in continuous wave (see CW) NMR. But, in pulse FOURIER transform (see PFT) NMR the above mentioned effects show a quite different dependence on the radio-frequency flip angle [9,10] and therefore, a linear combination of two experiments with different flip angles should separate them. A disadvantage of such a procedure is that it will strongly depend on the accuracy of the radio-frequency flip angles. In particular, the multiplet effect should be unobservable after a 90° pulse in a non-equilibrium state, but in practice this is not true because of the radio-frequency inhomogeneity. Recently, both Bodenhausen [11] and Allouche et al. [12] have suggested that these effects could be observed using 2D-NMR.

Kaptein et al. [13] have recently shown the separation of net polarization and multiplet effect in coupled spin systems by two-dimensional CIDNP. With a 90° radio-frequency pulse the multiplet effect is transformed into a mixture of multi-quantum coherences, which evolves during the t_1 period of a 2D experiment and can be detected by a second 90° pulse. After FOURIER transformation (see FT), net and multiplet effects occur at different ω_1 frequencies.

The method has been applied to the photoreaction of flavin I with *N*-acetyl tyrosine and *p*-methoxyphenol [13] and can also abbreviated as CIDNP-COSY (see p 48).

References

1. Closs GL (1974) In: Waugh JS (ed) Advances in magnetic resonance, vol 7, Academic, New York, p 157
2. Kaptein R (1975) In: Williams GH (ed) Advance in free radicals chemistry, vol 5, Elek Science, London, p 319
3. Kaptein R: Chem. Commun. 1971: 732

4. Kaptein R (1982) In: Berliner LG, Reuben J (eds) Biological magnetic resonance, vol 4, Plenum, New York, p 145
5. Closs GL, Czeropski MS (1977) Chem. Phys. Lett. 45: 115
6. de Kanter FGG, Kaptein R (1979) Chem. Phys. Lett. 65: 421
7. Scheek RM, Stob S, Boelens R, Dijkstra K, Kaptein R (1984) Faraday Discuss. Chem. Soc. 78: 245
8. Scheek RM, Stob S, Boelens R, Dijkstra K, Kaptein R (1985) J. Am. Chem. Soc. 107: 705
9. Schäublin S, Hochener A, Ernst RR (1974) J. Magn. Reson. 13: 196
10. Schäublin S, Wokaun A, Ernst RR (1976) Chem. Phys. 14: 285
11. Bodenhausen G (1981) Progr. NMR Spectrosc. 14: 137
12. Allouche A, Marinelli F, Pouzard G (1983) J. Magn. Reson. 53: 65
13. Boelens R, Podoplelov A, Kaptein R (1986) J. Magn. Reson. 69: 116

2D-CRAMPS

Acronym for **Two-D**imensional **C**ombined **R**otation **A**nd **M**ultiple-**P**ulse Spectroscopy used in Nuclear Magnetic Resonance (see NMR)

2D-CRAMPS, a two-dimensional variant of the originally proposed CRAMPS (see p 58) experiment [1], has been used by Taylor to study *proton spin diffusion* in polycrystalline compounds [2] which results from the dipolar interactions of nuclear spins in their $1/r^6$ spatial dependence [3].

Spin diffusion measurements yield information on local order or structure in physical systems. For example, many interesting polymers have amorphous and crystalline domains with quite different mobilities on the molecular scale, resulting in bi-exponential spin-spin relaxation (T_2) decays. Spin diffusion rates in those polymers can be measured either by the well-known GOLDMAN-SHEN experiment [4] or by modifications of it, including *multiple-pulse line narrowing sequences* [5]. R.R. Ernst et al. have extended this approach to a 2D version and applied it to polymer blends [6] cast from different solvents. Such extensions do not necessarily dependent on differences in relaxation times, such as T_2 or $T_1\rho$, but rely simply on direct spectral resolution as shown by R.R. Ernst et al. [7].

That is, CRAMPS [1], used to suppress homonuclear dipolar and anisotropic chemical shift (see CSA) broadening can be implemented into an experiment of the GOLDMAN-SHEN type in order to study proton spin diffusion in systems where the free induction decay (see FID) follows a single exponential. Because, examples of this approach rely on spectral resolution as mentioned above, a *2D version of CRAMPS* has been used by Taylor [2] to look at the ^{1}H spin diffusion process in polycrystalline glycine and D,L-aspartic acid. This spin diffusion experiment can generally divided into three time intervals [2]:

a) the initial labeling of the spin order,
b) the actual spin diffusion process, and
c) the detection process.

The pulse sequence for 2D ^{1}H spin diffusion under *magic angle spinning* (see MAS, MASS, and MAR) typically consists of four time periods:

P   preparation by a radio-frequency pulse;
E   evolution for time t_1 under MREV-8 (see p 182) homonuclear decoupling [8,9];
M   spin diffusion during the mixing time t_m;
D   detection in the presence of MREV-8 multiple-pulse decoupling.

The MREV-8 pulse sequence (all pulses are 90° with X, Y, $\bar{Y}$, $\bar{X}$ indicating the radio-frequency phase in the rotating frame, with a cycle time $t_c = 12\tau$) creates an effective field along the [1, 0, 1] axis in the rotating frame about which nuclear magnetization will precess [1]. The 90_x° preparation pulse places magnetization perpendicular to the effective rotation axis and prevent spin-locking. Also, when the simple two-pulse dipolar echo sequence is short compared with the rotor period, synchronization of the MREV-8 sequence is not necessary as shown by Taylor [2].

The advantages of studying spin diffusion via protons is well-known in NMR: the larger magnetogyric ratio (γ) and higher natural abundance allow ^1H spin diffusion to occur several orders of magnitude faster than ^{13}C spin diffusion and to be detected with higher sensitivity. Even in cases of very short spin-lattice relaxation times (T_1), spin diffusion processes among protons can be easily measured.

References

1. Gerstein BC, Dybowski CR (1985) Transient techniques in NMR of solids, Academic, Orlando
2. Taylor RE: Bruker Report 1/1988: 16
3. Abragam A (1961) The principles of nuclear magnetism, Oxford University Press, Oxford
4. Goldman M, Shen L (1966) Phys. Rev. 144: 321
5. Cheung TPP, Gerstein BC, Ryan LM, Taylor RE, Dybowski CR (1980) J. Chem. Phys. 73: 6059
6. Caravatti P, Neuenschwander P, Ernst RR (1985) Macromolecules 18: 119
7. Caravatti P, Neuenschwander P, Ernst RR (1986) Macromolecules 19: 1889
8. Rhim WK, Elleman DD, Vaughan RW (1973) J. Chem. Phys. 58: 1772
9. Mansfield P (1971) J. Phys. C. C4: 1444

2D-HAHA
Acronym for **Two-D**imensional **HARTMANN–HAHN** Spectroscopy

D.G. Davis and A. Bax [1] proposed the use of a new type of two-dimensional nuclear magnetic resonance (see 2D-NMR) experiment for the determination of homonuclear scalar connectivity of protons in complex molecules. This new method relies on the principle of cross-polarization (see CP), first introduced by Hartmann and Hahn [2] (see HAHA) and commonly used for sensitivity enhancement in solid-state carbon-13 NMR [3].

In Davis and Bax's experiment, homonuclear cross-polarization is obtained by switching on a single coherent radio-frequency field. In cases where the effective radio-frequency field strengths experienced by two scalar coupled protons are identical, a perfect HAHA match is established and gives rise to oscillatory exchange-with period 1/J-of spin-locked magnetization.

A major advantage of homonuclear cross-polarization is that *net* magnetization transfer (see MT) occurs and a phase-sensitive 2D spectrum with all peaks in the near absorption mode can be obtained. This is in contrast with the normal COSY experiment [4–7] (see COSY) where diagonal and cross-peaks are 90° out of phase and individual cross-multiplet components are 180° out of phase relative to one another, leading to partial signal cancellation and important sensitivity loss. But, it is noteworthy that in the new 2D-HAHA experiment, there is also some dispersive character to the individual diagonal- and cross-multiplets.

However, their antiphase nature causes significant mutual cancellation while the absorptive components remain.

Another consequence of the *net* magnetization transfer is the introduction of relayed connectivity.

References

1. Davis DG, Bax A (1985) J. Am. Chem. Soc. 107: 2820
2. Hartmann SR, Hahn EL (1962) Phys. Rev. 128: 2042
3. Pines A, Gibby MG, Waugh JS (1973) J. Chem. Phys. 59: 569
4. Aue WP, Bartholdi E, Ernst RR (1976) J. Chem. Phys. 64: 2229
5. Bax A, Freeman R (1981) J. Magn. Reson. 44: 542
6. Nagayama K, Kumar A, Wüthrich K, Ernst RR (1980) J. Magn. Reson. 40: 321
7. Marion D, Wüthrich K (1984) Biochem. Biophys. Res. Commun. 56: 207

2D-INADEQUATE
Acronym for **Two-D**imensional INADEQUATE (see p 141)

The 2D-INADEQUATE experiment in nuclear magnetic resonance (see NMR) spectroscopy was introduced by A. Bax et al. some years ago [1–3] as an alternative procedure to the original one-dimensional INADEQUATE.

This experiment has become a valuable tool for obtaining carbon-13 NMR assignments for structure elucidation [4] as well as carbon-13 multiple labeling studies [5].

2D-INADEQUATE allows the determination of ^{13}C–^{13}C double-quantum precessional frequencies for samples containing carbon-13 in natural abundance. This information can be used to identify carbon-13 NMR resonances arising from adjacent (or connected) carbon atoms of a molecule under investigation.

The most important limitation of this method is its low sensitivity, due to both the very low probability (1 to 10^4) of isotopomers bearing connected carbon-13 atoms and the well-known low gyromagnetic ratio (γ) of carbon-13.

This somewhat frustrating situation of having an important useful but insensitive procedure has prompted several proposals for increasing the sensitivity of such an experiment [6–11].

These suggestions include experiments named INEPT-INADEQUATE (see p 144) [6], DEPT-INADEQUATE (see p 69) [7], and heteronuclear relay techniques [8].

The sensitivity advantage [12] over INADEQUATE should be of the order of 64 ($[\gamma_H/\gamma_C^3]$).

But, the proton-monitored INADEQUATE pulse sequence, which has been dubbed, by Keller and Vogele [13], INSIPID (standing for *IN*adequate *S*ensitivity *I*mprovement by *P*roton *I*ndirect *D*etection) (see p 145) seems to be more effective than former approaches.

References

1. Bax A, Freeman R, Frenkiel TA (1981) J. Am. Chem. Soc. 103: 2102
2. Bax A, Freeman R (1980) J. Magn. Reson. 41: 507
3. Bax A (1982) Two-dimensional nuclear magnetic resonance in liquids, Delft Univ. Press, Delft/NL
4. Benn R, Mynett R (1985) Angew. Chem. 97: 330

5. Keller PJ, Le Van Q, Bacher A, Kozlowski JF, Floss HG (1983) J. Am. Chem. Soc. 105: 2502
6. Sørensen OW, Freeman R, Frenkiel T, Mareci TH, Schuck R (1982) J. Magn. Reson. 46: 180
7. Sparks SW, Ellis PD (1985) J. Magn. Reson. 62: 1
8. Bolton PH (1982) J. Magn. Reson. 48: 336
9. Unkefer CJ, Earl WL (1985) J. Magn. Reson. 61: 343
10. Piveteau D, Delsuc MA, Guittet E, Lallemand JY (1985) Magn. Reson. Chem. 23: 127
11. Kessler H, Bermel W, Griesinger C (1985) J. Magn. Reson. 62: 573
12. Martin ML, Delpuech JJ, Martin GJ (1980) Practical NMR spectroscopy, Heyden, Philadelphia, p 8
13. Keller PJ, Vogele KE (1986) J. Magn. Reson. 68: 389

2D

2D-MAS-NMR
Abbreviation of **Two-**Dimensional **M**agic **A**ngle **S**pinning **N**uclear **M**agnetic **R**esonance

This method is one of the various techniques used in NMR of solids. Theory and applications of these new two-dimensional NMR techniques (see 2D-NMR) for characterizing molecular structure and order of partially oriented solids have been described in detail by Spiess et al. recently [1].

After deriving a general expression for the signal intensities in the two dimensions in the case of perfect order the influence of disordering on the NMR spectra have been discussed carefully [1].

Spiess and his group [1] have demonstrated that an expansion of the orientational distribution in terms of the well-known WIGNER's rotation matrices allows NMR spectra from residues with an arbitrary degree of order and an arbitrary chemical shielding tensor orientation to be fully analyzed.

The new technique has been illustrated by obtaining the orientational distribution function for the crystalline and amorphous regions of semi-crystalline polyethylene terephthalate, showing quite different molecular orientation in the two phases [1].

References
1. Harbison GS, Vogt V-D, Spiess HW (1987) J. Chem. Phys. 86: 1206

2D-NMR
Abbreviation of **Two-**Dimensional **NMR** spectroscopy

General aspects of the new method in nuclear magnetic resonance have been presented by A. Bax [1] in a clear mathematical way. This paper has encouraged a lot of NMR spectroscopists to find out the scope and limitations of this method for structure elucidation [2,3]. J. Jeener was the first, in 1971, to propose the idea of two-dimensional Fourier transformation (see FOURIER Transformation) as a function of two time variables, yielding a NMR spectrum as a function of two frequency variables [4]. At the time, Jeener proposed such an experiment as an alternative for common homonuclear double resonance. R. Ernst et al. [5] presented, much later, a detailed and rather complex theoretical description. Extension of the procedure to the detection of multiple quantum transitions [6–8] was a direct consequence of this concept.

2D

Three main categories of 2D-NMR can be now distinguished:

a) *Shift correlation spectroscopy*; a class of experiments which can be considered as an alternative for double resonance, correlating shifts of coupled nuclei, exchanging nuclei or nuclei with cross-relaxation interaction.
b) *J-spectroscopy*; in the two-dimensional way one can determine chemical shift and scalar coupling by separating each other.
c) *Mutiple quantum spectroscopy*; two dimensional FT experiments (see FT) facilitate the detection and enlarge the applicability of multiple quantum transitions.

R.R. Ernst [9] has designed a hierarchical tree of 2D correlation techniques in which the various methods are grouped according to the degree of complexity of the resulting 2D spectra.

All the two-dimensional NMR spectra (see 2D-NMR spectra) can be obtained principally by the same following experiment consisting of four periods:

1) preparation,
2) evolution,
3) mixing, and
4) detection.

The evolution period t_1 is varied systematically in its duration in a series of experiments. The signal $s(t_1, t_2)$ is measured in the detection period as a function of t_2. The nuclear precession during the evolution period causes an oscillatory dependence on t_1 of the initial signal amplitude $s(t_1,0)$. A 2D FOURIER transformation delivers the 2D spectrum, $S(\omega_1, \omega_2)$, which can be perceived as a representation of the mixing process that transfers coherence (see CT) between different transitions [9]. This type of time-domain FT spectroscopy is the most frequently used approach to 2D-NMR spectroscopy although further procedures are conceivable [10]. It is, for example, possible to compute 2D spectra from the stochastic response (see STNMR) of non-linear systems [11].

Almost all 2D techniques will be explained under their individual acronyms or abbreviations in this dictionary, but the interested reader is also referred to some recently published monographs [10,12] and review articles [9,13].

In the following section we will give some remarks about the analysis of 2D-NMR experiments as an addendum.

References

1. Bax A (1982) 'Two-dimensional nuclear magnetic resonance in liquids', Delft University Press and Reidel, Delft and Dordrecht
2. a) Morris GA (1982) In: Marshall AG (ed) Fourier, Hadamard and Hilbert transforms in chemistry, Plenum, New York
   b) Morris GA (1984) In: Levy GC (ed) Topics in carbon-13 NMR spectroscopy, vol 4, chap 7, Wiley, New York
3. a) Bax A (1985) Bull. Magn. Reson. 7: 167
   b) Nagayama K (1986) In: Marshall J L (ed) NMR in stereochemical analysis, vol 6, chap 5, VCH, Deerfield Beach
   c) Morris GA (1986) Magn. Reson. Chem. 24: 371
4. Jeener J (1971) Ampère International Summer School, Basko Polje, Yugoslavia
5. Aue WP, Bartholdi E, Ernst RR (1976) J. Chem. Phys. 64: 2229
6. Anderson WA (1956) Phys. Rev. 104: 850
7. Anderson WA, Freeman R, Reilly CA (1963) J. Chem. Phys. 39: 1518

8. a) Bucci P, Martinelli M, Santucci S (1970) J. Chem. Phys. 52: 4041
   b) Bucci P, Martinelli M, Santucci S (1970) J. Chem. Phys. 53: 4254
9. Ernst RR (1987) Chimia 41: 323; R.R. Ernst pointed out at the end of this article that the very first paper ever published on 2D spectroscopy appeared in Chimia in 1975 (see Ref 14); a comparison with the present paper will show the interested reader the enormous progress that has been achieved since 1975.
10. Ernst RR, Bodenhausen F, Wokaun A (1987) Principles of nuclear magnetic resonance in one and two-dimensions, Clarendon, Oxford
11. Blümich B (1987) Prog. NMR Spectrosc. 19: 331
12. Carlson RM, Croasmun WR (eds) (1987) Two-dimensional NMR for chemists and bio-chemists, Verlag Chemie International, Deerfield Beach (Methods in stereochemical analysis vol 9)
13. Kessler H, Gehrke M, Griesinger C (1988) Angew. Chem. 100: 507; (1988) Angew. Chem. Ed. Engl. 27: 460
14. Ernst RR (1975) Chimia 29: 179

Addendum

Remarks about the analysis of 2D-NMR experiments.
The first expositions of two-dimensional NMR have used, among others,

– the density matrix formalism [1],
– the HEISENBERG vector picture [2],
– the semiclassical vector model [3],
– the single-transition operator formalism [4],
– and a spherical tensor expansion of the density operator [5].

More recently, it has been demonstrated, following the spirit of Fano's interesting review [6], that the density operator could be easily and conveniently expanded in terms of products of single-spin operators [7–9]. In this last formalism, the analysis of a two-dimensional NMR experiment reduces to a computation of the effects of various pulses and delays on the basis operators. A final trace operation yields the signal.

Fano [6] also suggested another approach to the solution of a quantum-mechanical problem. It consists of writing down and solving the equations of motion for the expectation values of the basis operators, much as is done in the well-known theory of relaxation in liquids [10].

J.-Ph. Grivet [11] has presented an alternative formalism for 2D-NMR which takes use of Fano's original ideas (6). The density operator is here expanded on a complete basis of spin operator products. But, the principal variables are the expectation values of these basis operators. Systems of differential equations governing these expectation values are set up for each HAMILTONian (see p 122) of interest, and then solved, giving directly the time dependence of physical relevant quantities like the transverse magnetization. This procedure has been demonstrated by computing in detail the effect of various pulse sequences on a system of two weakly coupled spins 1/2 [11].

References

1. Aue WP, Bartholdi E, Ernst RR (1976) J. Chem. Phys. 64: 2229
2. Pegg DT, Bandall MR, Doddrell DM (1981) J. Magn. Reson. 44: 238
3. Bodenhausen G, Freeman R (1977) J. Magn. Reson. 28: 471
4. Wokaun A, Ernst RR (1977) J. Chem. Phys. 67: 1752
5. Bain AD, Brownstein S (1982) J. Magn. Reson. 47: 409
6. Fano U (1957) Rev. Mod. Phys. 29: 74

2D

7. van de Ven FJM, Hilbers CW (1983) J. Magn. Reson. 54: 512
8. Packer KJ, Wright KM (1983) Mol. Phys. 50: 797
9. Sørensen OW, Eich GW, Levitt MH, Bodenhausen G, Ernst RR (1983) Prog. NMR Spectrosc. 16: 163
10. Abragam A (1961) Principles of nuclear magnetism, chap 8, Oxford University Press, London
11. Grivet J-Ph (1985) J. Magn. Reson. 62: 269

2D-NMR spectra

are the products of two-dimensional experiments in nuclear magnetic resonance (see 2D-NMR). The different experiments themselves have been explained under their individual acronyms, abbreviations, and symbols in this dictionary.

We will discuss some features of 2D-NMR spectra here.

It has become apparent that *2D-NMR spectra* are very rich in information content, especially for large molecules, and their *analysis* can become quite demanding and very time-consuming.

On the other hand, the structure of the 2D spectra follows well defined rules [1] and it should therefore be possible in principle to perform a *spin-system analysis* (as in the case of a one-dimensional NMR spectrum) at least partially by a computer program with or without operator interaction.

Several attempts [2–6] have been made in this direction so far. It is possible to start with a detailed analysis, searching for the *basic square patterns*, mentioned by R.R. Ernst [7], and to build up successively *cross-peaks* and *coupling networks* [2,3]. Alternatively, one can employ *cluster analysis* procedures to localize cross-peaks or cross-peak clusters which are then further analyzed using local symmetry analysis methods [4,5]. A recently developed *local symmetry algorithm* has been applied to the E. COSY (see p 84) spectrum of a cyclic decapeptide [5].

Multiple quantum filtering [1,8] and spin topology filtering [7,9] are important concepts which allow us to *simplify* 2D spectra. Recently [10] a multivariate data analysis for *pattern recognition* has been proposed for 2D-NMR spectra. This new concept for spectral interpretation relies on multivariate analysis, abbreviated to MVA. The published results show that MVA can be used to separate mixtures of spin-systems, regardless of coupling approximation, and to classify them. Within this concept principal component analysis (PCA) and its possibilities has also been discussed [10].

M.A. Delsuc [11] has presented a paper dealing with the spectral *representation of 2D-NMR spectra* by hypercomplex numbers. With this representation, it has been shown that the FOURIER transformation (see FT and FFT) and the phasing of phase-sensitive 2D-NMR spectra can take relatively simple expressions.

References

1. Ernst RR, Bodenhausen G, Wokaun A (1987) Principles of nuclear magnetic resonance in one and two dimensions, Clarendon, Oxford.
2. Meier BU, Bodenhausen G, Ernst RR (1984) J. Magn. Reson. 60: 161
3. Pfändler P, Bodenhausen G, Meier BU, Ernst RR (1985) Anal. Chem. 57: 2510
4. Mádi Z, Meier BU, Ernst RR (1987) J. Magn. Reson. 72: 584
5. Meier BU, Mádi Z, Ernst RR (198) J. Magn. Reson.

6. a) Pfändler P, Bodenhausen G (1986) J. Magn. Reson. 70: 71
   b) Nović M, Oschkinat H, Pfändler P, Bodenhausen G (1987) J. Magn. Reson. 73: 493
7. Ernst RR (1987) Chimia 41: 323
8. Piantini U, Sørensen OW, Ernst RR (1982) J. Am. Chem. Soc. 104: 6800
9. Levitt MH, Ernst RR (1985) J. Chem. Phys. 83: 3297
10. Grahn H, Delaglio F, Delsuc MA, Levy GC (1988) J. Magn. Reson. 77: 294
11. Delsuc MA (1988) J. Magn. Reson. 77: 119

2D

2D-NOE
Acronym for **Two-D**imensional Nuclear **OVERHAUSER E**nhancement or Effect

The basic 2D experiment used for correlations via the dynamic NOE, known by analogy with COSY (see p 53) as NOESY (see p 200) [1–3], is identical with that used for investigating chemical exchange (see EXSY):

$$90_{\phi_1} - t_1 - 90_{\phi_2} - \tau - 90_0 - \text{acquire} \tag{1}$$

The first two 90° pulses impart a simple cosine dependence on the product of t_1 and chemical shift to the longitudinal magnetizations of different spins, which effectively "labels" their z-magnetization with their chemical shifts. During the period τ these magnetizations can mix, either by direct chemical exchange or, in the context of the OVERHAUSER experiment, through cross-relaxation. Any exchange of the z-magnetization during τ will lead to signals detected during t_2 which have f_1 modulation frequencies different from their precession frequencies during t_2 and hence after double FOURIER transformation (see FT) to the cross-peaks in the 2D-NOE spectrum. Relaxation delays between the individual measurements of 2 to 5 times T_1 (spin-lattice relaxation time) may be sufficient, depending on the dynamic range of the spectrum and on the accuracy of results required.

The resemblance between Eq. (1) and Eq. (2)

$$90 - t_1 - 90_{\phi_1} - 90_{\phi_2} - \text{acquire} \tag{2}$$

is a reminder that the choices of phase-cycling and delay times are crucial for determining the form of a 2D spectrum. Equation (2) sets out to show cross-peaks due to scalar couplings, while the 2D-NOE experiment needs to exclude such peaks while slowing cross-peaks from z-magnetization exchange. The phase-cycling for Eq. (1) is therefore required to ensure that only coherence which is of the order zero during τ is observed:

$$\phi = (0123 \ 1230 \ 2310 \ 3012)$$
$$\phi = (0_4 1_4 2_4 3_4) \tag{3}$$

But, this does not rule out the possibility of seeing cross-peaks which stem from scalar couplings rather than from the dynamic OVERHAUSER effect or from chemical exchange, since longitudinal magnetization and zero-quantum coherence (see ZQC) are both of the order zero and, therefore, show the same behavior when subjected to pulses trains.

Considerable efforts have been made to exclude ZQS effects from 2D-NOE spectra. One of the most effective attempts [1,4] is to insert an extra 180° pulse into the delay τ, which is moved through τ on successive measurements in sympathy with t_1 as shown with Eq. (4).

$$90_{\phi_1} - t_1 - 90_{\phi_2} - (\tau/2 - kt_1) - 180 - (\tau/2 + kt_1) - 90_0 - \text{acquire} \qquad (4)$$

In Eq. (4) k is a constant, typically around 0.25 [5]. The additional 180° pulse has no significant effect on the dynamic NOE, but does affect ZQC with the result that the effective t_1 modulation frequency is made to vary during the experiment, so that these cross-peaks become "smeared out" in f_1. The OVERHAUSER cross-peaks are unaffected and easily recognized.

It has been pointed out by G.A. Morris [5], that much early work on relatively small molecules using Eq. (1) failed to take proper account of the dangers of assigning cross-peaks to OVERHAUSER effects without checking for ZQC.

2D-NOE or NOESY (see p 200), in the meantime, has become an indispensible tool for resonance assignments on complex macromolecules such as proteins [6–9], and the spatial structures in solution of some of them have already been determined on the basis of the NOE data [10–13]. For a comprehensive overview the interested reader is referred to a recently published monograph written by K. Wüthrich [14].

The quality of such spatial structures determined depends essentially on the number of the NOE cross-peaks identified [15], and also on the accurate determination of distances from the cross-peak intensities. Recently [16] a computer procedure which yields cross-peak intensities via matrix operations has been published. This procedure is a projection of the 2D-NOE spectrum on a linear space spanned by a set of reference lines and is equivalent to a linear least-squares fit of the 2D spectrum to a set of reference lineshapes with the line intensities as fitting parameters. The output of the program consists of a matrix of volumes of cross-peaks between all signals represented by reference lines. The method can readily be applied to "integrate" overlapping cross-peaks in the spectra, too [16].

It has been demonstrated [17] that direct and relayed NOE's can be distinguished by using the "spin-locked" NOE spectroscopy [18,19]. In the spin-locked NOE experiment the NOE is always positive in sign and increases with slower molecular tumbling. Thus, cross-peaks arising from direct NOEs are always opposite in sign relative to the diagonal and consequently, cross-peaks relayed via an intermediate nucleus will be in phase with the resonances on the diagonal. Of course, doubly relayed signals would be of opposite sign again but are usually too weak to be detected unless very long mixing times are used [17].

Problems arising from cross-correlation in 2D-NOE spectra were discussed in detail by Bull [20] in 1987. Table 1 shows values of the maximum NOE for six different nuclei. In all cases the irradiated nucleus is the proton. Please note, η will have a negative value if a nucleus with a negative γ (the magnetogyric ratio) is observed while the proton is decoupled.

Table 1. Data of different nuclei in NOE spectroscopy

Observed nucleus:	^1H	^{13}C	^{15}N	^{19}F	^{29}Si	^{31}P
γ [rad $T^{-1} s^{-1}$] 10^7:	26.75	6.726	−2.711	25.16	−5.314	10.83
η max:	0.5	1.99	−4.93	0.53	−2.52	1.24

Much more recently [21] strong coupling effects and their suppression in two-dimensional *heteronuclear* NOE experiments have been analyzed experimentally. A simple method has been proposed for the suppression of "ghost" peaks resulting from second-order effects. But, this procedure decreases the sensitivity.

References

1. Wider G, Macura S, Kumar A, Ernst RR, Wüthrich K (1984) J. Magn. Reson. 56: 207
2. Jeener J, Meier BH, Bachmann P, Ernst RR (1979) J. Chem. Phys. 71: 4546
3. Kumar A, Ernst RR, Wüthrich K (1980) Biochem. Biophys. Res. Commun. 95: 1
4. Macura S, Wüthrich K, Ernst RR (1982) J. Magn. Reson. 46: 269
5. Morris GA (1986) Magn. Reson. Chem. 24: 371
6. Wagner G, Wüthrich K (1982) J. Mol. Biol. 155: 347
7. Wider G, Lee KH, Wüthrich K (1982) J. Mol. Biol. 155: 367
8. Williamson MP, Marion D, Wüthrich K (1984) J. Mol. Biol. 173: 341
9. Wemmer D, Kallenbach NR (1983) Biochemistry 22: 1901
10. Braun W, Wider G, Lee KH, Wüthrich K (1983) J. Mol. Biol. 169: 921
11. Arseniev AS, Kondakov VI, Maiorov VN, Bystrov VF (1984) FEBS Lett. 165: 57
12. Williamson MP, Havel TF, Wüthrich K (1985) J. Mol. Biol. 182: 295
13. Zuiderweg ERP, Billeter M, Boelens R, Scheek RM, Wüthrich K, Kaptein R (1984) FEBS Lett. 174: 243
14. Wüthrich K (1986) NMR of proteins and nucleic acids, Wiley, New York
15. Havel TF, Wüthrich K (1985) J. Mol. Biol. 182: 281
16. Denk W, Baumann R, Wagner G (1986) J. Magn. Reson. 67: 386
17. Bax A, Sklenár V, Summers MF (1986) J. Magn. Reson. 70: 327
18. Bothner-By AA, Stephens RL, Lee JT, Warren CD, Jeanloz RW (1984) J. Am. Chem. Soc. 106: 811
19. Bax A, Davis DG (1985) J. Magn. Reson. 63: 207
20. Bull TE (1987) J. Magn. Reson. 72: 397
21. Köver KE, Batta G (1987) J. Magn. Reson. 74: 397

In the following *appendix* we want to discuss an alternative approach to NOE measurements by changing solvent viscosity and guidelines for kinetic NOE experiments.

Appendix

An *alternative approach* to successful NOE studies of *small molecules* is to create experimental conditions so that $\omega\tau_c$ is no longer close to unity.

Varying the field strength seems to be impractical, since the highest field is preferred for greater sensitivity and shift separation.

A more attractive procedure is to alter the frequency of motion, τ_c, by *changing the solvent viscosity*.

Indeed, several solvents such as tetramethylene-d_8 sulfone [1], ethylene glycol-d_6 [2], phosphoric acid-d_3 [3], acetic acid-d_4 [4], and "oil voltale 10 S" [5] have been used so far for this purpose in NOE studies of relatively small molecules.

Recently a simple procedure has been proposed [6] that involves the use of the mixed solvent $(CH_3)_2SO$-d_6/D_2O to increase the viscosity of the solution such that $\omega\tau_c > 1$ because of the well-known physical properties of the dimethyl sulfoxide/water mixture studied over many years [7–11]. Using this solvent system more NOE's can be observed, and these NOE's can be more accurately quantified in terms of proton-proton distances [12]. The ability to control size and number of the NOE's observable using this method has proven useful in

2D

conformational investigations [13] where many NMR distance constraints are necessary to define precisely three-dimensional structures in solution.

In 1987 [14] *guidelines* for the design of kinetic *NOE experiments* from computer simulation were published. A FORTRAN computer program (see FORTRAN) has been written that calculates the time course of NOE's in spin systems of arbitrary complexity. The program starts with calculations of the spin-lattice and cross-relaxation rates from known molecular parameters, and then calculates all NOE's of interest simultaneously via a numerical integration of SALOMON's equations [15]. This program may be applied for the calculation of truncated drive NOE's or transient NOE's to peptide and protein structures using various correlation times, in order to construct a picture of the behaviour of the NOE under different experimental conditions. The results obtained have been used to suggest guidelines for the design of NOE experiments to ensure that structural information is recovered in an efficient way, and that experiments are not misinterpreted. The distinction of an α helix from the 3_{10} helix by NOE's has been discussed in detail by the author [14].

References

1. Gierasch LM, Rockwell AL, Thompson KF, Briggs MS (1985) Biopolymers 24: 117
2. Zhu PP (1985) Ph.D. thesis, Illinois Institute of Technology, Chicago
3. Szeverenyi NM, Bothner-By AA, Bittner R (1980) J. Phys. Chem. 84: 2880
4. Neuhaus D, Rzepa HS, Sheppard RN, Bick IRC: Tetrahedron Lett. 1981: 2933
5. Williamson MP, Williams DH: J. Chem. Soc., Chem. Commun. 1981: 165
6. Fesik SW, Olejniczak ET (1987) Magn. Reson. Chem. 25: 1046
7. Cowie JMG, Toporowski PM (1961) Can. J. Chem. 39: 2240
8. Packer KJ, Tomlinson DL (1971) Trans. Faraday Soc. 67: 1302
9. Fox MF, Whittingham KP (1975) J. Chem. Soc., Faraday Trans. I, 71: 1407
10. Schott H (1980) J. Pharm. Sci. 69: 369
11. Madigosky WM, Warfield RW (1983) J. Chem. Phys. 78: 1912
12. Olejniczak ET, Gampe RT, Fesik SW (1986) J. Magn. Reson. 67: 28
13. Fesik SW, Bolis G, Sham HL, Olejniczak ET (1987) Biochemistry 26: 1851
14. Williamson M P (1987) Magn. Reson. Chem. 25: 356
15. Noggle JH, Schirmer R E (1971) In: The nuclear Overhauser effect, Academic, New York

2QT-ENDOR
Abbreviation of **D**ouble **Q**uantum **T**ransitions (in) **E**lectron **N**uclear **D**ouble **R**esonance

A new kind of ENDOR spectroscopy (see ENDOR) in electron spin or para-magnetic resonance (see ESR or EPR). Double quantum transitions (2 QTs) in ENDOR spectroscopy may be generated by one or two intensive radio-frequency fields [1]. Observation of 2 QTs (other name: double quantum coherence) in ENDOR-ESR under elimination of 1 QTs by suitable modulation procedures [2] will be helpful in elucidation of connectivities in the case of very complicated energy niveau diagrams such as the carbon-carbon-connectivity test in nuclear magnetic resonance (see for example COSMIC). The concept of double quantum coherence became very important in both magnetic resonance spectroscopies.

References
1. Rudin M, Schweiger A, Günthard HH (1983) J. Magn. 51: 278
2. Schweiger A (1986) Chimia 40: 111

2QT

3D J-NMR
Abbreviation of **Three-D**imensional **J** Resolved-**N**uclear **M**agnetic Resonance Spectroscopy

3D

Sometimes in multidimensional NMR spectroscopy a third time domain has been introduced, e.g. in nuclear OVERHAUSER enhancement buildup (see NOE) experiments [1] and the so-called accordian technique [2,3], in order to quantify the NOE effect. The concept of a third domain and FOURIER transformation (see FT) in three dimensions has been described for NMR FOURIER zeugmatography experiments [4]. On the other hand, three-dimensional methods have been used in the past to measure carbon-13 spectra in which ^{13}C signals are modulated by both chemical shift and spin-spin coupling [5].

Recently [6] a homonuclear three-dimensional J-resolved experiment has been reported in which the intensities and the phases of the cross peaks of a COSY (see p 53) experiment are modulated by J couplings, thus allowing determination of spin multiplets [7].

References
1. Kumar A, Wagner G, Ernst RR, Wüthrich K (1981) J. Am. Chem. Soc. 103: 3654
2. Bodenhausen G, Ernst RR (1981) J. Magn. Reson. 45: 367
3. Bodenhausen G, Ernst RR (1982) J. Am. Chem. Soc. 104: 1304
4. Kumar A, Welti D, Ernst RR (1975) J. Magn. Reson. 18: 69
5. Bolton PH (1982) J. Magn. Reson. 46: 343
6. Vuister GW, Boelens R (1987) J. Magn. Reson. 73: 328
7. A similar experiment was reported by H.D. Plant et al. at the 27th Experimental NMR Conference in Baltimore/USA, April 1986.

3D-NMR
Acronym for **Three-D**imensional **N**uclear **M**agnetic **R**esonance

After the publication of the first two-dimensional nuclear magnetic resonance (see 2D-NMR) experiments [1,2] it seemed clear that the principles of two-dimensional spectroscopy [3] can be extended to three or even higher dimensions. Such experiments have indeed been performed in the context of nuclear magnetic resonance imaging (see NMRI), displaying all the three spatial coordinates [4] or mapping spatially resolved chemical shift information [5]. In addition, a three-dimensional J-resolved NMR method (see 3D J-NMR) has been published recently [6]. On the other hand, it has been thought sometimes that analogous high-resolution 3D-NMR experiments are impracticable because of large data matrices and extremely long measurement times. Also the potential usefulness of 3D-NMR experiments has been questioned as the information of a 3D spectrum can sometimes be evaluated from two separate 2D-NMR spectra.

Suggestions for a much more practical approach to 3D-NMR spectroscopy have been given by R.R. Ernst et al. [7] including a technique for reduction of data matrices and application possibilities in those cases where the information extraction from 2D spectra or series of them is difficult or impossible.

Three-dimensional pulse sequences can be derived by a combination of two 2D pulse sequences. These two experiments are linked together leaving out the detection period of the first and the preparation period of the second experiment.

Analogous to this construction principle, R.R. Ernst et al. [7] named 3D experiments according to the well-known 2D-NMR experiments from which they arise: COSY-COSY, COSY-NOESY and NOESY-COSY (see these two-dimensional techniques under their individual acronyms or abbreviations). This proposed nomenclature is in accordance with the 3D experiment COSY-J published by Plant and co-workers [6].

3D

R.M. Scheek and his group has given [8] an interesting comparison of different algorithms for *generating 3D structures* of molecules. Several approaches exist for the generation of molecular structures that satisfy *experimental distance constraints*. Scheek et al. reported their own experiences with the metric matrix distance geometry approach (both Havel's implementation, known as DISGEO, and the implementation by Thomason, Scheek and Kuntz), with DISMAN (Braun and Go's procedure for minimizing a variable target function in dihedral angle space), with the ellipsoid algorithm (Billetter and Havel's algorithm for global minimization of a standard error function, again in dihedral angle space) and with restrained *molecular dynamics* (Van Gunsteren and Kaptein, Clore and Karplus). In addition a new technique, tentatively called *distance-bounds-driven dynamics*, was tested as a means of improving the sampling behaviour of distance geometry algorithms.

References

1. Aue WP, Bartholdi E, Ernst RR (1976) J. Chem. Phys. 64: 2229
2. Kumar A, Welti D, Ernst RR (1975) J. Magn. Reson. 18: 69
3. Ernst RR, Bodenhausen G, Wokaun A (1987) Principles of nuclear magnetic resonance in one and two dimensions, Clarendon, Oxford
4. Lai CM (1981) J. Appl. Phys. 52: 1141
5. Maudsley AA, Hilal SK, Perman WH, Simon HE (1983) J. Magn. Reson. 51: 147
6. Plant HD, Mareci TH, Cockman MD, Brey WS (1986) paper presented at the 27th ENC, Baltimore
7. Griesinger C, Sørensen OW, Ernst RR (1987) J. Magn. Reson. 73: 574
8. Scheek RM, van Hoesel F, de Vlieg J, van Gunsteren WF (1986) paper presented at the 27th ENC, Baltimore

ABC

Notation for a very complex three-spin system in nuclear magnetic resonance spectroscopy (see NMR).

The first capitals of the alphabet symbolize, in NMR, a very complex three-spin system: there are small differences in the relative chemical shifts of these three nuclei, whilst the involved coupling constants may be in the same order of magnitude. It has been shown, that factorization of the representative matrix of the HAMILTONian is impossible. Therefore such ABC spin systems can be analyzed only via iterative computer programs such as LAOCOON (see p 155), or DAVINS (see p 65). On the other hand, in the past there were a lot of attempts, to analyze such complex spin systems by *"more direct solutions"* [1–7].

References

1. Brügel W, Ankel T, Krückeberg F (1960) Z. Elektrochem. 64: 1121
2. Corio PL (1966) Structure of high resolution NMR spectra, Academic, New York
3. Cavanaugh JR (1963) J. Chem. Phys. 39: 2378; (1964) J. Chem. Phys. 40: 248
4. Whitman DR (1963) J. Mol. Spectrosc. 10: 250
5. Kellerhals H (1970) Dissertation, Basel
6. Diehl P, Kellerhals H, Lustig E (1972) Computer assistance in the analysis of high resolution NMR spectra, Springer, Berlin Heidelberg New York. (NMR basic principles and progress, vol 6)
7. Kowalewski V J (1986) J. Magn. Reson 67: 362

ACCORDION

Synonym for a special kind of spectroscopy in nuclear magnetic resonance (see NMR)

This spectroscopy is so called because "the concerted stretching of the evolution and mixing times recalls the motion of an accordion" [1] and is a modification of the NOESY experiment (see NOESY).

The necessity of choosing a certain mixing time, t_m, in a NOESY experiment can be cumbersome when studying chemical exchange or more, in general, when cross-relaxation rates are unknown and the optimum t_m is uncertain. The ACCORDION experiment was developed [1,2] to allow for a systematic variation of t_m synchronous with t_1 incrementation, i.e. $t_m = t_m^0 + kt_1$, where k is a constant defining the range over which t_m varies. Variations in the intensity of correlation (or diagonal) signals as a function of t_m then appear as line shape effects in the frequency domain, F_1, which can be used to get kinetic information.

The ACCORDION spectroscopy of proportional incrementation of a mixing period is also of interest in homonuclear [3] and heteronuclear [4] RELAY experiments (see RELAY) to allow for a range of coupling constants.

References

1. Bodenhausen G, Ernst RR (1982) J. Am. Chem. Soc. 104: 1304
2. Bodenhausen G, Ernst RR (1981) J. Magn. Reson. 45: 367
3. Homans SW, Dwek RA, Fernandes DL, Rademacher TW (1984) Proc. Natl. Acad. Sci. USA 81: 6286
4. Bolton PH, Bodenhausen G (1982) Chem. Phys. Lett. 89: 139

Appendix

Recently, [1] an ACCORDION version of the 2D-NOE (see p 11) experiment has been applied to study the structure of micelles. In aqueous solution, the monomeric surfactant 1-methyl-4-dodecylpyridinium iodide [2] exhibits *positive NOE's* whereas *negative NOE's* has been observed for the micellar aggregate. In contrast, 1-methyl-4-*n*-butylpyridinium iodide, which does not form micelles, *only* shows *positive NOE's*. The authors have argued that the observation of *negative NOE's* for micelles is reconcilable with quantitative data for the molecular dynamics of these surfactant aggregates.

Previously, 2D-NOE experiments yielded important information about the structure of vesicles [3], which are surfactant bilayer membranes of a less dynamic nature as compared with micellar aggregates.

References

1. Nusselder JJH, Engberts JBFN, Boelens R, Kaptein R (1988) Recl. Trav. Chim. Pays-Bas 107: 105
2. Sudhölter EJR, Engberts JBFN (1979) J. Phys. Chem. 83: 1854
3. Ellena JF, Hutton WC, Cafiso DS (1985) J. Am. Chem. Soc. 107: 1530

ADC
Abbreviation of **A**nalog-to-**D**igital **C**onverter

ADCs are most important elements used in both nuclear magnetic resonance (see NMR) and infrared (see IR) spectroscopy in the case of FOURIER transform (see FT, FFT, etc.) mode.

In the following part we will discuss at first the relevance of the ADC in the field of *FT-IR spectroscopy* in connection with the so-called *dynamic range*. Besides the question, how often the interferogram should be *digitized*, we have to discuss to what accuracy the amplitude of the signal should be *sampled*.

To give an example of the range of signals in IR interferometry, let us consider the case of the interferogram of an incandescent blackbody-source generated by a rapid-scanning interferometer and detected by a mid-infrared bolometer. The ratio of the intensity of the observed signal to zero-retardation to the root-mean-square (see RMS) noise level (often called the *dynamic range*) can often be higher than $10:1$. State-of-the-art ADCs have a resolution of approximately 16 bits, which means that the signal itself can be divided up into a maximum of 2^{16} levels. Thus, if the *dynamic range* of the interferogram is 20,000, only the two least significant bits of the ADC would be used to digitize the noise level. If the *dynamic range* were an order of magnitude higher, the noise level would fall below the least significant bit of the ADC and consequently real information would be lost from the interferogram.

For further details the interested reader is referred to the FT-IR Reference [1].

Reference

1. Griffiths PR, de Haset JA (1986) Fourier transform infrared spectrometry, Wiley, New York

Appendix

When the *FT method* was first introduced into high-resolution *NMR spectroscopy* by Ernst and Anderson [1] it was difficult to find any negative aspects of this most important technique. FT provides a dramatic improvement in sensitivity and greatly facilitates time-resolved experiments, without any concomitant disadvantages compared with slow-passage continuous-wave (see CW) spectroscopy.

The sole exception so far is the *dynamic range* problem. This problem arises when weak signals are examined in the presence of a very strong solvent peak. Three different sections of the spectrometer are involved: the receiver, which can exhibit nonlinear mixing effects, the ADC, which cannot properly digitize very weak signals without running into overflow on the very strong signal, and the computer, where the wordlength restricts the dynamic range that can be handled, limiting the extent of time averaging that can be used [2].

The first and last problems are not too serious. Spurious harmonics of the solvent peak, and sum and difference frequencies are normally weak and will appear at readily predictable positions. In the computer, double-precision arithmetic for FT and storage accommodates signals with a very large dynamic range. There remains the *serious problem of the ADC itself*, not easily circumvented by hardware improvements since lower digitization rates tend to go hand-in-hand with improved dynamic range.

It seems to be probably symptomatic of the current trends in the so-called solvent-suppression techniques (see SST) *to negotiate the ADC "bottleneck"*.

But, recently R. Freeman and co-workers [2] have proposed an approach which attacks the problem at its source, the conversion of the NMR signal into its digital form, a process most essential for computer methods of FT. In this approach, the transmitter frequency is set close to the water (or solvent) resonance so that this signal is carried at a very low frequency in the free induction decay (see FID). This is facilitated if quadrature phase detection (see QPD) is used, but even on spectrometers without this feature quite the same effect can be achieved by a method suggested by Redfield and Kunz [3]. The FID is not digitized in the common manner but in a differential mode where each coordinate represents the *increment* in signal intensity with respect to the previous ordinate. This can easily implemented under using an additional sample-and-hold circuit which "remembers" the signal ordinate from the previous sampling operation and compares it with the current signal ordinate in a differential amplifier feeding the ADC.

For further details the reader is referred to the original paper of R. Freeman and his group [2].

References

1. Ernst RR, Anderson WA (1966) Rev. Sci. Instrum. 37: 93
2. Davies S, Bauer C, Barker P, Freeman R (1985) J. Magn. Reson. 64: 155
3. Redfield AG, Kunz SD (1975) J. Magn. Reson. 19: 250

ADRF
Acronym for **A**diabatic **D**emagnetization in the **R**otating **F**rame

Using the pulse sequence as shown with Fig. 1 in the field of solid-state nuclear magnetic resonance (see NMR), a special state of proton dipolar order is created by a process called ADRF.

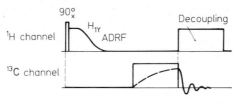

Fig. 1

This dipolar state relaxes with the time constant T_1^D, the so-called dipolar spin-lattice relaxation time. Cross-polarization (see CP) then occurs from this proton dipolar state to the carbon-13 spin-lock state at a rate $1/T_{CH}^D$. This process can compete with motional or spin-lattice processes in the so-called $T_1\,\rho$ (spin-lattice relaxation time measurement in the rotating frame) experiment [1]. It is much more efficient if the carbon H_1 field is less than the local dipolar field arising from the protons. Note, the quantity T_{CH}^D is not to be confused with T_{CH} the cross-relaxation time under proton spin-lock conditions!

The HARTMANN-HAHN condition (see HAHA) is not satisfied for T_{CH}^D, and therefore this quantity is much longer than T_{CH}.

For further details see the literature dealing with principles, general aspects, and "state of the art" in solid-state NMR [2].

References

1. Schaefer J, Stejskal EO (1979) Top. Carbon-13 NMR Spectrosc. 3: 283
2. Komoroski RA (1986) Principles and general aspects of high-resolution NMR of bulk polymers. In: Komoroski RA (ed) High-resolution NMR spectroscopy of synthetic polymers in bulk. VCH Publisher, Deerfield Beech (Methods in stereochemical analysis, vol 7)

AM and PM
Abbreviation of **A**mplitude **M**odulation and of **P**hase **M**odulation

In infrared spectroscopy (see IR and FT-IR) one disadvantage of slow-scanning interferometers is that slow variations in the intensity of the source can result in variations of the baseline of the interferogram. One method of getting around the problem of slow varying sources used in conjunction with a slow scanning interferometer, while at the same time picking up approximately a factor of 2 in signal-to-noise ratio (see SNR) by eliminating the chopper, is to apply a technique known as *phase modulation* (PM).

This technique has been independently developed by Chamberlain and his group [1,2] for far-infrared (see FIR) and by Connes et al. [3,4] for near-infrared (see NIR) measurements. The method (called internal modulation by Connes et al.) requires the use of a slow-scanning interferometer often employing the step-and-integrate method of traversing the moving mirror. At each step, the beam is modulated not by a chopper (called *amplitude modulation*, or AM, by the Chamberlain's group) but by periodically varying the retardation by a low-amplitude oscillation of the fixed mirror.

For comparison of AM and PM the interested reader is referred to the literature [5]. But, it should be pointed out here that the efficiency of PM measurements, relative to those of AM, falls off at the extremes of the spectral region being measured. This situation is predominantly due to the fact that modulation of a cosine function (here the interferogram) by another cosine function (the modulation of the mirror) will generate BESSEL's coefficients. Since only the lowest-order coefficient is detected, energy is lost in the higher-order coefficients.

References

1. Chamberlain J (1971) Infrared Phys. 11: 25
2. Chamberlain J, Gebbie HA (1971) Infrared Phys. 11: 57
3. Connes J, Connes P, Maillard JP (1967) J. Phys. 28: 120
4. Connes J, Delouis H, Connes P, Guelachvili G, Maillard JP, Michel G (1970) Nouv. Rev. Opt. Appl. 1: 3
5. see Griffiths PR, de Haseth JA, (1986) Fourier transform infrared spectrometry, Wiley, New York (Chemical analysis, vol 83)

Apodization
This word refers to the suppression of the side lobes or feet of the (instrument) line shape (see ILS), important both in infrared (see IR) and nuclear magnetic resonance (see NMR) spectroscopy. The word itself is apocryphally derived from Greek, 'α πòδos' (without feet).

Functions which weight the interferogram or the free induction decay (see FID) are known as apodization functions, sometimes called truncation functions.

The function $A_1(\delta)$ given with Eq. (1) and their limitating conditions is called a triangular apodization function and is the most common apodization function used in infrared FOURIER transform spectrometry (see FT-IR) [1].

$$A_1(\delta) = 1 - |\delta/\Delta| \quad \text{for} -\Delta \leq \delta \leq \Delta$$

$$A_1(\delta) = 0 \qquad \text{for } \delta > |-\Delta| \tag{1}$$

Several general studies on apodization functions have been carried out. For example, Filler [2] investigated a variety of trigonometric functions, and Norton and Beer [3] tested more than a thousand functions of the general form

$$A(\delta) = \sum_{i=0}^{n} C_i [1 - (\delta/A)^2]^i .$$

These authors found that there is a distinct empirical boundary relation between FWHH (see p 118) and the amplitude of the largest side lobe. Series of different apodization functions and their corresponding instrument line shape (see ILS) functions have been well documented [4] in the past. Among these we found the so-called boxcar truncation, trapezoidal, triangular (see above), and the triangular squared apodizations on one hand, and BESSEL's function, cosine and sine^2 functions as well as GAUSSian apodization on the other hand.

One other popular function for apodization purposes is the HAPP/GENZEL function given by Eq. (2).

$$A(\delta) = 0.54 + 0.46 \cos \pi(\delta/\Delta) . \tag{2}$$

Although the HAPP/GENZEL function has been recommended by the manufacturers of several commercial FT-IR spectrometers, the FWHH of the ILS functions given by the HAPP/GENZEL and the triangular apodization procedure are quite similar [1]. It is noteworthy that in some cases both methods are far from the boundary conditions given by Norton and Beer [3].

References

1. Griffiths PR, de Haseth JA (1986) Fourier transfer infrared spectrometry, Wiley, New York (Chemical analysis, vol 83)
2. Filler AH (1964) J. Opt. Soc. Am. 54: 762
3. Norton RH, Beer R (1976) J. Opt. Soc. Am. 66: 259; for an erratum see: (1977) J. Opt. Soc. Am. 67: 419
4. Kauppinen JK, Moffat DJ, Cameron DG, Mantsch HH (1981) Appl. Opt. 20: 1866

APT
Abbreviation for **A**ttached **P**roton **T**est

This method—working with typical values for $^1J_{C,H}$ in a modulation sequence—provides definite information about typical sp^2- and sp^3-hybridized carbons with different numbers of attached protons. As a typical experiment in the task of discriminating different carbon-13 species it is an alternative to DEPT (see p 69). Figure 2 explains APT which was introduced in carbon-13 nuclear magnetic resonance (see NMR) in 1982 [1].

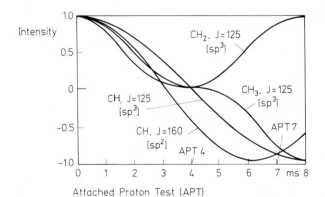

Attached Proton Test (APT)
for selection of carbon species in C-13nmr spectra

Fig. 2

Reference

1. Patt SL, Shoolery J (1982) J. Magn. Reson. 46: 535

AR
Abbreviation of Acousting Ringing

AR [1,2] is a problem in pulsed FOURIER transform (see PFT) experiments in nuclear magnetic resonance (see NMR) spectroscopy and has attracted considerable interest in the last ten years.

Several techniques and attempts of avoiding these AR effects have been suggested, including new technologies in probe constructions [1–3], computer approaches [3], and multi-pulse methods [3–9]. Multi-pulse sequences are widely applied since they are readily applicable routinely and they will result, if certain conditions are fulfilled, in a significant alleviation of the acoustic distortions [3].

Recently Goc and Fiat [10] published experimental results on a simple pulse sequence which does not involve phase-cycling. These authors have concluded that this simple pulse sequence results in a significant alleviation of baseline distortions similar to that realized by using variants of the three-pulse sequence and the extended spin-echo sequence [7], which can be only performed with spectrometers capable of phase-cycling.

Gerothanassis [11] believed, however, that these statement deserve clarification in view of the fact that, according to Canet and co-workers [12], the originally suggested pulse sequence [10] does not seem to offer important advantages in eliminating the unwanted acoustic distortions. He therefore studied the effectiveness of pulse sequences for suppression of AR in FT NMR much more carefully [11] and found out their scope and limitations.

References
1. Buess ML, Petersen GL (1978) Rev. Sci. Instrum. 49: 1151
2. Fukushima E, Roeder SBW (1979) J. Magn. Reson. 33: 199
3. Gerothanassis IP (1987) Prog. NMR Spectrosc. 19: 267
4. Patt SL (1982) J. Magn. Reson. 49: 161
5. Belton PS, Cox IJ, Harris RK (1985) J. Chem. Soc., Faraday Trans. II 81: 63
6. Morris GA, Toohey MJ (1985) J. Magn. Reson. 63: 629
7. Gerothanassis IP, Lauterwein J (1986) J. Magn. Reson. 66: 32
8. Kunwar AC, Turner GL, Oldfield E (1986) J. Magn. Reson. 69: 124
9. Gerothanassis IP (1986) Magn. Reson. Chem. 24: 428
10. Goc R, Fiat D (1986) J. Magn. Reson. 70: 295
11. Gerothanassis IP (1987) J. Magn. Reson. 75: 361
12. Canet D, Brondeau J, Marchal JP, Lherbier BR (1982) Org. Mag. Reson. 20: 51

AR model
Abbreviation of Autoregression Model

AR models have been used in non-linear spectral analysis in nuclear magnetic resonance (see NMR) spectroscopy [1–3].

One can explain the AR model with the following three statements:

a) The system model is a finite sum of non-interacting, damped simple harmonic oscillators (see SHO) at arbitrary (complex) frequencies. The important parameters are the frequencies and the complex amplitudes of the oscillators. Data are precise samples obtained from a time record of the system response to noise or pulse excitation.

b) Frequency parameters are obtained from a linear autoregression fit to match with the least square residue. The number of oscillators is subjectively estimated from the error residue. Two algorithms have been described for this linear least squares fit:

the "singular value decomposition" (see SVD) and

the "LEVINSON-BERG recursion method".

The amplitude parameters are then obtained from a second linear least squares fit of the model output to the data.

c) The spectrum itself is the superposition of LORENTZians of the model simple harmonic oscillators.

References

1. Barkhuijsen H, de Beer R, Bovée WNH, Van Ormondt D (1985) J. Magn. Reson. 61: 465
2. Barkhuijsen H, de Beer R, Van Ormondt D (1985) J. Magn. Reson. 64: 343
3. Tang J, Lin CP, Bowman MK, Norris JR (1985) J. Magn. Reson. 62: 167

ARMA model
Abbreviation of Autoregression Moving Average **Model**

This model can be used in non-linear spectral analysis in nuclear magnetic resonance (see NMR) spectroscopy instead of the AR model (see p 24).

The AR model does not apply to the usual NMR situation where the measured data are not precise enough as consequence of radio-frequency noise. Addition of output noise to the AR model yields an ARMA process. This process is characterized by the same filter in the AR feedback and MA feedforward path. The frequency parameters of the model are taken from the autoregression filter coefficient that form an eigenvector of the autocorrelation data matrix. A special eigenvalue algorithm for the ARMA method is necessary. But, all other procedures are quite similar to those used in the AR model.

Reference

1. Kaiser R (1986) paper presented at the 8th European ENC, June 3–6, 1986, Spa/Belgium

ASIS
Acronym for Aromatic Solvent Induced Shifts

In nuclear magnetic resonance (see NMR) spectroscopy one can use the specific solvent effects of benzene and its derivatives on different protons of a molecule under investigation. Particularly valuable for this purpose is benzene itself because of its high magnetic anisotropy and its well-known tendency to form specific complexes with the solute [1].

Especially in steroid chemistry, ASIS have been systematically studied and used for better shift separation and assignment. For example, if benzene is used instead of chloroform, the proton resonance signals of the four methyl groups in 3-oxo-4,4-dimethyl-5α-androstane could be differentiated [2].

ASIS have been found to be helpful in problems of configurational and conformational assignment of small molecules, too [3].

Sometimes ASIS have been used to simplify complex proton spin systems and their analysis.

References

1. Günther H (1980) NMR spectroscopy. An introduction, John Wiley, Chichester
2. Bhacca NS, Williams DH (1965) Application of NMR spectroscopy in organic chemistry, Holden-Day, San Francisco
3. Wendisch D (1966) Dissertation, University of Cologne

A

ATR
Acronym for **A**ttenuated **T**otal **R**eflection

ATR is a very important method used in infrared (see IR) spectroscopy. In internal reflection spectroscopy (see IRS) the sample is in a direct optical contact with a prism, optically denser than the sample. The incoming light produces a standing wave pattern at the interface within the prism, whereas in the rare medium of the sample the amplitude of the electric field falls off exponentially with the distance from the phase boundary. If the rare medium exhibits absorption, the penetrating wave becomes attenuated. The resulting energy loss in the reflected wave is referred to as attenuated total reflection (ART). Multiple reflections are used for sensitivity enhancement, therefore this technique is often called multiple internal reflection (MIR). The usefullnes of the phenomenon of ATR was first demonstrated by Fahrenfort [1,2], while Harrick developed the technique [3,4] and designed ATR cells for commercial purposes. ATR measurements became very popular in the field of IR studies on polymers [5].

References

1. Fahrenfort J (1961) Spectrochim. Acta 17: 698
2. Fahrenfort J, Visser WM (1962) Spectrochim. Acta 18: 1103
3. Harrick NJ (1962) Phys. Rev. 125: 1165
4. Harrick NJ (1967) Internal reflection spectroscopy, Wiley, New York
5. Koenig JL (1984) Fourier transform infrared spectroscopy, Springer, Berlin Heidelberg New York (Advances in Polymer Science, vol 5)

BB
Abbreviation of **B**road-**B**and Decoupling

BB decoupling or proton noise decoupling (see PND) became very important in nuclear magnetic resonance (see NMR) of nuclei with low natural abundance and low magnetogyric ratio (γ) such as carbon-13, etc.

BEER's law
Abbreviation of the well-known BOUGUER-BEER-LAMBERT law used in absorption spectroscopy, especially in infrared (see IR) spectrometry

B

This abbreviation was introduced recently by P.R. Griffiths [1]. For a single solute in a nonabsorbing solvent, BEER's law gives the absorbance at any wavenumber $\bar{v}$ as

$$A(\bar{v}) = -\log_{10}\tau(\bar{v}) = a(\bar{v})bc \tag{1}$$

where $\tau(\bar{v})$ and $a(\bar{v})$ are the transmittance and absorptivity at $\bar{v}$, respectively, b is the pathlength, and c is the concentration of the sample under investigation.
For a mixture of n components, the total absorbance at $\bar{v}$ is given by

$$A(\bar{v}) = \sum_{i=1}^{n} a_i(\bar{v})bc_i \tag{2}$$

Deviations from BEER's law are known to occur for many reasons. Suffice it to say [1] that deviations from linearity for plots of $A(\bar{v})$ versus c_i can be caused by stray radiation, insufficient resolution, and chemical effects. The stray light levels in FOURIER transform-infrared spectra (see FT-IR) are normally very small. For modern FT-IR spectrometers, the most important factor leading to deviations from BEER's law is the effect of insufficient resolution. The shape of each band in an absorption spectrum of samples in the condensed phase is given—to a fairly good approximation—by the LORENTZian (or CAUCHY) function, here expressed with Eq. (3).

$$A(\bar{v}) = A_{peak}^{t} \frac{\gamma^2}{\gamma^2 + (\bar{v} - \bar{v}_0)^2} \tag{3}$$

In Eq. [3] A_{peak}^{t} stands for the true absorbance at the wavenumber of maximum of absorption, $\bar{v}_0$, and γ is the half-width at half-height. Note please, that FWHH (see p 118) will be equal to 2γ!

Reference
1. Griffiths PR, de Haseth JA (1986) Fourier transform infrared spectrometry. John Wiley, New York. chap 10 (Chemical analysis, vol 83)

BIRD
Acronym for **Bi**linear **R**otation **D**ecoupling

The BIRD pulse sequence [1,2] can be used in various experiments dealing with heteronuclear magnetic resonance of nuclei with a low gyromagnetic ratio γ.

BIRD has been incorporated in SHARP [3] experiments (see SHARP) on complex systems where it is desirable to collapse all multiplet structures to improve the resolution of the chemical shift information [3]. Recently BIRD has been used for sensitivity-enhanced two-dimensional heteronuclear shift correlation NMR spectroscopy (see COSY) in the following way [4]: All protons not coupled to carbon-13 (low-γ nucleus) are inverted by the BIRD pulse sequence [1,2]: 90°_x (^{1}H) $- 1/(2\,\mathrm{J}) - 180^\circ_x$ (^{1}H, ^{13}C) $- 1/(2\,\mathrm{J}) - 90^\circ_{-x}$ (^{1}H), where J is the one-bond ^{13}C–^{1}H-scalar coupling constant. Protons coupled to ^{13}C are not affected by this bilinear pulse, whereas the magnetization of all other protons is inverted. At the time, τ, when the inverted magnetization changes from negative to positive (i.e. when protons not coupled to carbon-13 are nearly saturated), the first 90° pulse of a heteronuclear multiple-quantum coherence (see HMQC) experiment is applied. In practice, the T_1's of the different protons in the molecule will vary, which at first sight makes it impossible to chose a single τ value for which all protons involved not coupled to carbon-13 are near saturation. But, by keeping the delay time, T, between experiments, one can largely circumvent this complication [4].

References

1. Garbow JR, Weitekamp DP, Pines A (1982) Chem. Phys. Lett. 93; 504
2. Bax A (1983) J. Magn. Reson. 53: 517
3. Gochin M, Weitekamp DP, Pines A (1985) J. Magn. Reson. 63: 431
4. Bax A, Subramanian S (1986) J. Magn. Reson. 67: 565

BLEW-48
Abbreviation of a 48-pulse sequence used in nuclear magnetic resonance (see NMR): **B**urum, **L**inder, **E**rnst **W**indowless

Like BLEW-12, the simplest *fully windowless* pulse cycle, BLEW-48 [1,2] was originally developed as special decoupling sequence in solid-state NMR in competition with other techniques (see MREV-8, BR-24, and BR-52).

Recently [3] BLEW-48 found a new application: the ^{13}C-spectrum of a liquid crystal is simplified and exhibits "first-order" coupling patterns if the proton-proton dipolar couplings are removed by the special decoupling sequence BLEW-48. The splitting between adjacent peaks in a multiplet is then given by

$$\Delta v = f/(3\cos^2\theta - 1)\cdot D + J/ \ ,$$

where f is a scaling factor characteristic of this dipolar decoupling sequence, D corresponds to the dipolar coupling constant, and J is the scalar coupling constant.

References

1. Burum DP, Linder M, Ernst RR (1981) Joint ISMAR-Ampère Intl. Conference on Magnetic Resonance, Aug. 25–29, 1980, Delft/NL; Bull. Magn. Reson. 2: 413; note, that this preliminary report contained incorrect and misleading statements (see Ref [2])
2. Burum DP, Linder M, Ernst RR (1981) J. Magn. Reson. 44: 173
3. Fung BM (1986) J. Magn. Reson. 66: 529

BLOCH's

equations are the background to a classical description of nuclear magnetic resonance (see NMR) phenomenon.

In his original formulation of the behaviour of nuclear magnetic moments in variable magnetic fields, F. BLOCH [1] used a set of macroscopic or *phenomenological equations* for the variation of the components of the total nuclear magnetic moment per unit volume.

Although these equations were introduced originally as postulates, it was later found that their solutions reproduce many of the properties of NMR experiments. They have the further advantage of being suitable for the explanation of various *time-dependent phenomena* and *transient effects*. For a discussion of detailed conditions under which they are strictly valid, the interested reader may consult two papers in the literature [2,3].

In deriving BLOCH's equations, we are following here the didactic concept of H. Günther [4] and his choice of signs and arrangement of certain quantities.

According to the classical description, the behaviour of a magnetic moment M in a magnetic field B is given by Eq. (1), the so-called *equation of motion*.

$$\frac{dM}{dt} = \gamma\,[M \times B] \tag{1}$$

In Eq. (1), γ stands for the magnetogyric ratio. In an NMR experiment the static magnetic field B_0 and the rotating field B_1 contribute both to the field B. B_0 coincides with the z-axis in CARTESian coordinates while B_1 rotates, with the frequency ω, clockwise in the x–y plane. Thus the components of B are·

$$B_x = B_1 \cdot \cos \omega t, \quad B_y = -B_1 \cdot \sin \omega t, \quad \text{and} \quad B_z = 0 \ .$$

The vector product of Eq. [1] can be solved as follows:[1]

$$\frac{dM_x}{dt} = \gamma\,(M_y B_0 + M_z B_1 \sin \omega t) \tag{2a}$$

$$\frac{dM_x}{dt} = \gamma\,(M_z B_1 \cos \omega t - M_x B_0) \tag{2b}$$

$$\frac{dM_z}{dt} = \gamma\,(M_x B_1 \sin \omega t - M_y B_1 \cos \omega t) \tag{2c}$$

For the time-dependence of magnetization we must, in addition, consider *relaxation effects*. These effects have been included in Eq. (2) by BLOCH phenomenologically [3]. The relaxation time T_2 (or spin-spin relaxation time) is characteristic for the *transverse* magnetization in the x-y plane while variation of the *longitudinal* magnetization along the z-axis is a function of T_1 (or spin-lattice relaxation time). The complete form of BLOCH's equations is given with Eqs. (3a), (3b), and (3c) where M_0 means the equilibrium magnetization that is present in the

---

[1] Here we applied the general relation for the product of two vectors A and B: $[A \times B] = (A_y B_z - A_z B_y)i + (A_z B_x - A_x B_z)j + (A_x B_y - A_y B_x)k$ where i, j, and k are the coordinate unit vectors.

magnetic field B_0 at the beginning of the NMR experiment. For further presentations and/or transformation of Eq. (3) see the literature [4].

$$\frac{dM_x}{dt} = \gamma(M_y B_0 + M_z B_1 \sin \omega t) - \frac{M_x}{T_2} \tag{3a}$$

$$\frac{dM_y}{dt} = \gamma(M_z B_1 \cos \omega t - M_x B_0) - \frac{M_y}{T_2} \tag{3b}$$

$$\frac{dM_z}{dt} = \gamma(-M_x B_1 \sin \omega t - M_y B_1 \cos \omega t) - \frac{M_z - M_0}{T} \tag{3c}$$

For the case in which two (or more) nuclei, as a result of a fast equilibrium, periodically *exchange* their original chemical environment and thus their LARMOR frequencies, BLOCH's equations for NMR line shape calculations or anlyses must be modified. This is done most easily by combining Eqs. (3a) and (3b) if they are transformed into the rotating frame and under defining a complex x-y magnetization, $G = M_{x'} + i M_{y'}$. But, in the case of chemical exchange where more than two positions with different LARMOR frequencies are involved, the modification of BLOCH's equations yield a very complex form, normally written in matrix notation. For solution an appropriately programmed computer is necessary. The final form of such modified BLOCH's equations leads to quite the same results as obtained on the basis of the *quantum-mechanical* treatment given in the theory of *Anderson, Kubo* and *Sack* [5,6].

Up to this point we have been concerned with the application of a continuous radio-frequency magnetic field B_1 to a sample in an uniform magnetic field B_0. Rather different phenomena of a transient nature can be observed if the radio-frequency field is changed discontinuously. Useful NMR experiments of this kind are possible because electronic response times are short compared with the time decay T_2 of a LARMOR precession.

Torrey [7] made a detailed study of solutions of BLOCH's equations if the system starts with the magnetization vector along the direction of the static field B_0 and the radio-frequency field is suddenly switched on. The approach to the steady-state solution is oscillatory in character and got the name *TORREY's oscillations*.

Hahn [8,9] has developed an important method based on *electronic pulses* of the field B_1 which aims at registering the nuclear induction signal after the radio-frequency field is cut off. From the theoretical point of view, this is rather simpler than the experiment considered by Torrey [7] for after removing B_1, the nuclear magnetization vector only precesses in the static field B_0, and BLOCH's equations take a simpler form. Hahn [8] first examined the response of a spin system to a *single radio-frequency pulse*. The induction signal produced in the *off* period is the so-called *free induction decay* (see FID), previously discussed by BLOCH [1]. These basic experiments led to the commonly used pulse FOURIER transform (see PFT) mode in modern NMR. After discovering the existence of the so-called *spin echoes* (see SE) by Hahn [9] a very fruitful development of different *pulse techniques* was initiated. These pulse techniques and *pulse sequences* will be explained under their individual acronyms or abbreviations in this dictionary.

Recently approximate solutions of BLOCH's equations for selective excitation in NMR spectroscopy has been proposed [10] by using a perturbation method. High-order terms have been analyzed, employing some special properties of the corresponding multi-dimensional integrals in the spectra. It has been found that for excitation by a symmetric radio-frequency pulse applied in the presence of a field-gradient, the spin nutation angle as a function of the off-resonance frequency will be approximately proportional to the FOURIER cosine transform (see FT) of the radio-frequency function.

This fundamental analysis will be helpful for understanding the different behaviours of both spin-inversion and phase-reversal 180° selection radio-frequency pulses as well as selective-excitation pulses.

In addition, this approach will also offer more insight to the problems of designing better selective radio-frequency pulses (for comparison see DMRGE) A. Hasenfeld [11] has formulated a connection between BLOCH's equations and the KORTENWEG/DE VRIES equation which may be useful in the field of NMRI. By formulating the physical problem of NMR radio-frequency excitation as an typical inverse problem, he has been led to a family of radio-frequency amplitude modulations that realize a particular response through BLOCH's temporal evolution.

In the treatment of the NMR spectra of *multi-site exchanging systems*, it is sometimes possible to reduce the dimension of matrices due to the fact that some of the sites are equivalent. However, the existence of identical spin HAMILTON-ians at some of the sites has been found [12] to be insufficient in general for simplification of the problem: Symmetry of exchange network linking the sites must be considered, too. Some cases in solid-state and liquid-state NMR have been examined [12]. The conditions for correct simplification of multi-site exchange problems have been compared with the well-known distinctions between chemical and magnetic equivalence in high-resolution NMR of coupled spin systems [12].

References

1. Bloch F (1946) Phys. Rev. 70: 460
2. Wangsness R, Bloch F (1953) Phys. Rev. 89: 728
3. Bloch F (1956) Phys. Rev. 102: 104
4. Günther H (1980) NMR spectroscopy. An introduction, appendix no. 8, John Wiley, Chichester/UK
5. Binsch G (1968) In: Allinger NL, Eliel EL (eds) Topics in Stereochemistry, vol 3, Interscience, New York
6. Sack RA (1958) Mol. Phys. 1: 163
7. Torrey HC (1949) Phys. Rev. 76: 1059
8. Hahn EL (1950) Phys. Rev. 77: 297
9. Hahn EL (1950) Phys. Rev. 80: 580
10. Yan H, Chen B, Gore JC (1987) J. Magn. Reson. 75: 83
11. Hasenfeld A (1987) J. Magn. Reson. 72: 509
12. Levitt MH, Beshah K (1987) J. Magn. Reson. 75: 222

Addendum

to the *equation of motion* (1): As we know, analysis of some NMR experiments requires the solution of the *equation of motion* of the relevant *density matrix*. The spin operator representation of the density matrix helps to derive a concise equation for the time development of the complete density operator, given the initial state of the system. This approach corresponds to the correlation function method discussed first by Banwell and Primas [1]. The Baker-Campbell-Hausdorff formula may be employed to evaluate operator evolution. Since the relevant LIOUVILLE's space for NMR is finite dimensional, a finite number of commutators completely characterizes the evolution, the sets of operators produced constituting the commutator algebra (or the so-called LIE's algebra) generated by the HAMILTONian.

Operator equations have been published to characterize AX_n ($n = 1$ to 3) and A_2X_2 systems under the conditions of cross-polarization (see CP) [2–4] as well as isotropic mixing (see IM) [4,5], while the analysis of AX_n systems evolving under CP [4,6,7] and IM [4] conditions has been reported under using zero-quantum frame techniques for arbitrary n.

Recently [8] a computer program has been written for an automated generation of LIE's algebra of any general spin HAMILTONian.

References

1. Banwell CN, Primas H (1963) Mol. Phys. 6: 225
2. Chandrakumar N (1985) J. Magn. Reson. 63: 202
3. Chandrakumar N (1986) J. Magn. Reson. 67: 457
4. Chandrakumar N, Visalakshi GV, Ramaswamy D, Subramanian S (1986) J. Magn. Reson. 67: 307
5. Chandrakumar N, Subramanian S (1985) J. Magn. Reson. 62: 346
6. Chingas GC, Garroway AN, Bertrand RD, Moniz WB (1981) J. Chem. Phys. 74: 127
7. Müller L, Ernst RR (1979) Mol. Phys. 38: 963
8. Visalakshi GV, Chandrakumar N (1987) J. Magn. Reson. 75: 1

BOWMAN

Acronym for **B**idimensional **S**imulator **O**perating on **W**eakly Coupled Systems for **M**ultipulse Sequences on **A**ny **N**uclear Magnetic Resonance Software

BOWMAN is an computer-assisted approach [1] for generalizing the different kinds of software in nuclear magnetic resonance (see NMR) spectroscopy in its multiple resonance [2] methods.

References

1. Piveteau D, Delsuc MA, Lallemand J-Y (1986) J. Magn. Reson. 70: 290
2. McFarlane W, Rycraft DS (1985) Multiple resonance. In: Webb Ga (ed) Ann. rep. on NMR spectrosc., Academic, London, vol 16, p 293

BPP (approach)

Abbreviation of the names **B**loembergen, **P**urcell and **P**ound (and their elementary discussions of NMR relaxation)

The theory of relaxation in nuclear magnetic resonance is by now well developed and has been presented in textbooks and research monographs [1,2]. Many of these treatments are based on the original work of Bloembergen, Purcell and Pound [3] in which it is said relaxation is due to transitions between Zeeman levels that are induced by fluctuations in the interaction terms of the HAMILTONian (see p 122). On the other hand a phenomenological description of the relaxation process is contained in the Bloch equations [4]. Fundamental justifications of both of these approaches make use of density-matrix methods developed by Bloch [5] and Redfield [6]. In a new paper by Rorschach [7], it is shown that some of the simple results of the density-matrix theory can be obtained without the use of quantum mechanics. The classical equations of motion of a single spin in a time-varying magnetic field are solved, and expressions for the relaxation times are obtained in terms of the Fourier transforms of the correlation functions of the components of the magnetic field.

B

References
1. Abragam A (1961) The principles of nuclear magnetism, Clarendon, Oxford
2. Slichter CP (1978) Principles of magnetic resonance, 2nd edn, Springer, Berlin Heidelberg New York
3. Bloembergen N, Purcell EM, Pound RV (1948) Phys. Rev. 73: 679
4. Bloch F (1946) Phys. Rev. 70: 460
5. Wangsness RG, Bloch F (1953) Phys. Rev. 89: 728; Bloch F (1956) Phys. Rev. 102: 104; Bloch F (1957) Phys. Rev. 105: 1206
6. Redfield AG (1957) IBM J. Res. Dev. 1: 19
7. Rorschach HE (1986) J. Magn. Reson. 67: 519

BR-24 and -52
are the abbreviations of a 24- and 52- multi-pulse sequence introduced by Burum and Rhin [1] for dipolar decoupling in nuclear magnetic resonance (see NMR) of solid-state samples.

For comparison see: MREV-8 and BLEW-12 and -48 in this dictionary.

Reference
1. Burum DP, Rhin WK (1979) J. Chem. Phys. 71: 944

broadband-INEPT
Abbreviation of a new INEPT experiment in nuclear magnetic resonance

This new method is shown, both by theory and experiment, to have a sensitivity advantage over the normal INEPT sequence (see INEPT) in a variety of applications [1]. The procedure involves a heteronuclear coherence transfer (see CT) over a range of coupling constants. A method is described [1] whereby the performance of nuclear magnetic resonance pulse experiments which use fixed free precession periods with durations that should be matched to the reciprocal of a scalar, dipolar, or quadrupolar coupling constant or a resonance offset can be made less sensitive to the presence of a range of couplings or offsets. The example selected for full discussion is the INEPT experiment which uses fixed periods of free precession ("delays") with durations that should ideally be chosen in proportion to the reciprocal of the heteronuclear scalar coupling constant. By considering an anology with known composite pulses designed to compensate for radio-frequency field inhomogeneity (ΔB_1), a new experiment has been derived, called "broadband-INEPT", which is effective over a wider range of coupling constants.

Reference

1. Wimperis S, Bodenhausen G (1986) J. Magn. Reson. 69: 264

BSS
Acronym for **BLOCH-SIEGERT** Shift observed in Nuclear Magnetic Double Resonance (see NMDR)

In NMDR, for example, if one observes in the olefinic region the A resonance of an AX system, the X resonance of which is overlapped by absorptions of a series of CH_2 protons, selective irradiation with B_2 in the methylene region allows one to measure the frequency difference $v_A - v_X$ and from that the chemical shift of the X proton.

It is necessary to point out, however, that in the case of a large B_2 amplitude and a small frequency difference this method may lead to erroneous results, since according to the rigorous theory for NMDR, Eq. (1) holds for the proton A.

$$v'_A = v_A + \frac{\gamma^2 B_2^2}{8\pi^2 (v_A - v_2)} \approx v_A + \frac{\gamma^2 B_2^2}{8\pi^2 (v_A - v_X)} \tag{1}$$

The shift $v_A \to v'_A$ is the so-called the **BLOCH-SIEGERT** shift.

BURG ME (M)
Abbreviation of BURG's Maximum Entropy (Method)

A maximum entropy spectral analysis method, termed **BURG MEM** (or non-linear MEM) [1,2], has been widely applied in different spectroscopie methods and in various areas of research, and has recently been used for resolution enhancement of **RAMAN** spectra of biomolecules [3]. The BURG MEM provides better resolution than the conventional discrete **FOURIER** transform (see DFT) method, especially for data truncated in the time domain, and is therefore preferred for resolution enhancement of complex spectra of polypeptides and proteins [3]. It has been pointed out [4] that this MEM procedure can also be applied to reconstruct a phase-sensitive spectrum with arbitrary phase for each resonance, a situation that occurs frequently in FT NMR (FOURIER transform nuclear magnetic resonance spectroscopy; see p 102) especially in the new two-dimensional (see 2D-NMR) FT NMR spectroscopies [5]. There seems to be a common belief [6] that the BURG MEM algorithm is adequate only for calculating a power or magnitude spectrum from experimental data, but it has been demonstrated [4] that this procedure has a much wider applicability. The most important point is that the method of maximum entropy power spectral analysis has been shown [7] to correspond formally to an autoregressive (see AR) modeling of the underlying time series, which could represent values of the free induction decay (see FID) at various time in the case of FT NMR spectroscopy, and the AR coefficients are actually evaluated by means of the linear prediction— one step forward—equations in BURG's formulation. While this set of AR coefficients can be used for calculating a power/magnitude spectrum, as in the most common applications [3], they are equally applicable to evaluate the direct FT of time series that is extrapolated beyond the available data, as is done with the backward linear prediction under using the HOUSEHOLDER triangularization procedure, termed LPQRD (see p 163) [6].

References

1. Burg JP (1967) Proc. 37th Meeting Soc. Expl. Geophys.
2. Burg JP (1978) In: Childers DG (ed) Modern methods of spectral analysis. IEEE Press, New York, p 42
3. Ni F, Scheraga HA (1985) J. Raman Spectrosc. 16: 337
4. Ni F, Scheraga HA (1986) J. Magn. Reson. 70: 506
5. Sørensen OW, Eich GW, Levitt HM, Bodenhausen G, Ernst RR (1983) Prog. Nucl. Magn. Reson. Spectrosc. 16: 163
6. Tang J, Norris JR (1986) J. Chem. Phys. 84: 5210
7. Ulrych TJ, Bishop TN (1978) In: Childers DG (ed) Modern methods of spectral analysis. IEEE Press, New York, p 54

B

C4S
Acronym for the so-called Chemical Shift-Specific Slice Section (Method)

This new method has recently been proposed for chemical shift imaging (see NMRI) at high magnetic fields [1]. This procedure, especially suited to high-field imaging systems (4.7 Tesla or higher), combines chemical shift-selective excitation [2,3] with slice selection.

References

1. Volk A, Tiffon B, Mispelter J, Lhoste J-M (1987) J. Magn. Reson. 71: 168
2. Bottomley PA, Foster TH, Leue WM (1984) Proc. Natl. Acad. Sci. (USA) 81: 6856
3. Haase A, Frahm J (1985) J. Magn. Reson. 64: 94

C

CAMBRIDGE ME
Abbreviation for **Cambridge maximum entropy method**

Method is used in non-linear spectral analysis [3] in connection with incoherent excitation of a nuclear spin system in NMR spectroscopy [1,2]. The Cambridge ME method combines a loosely interpreted maximum entropy principle with a somewhat modified least square procedure in order to get an estimate of a "spectrum" from a noisy, truncated time record. The resulting digital "spectrum" is given by Eq. (1)

$$S_j \sim \exp\left[\lambda \sum_{k \in M} (\tilde{x}_k - x_k) e^{2\pi ijk/m} \right] \tag{1}$$

In Eq. (1) we have the following variables:

$x_1 \ldots x_m$: observed samples of the time record,

$\tilde{x}_1 \ldots \tilde{x}_m$: corresponding samples computed as inverse FT (see FT) of the spectrum S_j,

λ : parameter which adjusts the closeness of the fit of the $\tilde{x}$ to x data; λ is adjusted to make $\sum_{k \in M} (\tilde{x}_k - x_k)^2$ nearly equal $m\sigma^2$

where σ is the estimated noise amplitude in each data sample. The resulting "spectrum" is an amplitude spectrum, but, is necessarily positive at all frequencies. For the Cambridge algorithm see Refs. [4–7].

References

1. van Kampen NG (1981) J. Stat. Phys. 24: 175
2. Blümich B (1985) Bull. Magn. Reson. 7: 5, and 138 references contained within
3. Haykin S (ed) (1979) Nonlinear methods of spectral analysis, Springer, Berlin Heidelberg New York (Topics in Applied Physics vol 34); Kay SM, Marple SL (1981) Proc. IEEE 69: 1380, and 278 references contained within
4. Laue ED, Skilling J, Staunton J, Sibisi S, Brereton RG (1985) J. Magn. Reson. 62: 437
5. Laue ED, Skilling J, Staunton J (1985) J. Magn. Reson. 63: 418
6. Hore PJ (1985) J. Magn. Reson. 62: 561
7. Martin JF (1985) J. Magn. Reson. 65: 291

CAMELSPIN
Acronym for **C**ross-Relaxation **A**ppropriate for **M**inimolecules **E**mulated by **L**ocked **Spin**s

In cases where $w_0\tau_c \simeq 1$—as, for example, with medium-sized oligomers observed at very high field—both the traditional one- and two-dimensional NOE methods (see 1D-NOE, 2D-NOE, and NOESY) fail because the maximum observable nuclear OVERHAUSER enhancement of factor approaches zero. One alternative is to perform the magnetization transfer in the rotating reference frame, using a y-axis spin-locking field. This experiment has been dubbed CAMELSPIN by Bothner-By and co-workers [1] and has 2D counterparts [2].

The maximum attainable transient NOE's in the rotating reference frame when $w_0\tau_c \ll 1$ (the so-called "extreme narrowing limit") or $w_0\tau_c \gg 1$ (the so-called "spin-diffusion limit") are 38.5% and 67.5%, respectively. This is no improvement on classical procedures. Whether the two-dimensional CAMELSPIN produces better cross-peak intensity than NOESY in the "extreme narrowing limit" has not been reported so far.

In the meantime instead of CAMELSPIN the acronym ROESY (see there) seems to be almost accepted [3].

References
1. Bothner-By AA, Stephens RL, Lee J-M, Warren CD, Jeanloz RW (1984) J. Am. Chem. Soc. 106: 811
2. a) Bax A, Davis DG (1985) J. Magn. Reson. 63: 207;
   b) Neuhaus D, Keeler J (1986) J. Magn. Reson. 68: 568;
   c) Kessler H, Griesinger C, Kessebaum R, Wagner K, Ernst RR (1987) J. Am. Chem. Soc. 109: 607
   d) Griesinger C, Ernst RR (1987) J. Magn. Reson. 75: 261
3. Kessler H, Gehrke M, Griesinger C (1988) Angew. Chem. 100: 507; (1988) Angew. Chem. Int. Ed. Engl. 27: 490

Carbon-13 NMRI
Acronym for Carbon-13 **N**uclear **M**agnetic **R**esonance **I**maging (see NMRI)

NMRI experiments have so far been predominantly performed with protons (see NMRI). Although there is a widespread interest in imaging with the use of other nuclei than protons, only lithium-7 [1], fluorine-19 [2–4], sodium-23 [5–7], and phosphorous-31 [8] have been recorded.

Recently [9], the first NMR images of a truly dilute spin, namely carbon-13, have been published.

Although high-resolution carbon-13 NMR, both in liquids and solids, has generated numerous application possibilities in all branches of chemistry and material sciences for many years, *in vivo* carbon-13 NMR has just become feasible. A number of reasons were relevant. The availability of high-field imaging systems (say 4.7 Tesla or higher), which are required to improve the sensitivity of the carbon-13 signal, have been limited in the past. On the other hand, a number of inherent physical properties of the carbon-13 nucleus give rise to several undesirable features, as far as NMRI experiments are concerned. These will include long spin-lattice relaxation times (T_1), spin-spin couplings to protons, and large chemical shift dispersion, too.

The use of carbon-13 labeled compounds for *in vivo* imaging will be an exciting possibility in the future.

References

1. Renshaw PF, Haselgrove JC, Leigh JS, Chance B (1985) Magn. Reson. Med. 2: 512
2. Joseph PM, Fishman JE, Mukherji B, Sloviter HA (1985) Proceedings of the Society of Magnetic Resonance in Medicine, 4th Annual Meeting, London, p 995
3. Koutcher JA, Rosen BR, Metarland EW, Telcher BA, Brady TJ (1984) Proceedings of the Society of Magnetic Resonance in Medicine, 3rd Annual Meeting, New York, p 435
4. Nunnally RL, Babock EE, Horner SD, Peshock RM (1985) Magn. Reson. Imaging 3: 399
5. Hilal SK, Maudsley AA, Simon HE, Perman WH, Bonn J, Mawad ME, Silver AJ (1983) A.J.N.R. 4: 245
6. Moseley ME, Chen WM, Nishimura MC, Richards TL, Murphy-Boesch J, Young GB, Marschner TM, Pitts LH, James TL (1986) Magn. Reson. Imaging 3: 383
7. Joseph PM, Summers RM (1986) J. Magn. Reson. 69: 198
8. Maudsley AA, Hilal SK, Perman WH, Simon HE (1983) J. Magn. Reson. 51: 147
9. Kormos DW, Yeung HN, Gauss RC (1987) J. Magn. Reson. 71: 159

CARS
Acronym for **C**oherent **A**nti-**STOKES RAMAN** **S**cattering

CARS is a method used in RAMAN spectroscopy (see p 233). The known disturbance caused by fluorescence of the sample in RAMAN spectroscopy can be reduced by many orders of magnitude in CARS spectroscopy. In this technique, two laser beams (see LASER) are crossed at the sample with a small angle. One of the two laser beams is tunable in frequency. Strongly enhanced scattering is observed if the difference in frequency of the two lasers corresponds to a vibration of the sample under investigation [1,2]. The coherently scattered radiation lies in the neighborhood of the short-wavelength side of fluorescence and is partially directed; it can therefore be discriminated from fluorescence.

References

1. Bulkin BJ (1978) In: Brame EG Jr (ed) Applications of polymer spectroscopy, Academic, New York p 121
2. Lascombe J, Huong PV (eds) (1982) Raman spectroscopy, linear and nonlinear, Proc. of the 8th Internatl. Conf. on Raman Spectroscopy, Bordeaux/France, Wiley and Heyden, Chichester

CC
Acronym for **C**oupling **C**onstants used in Nuclear Magnetic Resonance (see NMR) Spectroscopy

The fine structure or the multiplet character of a resonance signal in NMR is caused by the so-called *spin-spin coupling*. Because of the magnetic field independence of this NMR parameter, we call it *coupling constant*, or CC. The symbol of CC is J, measured in Hz.

Besides the chemical shift (see CS) CC is the most important parameter of the NMR spectrum, and in the case of proton NMR CC give information about proton-proton connectivities.

^{2}J is the symbol for a geminal proton-proton CC, ^{3}J symbolizes a vicinal proton-proton CC, ^{4}J represents a CC over four bonds (often observed in allylic systems), and ^{5}J is a long-range CC which is, for example, found in homoallylic compounds.

In proton NMR, one can extract the CC directly from the multiplet if the so-called "rules of 1st-order analysis" are valid. In cases of higher-order spectra one has to analyze the total spin system or spectrum, if possible, via explicit mathematical equations, or by an interactive procedure using computer programs (see LAOCOON, NMRIT/NMREN, DAVINS, etc.) based on the HAMILTON-ian (see p 122) for the quantum-mechanical description.

Relative and absolute signs of CC can be determined experimentally via techniques of different complexity such as spin-tickling (see p 264).

The sign and magnitude of proton-proton CC give the most important information about constitution and conformation of a given molecule under investigation.

Following the CC over one bond in the hydrogen molecule, which amounts to 276 Hz, the geminal CC with values between -23 to $+42$ Hz forms the group with the largest spin-spin interaction between protons. Many factors are responsible for the sign and the magnitude of ^{2}J, some of them are:

a) dependence on the hybridization of the attached carbon,
b) influence of both α and β substituents, and
c) influence of adjacent π bonds.

There is extensive information on vicinal CC and their relation to chemical structure features. In agreement with results of theoretical calculations, it has been shown that the magnitude of ^{3}J, the sign of which was earlier found to be always positive, depends in essence upon the following four factors:

a) the dihedral angle, ϕ, between the C–H bonds under consideration,
b) the C–C bond length, $R_{\mu v}$,
c) the H–C–C valence angles, θ and θ', and
d) the electronegativity of the substituent on the H–C–C–H moiety.

For further details on proton-proton CC, the interested reader is referred to the literature [1–7].

Spin-spin interactions between protons and other nuclei than protons have been called heteronuclear CC. Among these heteronuclear CC spin-spin inter-actions with carbon-13, fluorine-19, and phosphorus-31 are the most important parameters for structure elucidation. The direct coupling of a carbon-13 nucleus with the attached proton has been correlated (see Eq. (1)) with the fractional s-character s(i) of C–H bonds involved in hydrocarbons.

$$^1J(^{13}C, {}^1H) = 500 \, s(i) \tag{1}$$

Since the s-character in turn is related by Eq. (2) to the hybridization parameter λ^2 of carbon orbital $sp^{\lambda 2}$, one can obtain information about the hybridization of a particular carbon atom via $^1J(^{13}C, {}^1H)$ measurements.

$$s(i) = 1/(1 + \lambda^2) \tag{2}$$

The homonuclear $^{13}C-^{13}C$–CC over one bond are sensitive parameters of the nature of the carbon-carbon bond involved. In the case of hydrocarbons, a dependence on the s-character product for the carbon orbitals ϕ_i and ϕ_j forming the C_i–C_j sigma bond has been observed and described with the empirical equation (3).

$$J(^{13}C, {}^{13}C) = 550\, s(i)\, s(j) \tag{3}$$

But, it must be pointed out that the basis of such a correlation is the assumption that only one of the various mechanisms that contribute to the spin-spin CC, the so-called FERMI contact term, dominates.

Information about carbon-carbon connectivities (see CCC) in a given molecule (that will mean carbon skeleton information) is obtainable via the measurement of ^{13}C–^{13}C–CC by techniques like INADEQUATE (see p 141).

For further details on heteronuclear CC the interested reader is referred to the literature [8,9].

References

1. Sternhell S (1969) Quart. Rev. 23: 236
2. Bothner-By AA (1965) In: Waugh JS (ed) Advances in magnetic resonance, vol 1, Academic, New York
3. Cookson RC, Crabb TA, Frankel JJ, Hudec J (1966) Tetrahedron, Suppl. 7: 355
4. Sternhell S (1964) Rev. Pure Appl. Chem. 14: 15
5. Barfield M, Chakrabarti B (1969) Chem. Rev. 69: 757
6. Hilton J, Sutcliffe LH (1975–1977) In: Emsley JW, Feeny J, Sutcliffe LH (eds) Progress in nuclear magnetic resonance spectroscopy, vol 10, Pergamon, Oxford
7. Kowalewski J (1977) In: Emsley JW, Feeny J, Sutcliffe LH (eds) Progress in nuclear magnetic resonance spectroscopy, vol 11, Pergamon, Oxford
8. Breitmaier E, Voelter W (1974) ^{13}C-NMR spectroscopy: methods and applications, Verlag Chemie, Weinheim, Sect. 3.2 (Monographs in modern chemistry, vol 5)
9. Marshall JL (1983) Carbon-carbon and carbon-proton NMR couplings: Applications to organic stereochemistry and conformational analysis, Verlag Chemie International, Deerfield Beach (Methods in stereochemical analysis, vol 2)

CD
Abbreviation of Convolution Difference

This method, used in nuclear magnetic resonance spectroscopy (see NMR), is an application of the exponential filtering concept [1]. It has been introduced as a resolution enhancement procedure [2].

References

1. see Ernst RR (1966) In: Waugh JS (ed) Sensitivity enhancement in magnetic resonance, Academic, New York, pp 1, 119, 123 (Advances in magnetic resonance, vol 2)
2. Campbell ID, Dobson CM, Williams RJP, Xavier AV (1973) J. Magn. Reson. 11: 172

CD
Acronym for Constrained Deconvolution

CD methods are used for nuclear magnetic resonance (see NMR) spectral enhancement, because a recurring problem in spectroscopy is how to improve signal-to-noise ratio (see SNR) and spectral resolution [1].

Using conventional apodization techniques (see Apodization), these two requirements are mutually exclusive [2]. The problem arises because the blanket application of an apodization function treats noise in the same way as the signal.

A more efficient approach is to treat signal and noise separately using a constrained deconvolution (CD) technique. To apply CD, it is necessary to measure the noise independently, using for example, the "tail" of the free induction decay (see FID). This information can then be used to construct the best (smoothest) spectrum consistent with the noise statistics.

A special case of CD is the maximum entropy method (see MEM). Although it has been claimed that MEM is the most unbiased or optimum version of CD [3] it has the disadvantage of requiring extensive computing time.

Belton and Wright have described a simpler and much more rapid version of CD [1]. Instead of the entropy they use a simple sum of squares of second derivatives of the spectrum as a measure of the smoothness of the reconstruction. In this procedure they are quite closely following Rosenfeld and Kak [4] who have discussed the method in considerable detail in the context of imaging via NMR.

C

References

1. Belton PS, Wright KM (1986) J. Magn. Reson. 68: 564
2. Martin ML, Delpeuch J-J, Martin GJ (1980) In: Practical NMR spectroscopy, Heyden, London, p 137
3. Laue ED, Skilling J, Staunton J, Sibisi S, Brereton RG (1985) J. Magn. Reson. 62: 437
4. Rosenfeld A, Kak AC (1982) In: Digital picture processing, 2nd edn, Academic, New York, vol 1, p 294

CD and ORD

CD stands for **C**ircular **D**ichroism and ORD for **O**ptical **R**otatory **D**ispersion

ORD and CD are the most important chiroptical methods. Variations, as a function of wavelength, of the optical rotation angle and the ellipticity angle as well as of the related quantities have been called *Optical Rotatory Dispersion* and *Circular Dichroism*, respectively (see ORD for definition).

ORD and CD curves are reported by plotting specific or molar optical rotation and molar ellipticity, or $\Delta\varepsilon$, the *molar coefficient of dichroic absorption*, against the wavelength or frequency.

An optically active chemical compound gives rise to a curve which shows a characteristic feature in the region of the isotropic absorption band known as the *COTTON effect* [1,2]. In the ideal case of an electronic transition well separated from all others of the same molecule, a *positive COTTON effect* will have a form as shown in Fig. 3.

A *positive COTTON effect* measured in the ORD has a positive optical rotation at longer wavelengths and a negative one at shorter wavelength values. $[\phi]_\lambda$, the *molar optical rotation*, decreases in absolute value with increase of the distance from the *peak* and from the so-called *trough* (see Fig. 3).

However, the effect of an optically active band is marked outside the region of the *COTTON effect*, thus producing overlapping contributions from different transitions in real cases [3]. Well outside the absorption region, the ORD curve

C

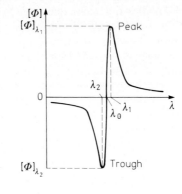

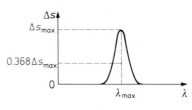

Fig. 3

can be described by the DRUDE equation [4,5] (1).

$$[\phi]_\lambda = \frac{K}{\lambda^2 - \lambda_0^2} \tag{1}$$

In Eq. (1), λ_0 stands for the *dispersion constant* being very close to the wavelength of the optically active transition under consideration and K for the so-called *rotation constant*. If several optically active transitions contribute to the optical rotation in a given spectral region, Eq. (1) becomes the more complex form Eq. (2).

$$[\phi]_\lambda = \sum_i K_i/(\lambda^2 - \lambda_i^2) \tag{2}$$

DRUDE's equation is not valid in the absorption region because $[\phi]_\lambda$, as observed, does not go to ∞ for $\lambda = \lambda_i$. In the absorption region therefore the modified equation (3) with a damping factor, G, is used.

$$[\phi] = \frac{K(\lambda^2 - \lambda_0^2)^2}{(\lambda^2 - \lambda_0^2)^2 + G\lambda^2} \tag{3}$$

The Eqs. (1), (2), and (3) have been commonly used in situations where the poor penetration of the commercially available instruments does not allow measurements in the *COTTON effect* region. But, the equations are becoming more and more obsolete, due to the considerable improvement in instrumentation [6].

In ORD curves the COTTON effect can be well characterized by the parameter a, the *amplitude*, and b, the *breath*:

$$a = \frac{[\phi]_{\lambda_1} - [\phi]_{\lambda_2}}{100} \qquad (4)$$

$$b = \lambda_1 - \lambda_2 \qquad (5)$$

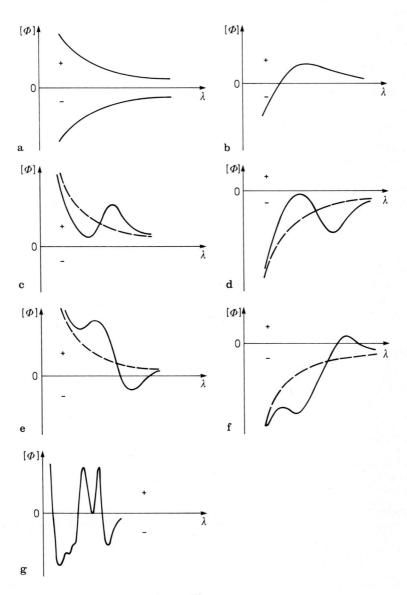

Fig. 4

The shape of a *positive COTTON effect* in the CD spectrum can be seen in Fig. 4. In order to avoid possible confusion with absorption spectra, the reader will note, that the term *peak* is commonly used for a maximum and the name *trough* for a minimum [7].

A CD band is usually described in terms of four parameters [3]:

– the wavelength of the maximum CD, λ_{max},
– the maximum value of dichroic absorption or of molar ellipticity ($\Delta\varepsilon_{max}$ or $[\theta]_{max}$, respectively),
– the half-width of the band, Δ, and
– the rotational strength, R, which derives from the area under the CD band [8].

For further details, the interested reader is referred to the literature [9].

Figure 4 shows seven ORD curves which are, in practice, more familiar to people dealing with *optical activity* problems. Here, we will divide ORD curves into the following groups:

– *Plain* (or normal) *curves* if the absolute value of the rotatory power increases regularly with decreasing wavelength, λ, as shown in Fig. 4a.
– *Complex* (or anomalous) *curves* if they exhibit changes in sign, maxima or minima (so-called pseudo-extrema), or inflection points arising from a superimposition of plain curves of different sign (see Fig. 4b).
– *COTTON effects curves* in cases of anomalies associated with the presence of optically active absorption bands. Curves of the type c–f in Fig. 4 have been called *single COTTON effect curves and their combinations multiple COTTON effect curves* as shown in Fig. 4g.

For a critical comparison between ORD and CD, the interested reader is referred to the literature [9]. Theoretical predictions and empirically found relationships for a lot of chromophores and their application possibilities have been very well documented [9–11].

References

1. Cotton A (1885) C.R.H. Acad. Sci. 120: 989 and 1044
2. Mitchell S (1933) The Cotton effect, Bell, London
3. Moscowitz A (1960) In: Djerassi C (ed) Optical rotatory dispersion, McGraw-Hill, New York, p 150
4. Velluz L, Legrand M, Grosjean M (1965) Optical circular dichroism, Verlag Chemie and Academic, London
5. Heller W (1958) J. Phys. Chem. 62: 1569
6. see 1. c. 9 and 12.
7. Djerassi C, Klyne W: Proc. Chem. Soc. 1957: 55
8. Moffitt W, Moscowitz A (1959) J. Phys. 30: 648
9. Ciardelli F, Salvadori P (eds) (1973) Fundamental aspects and recent developments in optical rotatory dispersion and circular dichroism, Heyden, London
10. Snatzke G (ed) (1965) Optical rotatory dispersion and circular dichroism in organic chemistry (Including applications from inorganic chemistry and biochemistry), Proceedings of NATO Summer School, Bonn/Germany
11. a) Minkin VI (1977) Dipole moments and stereochemistry of organic compounds—Selected applications. In: Kagan HB (ed) Stereochemistry: Fundamentals and methods, Georg Thieme, Stuttgart, vol 2 p 2
    b) Legrand M, Rougier MJ (1977) Application of the optical activity to stereochemical determinations. In Kagan HB (ed) Stereochemistry: Fundamentals and methods, Georg Thieme, Stuttgart, vol 2 p 33

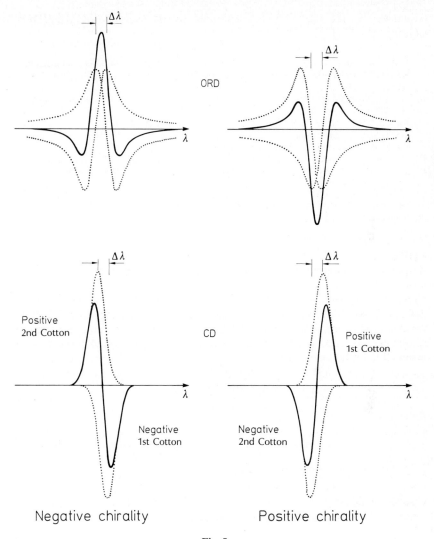

ORD

CD

Positive
2nd Cotton

Negative
1st Cotton

Positive
1st Cotton

Negative
2nd Cotton

Negative chirality

Positive chirality

Fig. 5

Further progress in the field was reported by Mason in 1979 [12]. Later on the use of the allylic benzoate method [13–15] or the application of a thiobenzoate chromophore [16] was described. Application of CD to oligosaccharides [17] and flower pigments [18] has been established. *Pars pro toto* we should remember here the determination of the absolute configuration of brevetoxin B, a red tide toxin, in 1981 [19].

Chiral exciton coupling and CD spectra or the so-called exciton chirality method (see ECM) brought new insights in the field of chiroptical properties [20].

Following this concept, Fig. 5 shows summation curves of ORD and CD spectra (symbolized by the solid lines) of two *COTTON effects* of opposite signs (visualized by the broken lines) separated by the well-known DAVIDOV's splittings, $\Delta\lambda$ [21].

12. Mason SF (ed) (1979) Optical activity and chiral discrimination, D. Reidel, Dordrecht
13. Harada N et al. (1978) J. Am. Chem. Soc. 100: 8022; Harada N et al. (1980) 102: 501, 506; (1981) 103: 5590
14. Naya Y, Yoshihara K, Iwashita T, Komura H, Nakanishi K, Hata Y (1981) J. Am. Chem. Soc. 103: 7009
15. Rastetter WH, Adams J, Bordner J (1982) Tetrahedron Lett. 23: 1319
16. Gawronski J, Gawronska K, Wynberg H: J. Chem. Soc., Chem. Commun. 1981: 307
17. Liu H-W, Nakanishi K (1981) J. Am. Chem. Soc. 103: 7005
18. Hoshino T, Matsumuto U, Harada N, Goto T (1981) Tetrahedron Lett. 22: 3621
19. Lin YY, Risk M, Ray SM, Engen DV, Clardy J, Golik J, James JC, Nakanishi K (1981) J. Am. Chem. Soc. 103: 6773
20. Harada N, Nakanishi K (1983) Circular dichroic spectroscopy-exciton coupling in organic chemistry, University Science Books, Mill Valley
21. Davidov AS (1962) Theory of molecular excitons, Mc Graw-Hill, New York

C

CDRE
Acronym for **C**onvolution **D**ifference **R**esolution **E**nhancement

CDRE is a special form of a method which has so far been called convolution difference (see CD) and used in nuclear magnetic resonance (see NMR) for resolution enhancement purposes. Practical guidelines for obtaining optimum results in *quantitative* FOURIER transform NMR (see FT-NMR) spectroscopy using CDRE [1,2] have been given recently [3]. Use of the optimum parameter set will diminish the errors in integral intensity and line position measurement caused by only poorly resolved NMR signals [3].

References
1. Campbell ID, Dobson CM, Williams RJ, Xavier A (1973) J. Magn. Reson. 11: 172
2. N.J. FT-80a Program, Research Programs for CFT-20, FT-80, and FT-80a NMR Spectrometer Systems, Varian Assoc. Palo Alto/USA, Pub. No. 87-172-617
3. Kupka T, Dziegielewski JO (1988) Magn. Reson. Chem. 26: 353

CHESS Imaging
Acronym for **C**hemical-**S**hift **S**elective **Imaging**

An important aim of spatially resolved biomedical nuclear magnetic resonance (see NMR and NMRI) is to produce spectroscopic images, i.e. to measure high-resolution spectra for each image element. Several attempts in this direction have been proposed in the literature. One class of experiments is based on CHESS imaging of predefined spectral components [1,2]. CHESS imaging has been shown to produce well-separated "fat" and "water" images useful in medical diagnostics [3]. Alternative methods have been published [4–10], some of them are explained under the acronyms FLASH and SPLASH Imaging in this dictionary.

References
1. Haase A, Frahm J, Hänicke W, Matthaei D (1985) Phys. Med. Biol. 30: 341
2. Haase A, Frahm J (1985) J. Magn. Reson. 64: 94

3. Matthaei D, Frahm J, Haase A, Schuster R, Bomsdorf H (1985) Lancet 2: 893
4. Maudsley AA, Hilal SK, Perman WH, Simon HE (1983) J. Magn. 51: 147
5. Dixon WT (1984) Radiology 153: 189
6. Park HW, Cho ZH (1986) Magn. Reson. Med. 3: 488
7. Haase A, Frahm J, Matthaei D, Merboldt KD, Hänicke W (1986) J. Magn. Reson. 67: 258
8. Ströbel B, Ratzel D (1986) Book of abstracts, 5th Annual Meeting of the Society of Magnetic Resonance in Medicine, Montreal, p 664
9. Haase A, Matthaei D (1987) J. Magn. Reson. 71: 550
10. Frahm J (1987) J. Magn. Reson. 71: 568

CHORTLE
Acronym for **C**arbon-**H**ydrogen Correlations from **O**ne-Dimensional Polarization-**T**ransfer Spectra by **Le**ast-Squares Analysis

The CHORTLE procedure has been introduced by G.A. Pearson [1] in order to get high-accuracy proton/carbon-13 chemical shift (see CS) correlations from one-dimensional polarization-transfer (see PT) carbon-13 nuclear magnetic resonance (see NMR) spectra. This technique has the speed of one-dimensional NMR experiments (see 1D-NMR), but yields information results equivalent to those of 2D-NMR (see p 7) experiments with a huge data matrix.

Reference
1. Pearson GA (1985) J. Magn. Reson. 64: 487

CIDEP
Acronym for **C**hemically **I**nduced **D**ynamic **E**lectron **P**olarization

A phenomenon in electron spin resonance (see ESR or EPR) like CIDNP in nuclear magnetic resonance (see CIDNP, CIDKP and PANIC). The effect was investigated first by Paul and Fischer [1] in the field of reduction and oxidation processes of acids.
    Interpretation of the phenomenon is possible using the Kaptein/Adrian theory or by the Glarum/Marshall model, a simple theory for the hydrogen atom alone.

Reference
1. Paul H, Fischer H (1970) Naturforsch Z. 25a: 443

CIDKP
German abbreviation for **C**hemisch **I**nduzierte **D**ynamische **K**ernpolarisation (see CIDNP), introduced by Bargon and Fischer [1].

Reference
1. Bargon J, Fischer H (1967) Naturforsch Z. 22a: 1556

CIDNP
Acronym for **C**hemically **I**nduced **D**ynamic **N**uclear **P**olarization

A phenomenon that was investigated originally by registering NMR spectra during radical reactions. In NMR one finds so-called emission lines or amplified absorption signals or a mixture of both. The effect was found in 1967 by Bargon, Fischer, and Johnson and Ward and Lawler independently. Interpretation of the effect was, in the beginning, a very rough analogy between dynamic nuclear polarization (see DNP) and the OVERHAUSER-effect (see OE). In the meantime new basic theoretical concepts (see CKO theory) have brought a more and more rational understanding of the phenomenon. Today CIDNP is used in studies of reaction mechanisms involving radicals. CIDNP-effects have been studied via ^{1}H-, ^{13}C-, ^{15}N- and ^{31}P-spectra [1].

Reference

1. Richard C, Granger P (1974) In: Diehl P, Fluck E, Kosfeld R (eds) Chemically induced dynamic nuclear and electron polarizations: CIDNP and CIDEP, Springer, Berlin Heidelberg New York (NMR-basic principles and progress, vol 8)

C

CIDNP-COSY

Acronym for **C**hemically **I**nduced **D**ynamic **N**uclear **P**olarization-**C**orrelation Spectroscopy

The CIDNP-COSY experiment was suggested [1,2] and later achieved [3] enabling the separation of net polarization and multiplet effect in coupled spin systems. For further details see the acronyms CIDNP and 2D-CIDNP in this dictionary.

References

1. Bodenhausen G (1981) Prog. NMR Spectrosc. 14: 137
2. Allouche A, Marinelli F, Pouzard G (1983) J. Magn. Reson. 53: 65
3. Boelens R, Podoplelov A, Kaptein R (1986) J. Magn. Reson. 69: 116

CJCP

Abbreviation of **C**orrected coupled **J** **C**ross-**P**olarization Spectra

In nuclear magnetic resonance (see NMR) a modified JCP (see p 152) pulse sequence has been presented [1], this purges coupled NMR spectra of both phase and multipled anomalies or artifacts by restoring genuine multiplet patterns. In this connection, it has been demonstrated [1] that a linear combination of the free induction decays (see FID's) resulting from the so-called PCJCP (see p 210) sequence and the modified JCP sequence, called *CJCP*, can lead to natural multiplet structures when spin-1 nuclei are the polarization source, as well.

In addition, it may be mentioned here that the author [1] has found a SHRIMP (see p 255) sequence [2,3] to respond to purging in a quite similar fashion.

References

1. Chandrakumar N (1985) J. Magn. Reson. 63: 202
2. Weitekamp DP, Garbow JR, Pines A (1982) J. Chem. Phys. 77: 2870
3. Braunschweiler L, Ernst RR (1983) J. Magn. Reson. 53: 521

CKO theory
Abbreviation for Closs-Kaptein-Oosterhoff theory

A theory involving radical pairs as important species in chemically induced dynamic nuclear polarization (see CIDNP or PANIC) which can explain qualitative and quantitative effects such as emission and enhanced absorption in NMR spectra registrated during radical reactions [1–2].

References
1. Closs GL (1969) J. Am. Chem. Soc. 91: 4552
2. Kaptein R, Oosterhoff JL (1969) Chem. Phys. Letters 4: 195; (1969) Chem. Phys. Letters 4: 214

C

COCONOSY
Acronym for Combined COSY-NOESY Experiments used in Nuclear Magnetic Resonance

A number of intriguing new pulse sequences require the acquisition of more than one free induction decay (see FID) during the course of the pulse sequence. The most well-known of these is the so-called COCONOSY experiment proposed by Haasnoot and co-workers in 1984 [1]. This experiment is simply a NOESY experiment (see NOESY) in which the first part of the mixing time is used to collect data for a normal COSY procedure (see COSY). Two separate sets of FID's are collected and are both processed individually to produce a COSY spectrum, showing the correlations between those protons which are spin-coupled to one another, and a NOESY spectrum, showing the correlations between hydrogens which are exchanging magnetization via either chemical exchange or nuclear OVERHAUSER enhancement (see NOE).

Reference
1. Haasnoot CAG, van de Ven FJM, Hilbers CW (1984) J. Magn. Reson. 56: 343

COLOC
Acronym for Correlation Spectroscopy for Long-Range Couplings

COLOC was introduced by Kessler and co-workers [1–3] as an important two-dimensional nuclear magnetic resonance technique (see 2D-NMR) for the elucidation of molecular constitution via long-range couplings to protons when signals overlap. But, in the meantime we have learned that this kind of technique has an application in the field of heteronuclear long-range couplings, too [4,5].

References
1. Kessler H, Griesinger C, Zarbock J, Loosli HR (1984) J. Magn. Reson. 57: 331
2. Kessler H, Bermel W, Griesinger C (1985) J. Am. Chem. Soc. 107: 1083
3. Kessler H, Griesinger C, Lautz J (1984) Angew. Chem. Int. Ed. Engl. 23: 444

4. Aue WP, Karhan J, Ernst RR (1976) J. Chem. Phys. 64: 4226
5. Wagner G, Wüthrich K, Tschesche H (1978) Eur. J. Biochem. 86: 67

COLOC-S

Acronym for a modified COLOC experiment (see COLOC), where S stands for Selective

The original COLOC experiment of Kessler and co-workers [1] modifies the basic C–H correlation sequence by introducing 180° pulses which are stepped incrementally through a fixed proton-evolution time, thus effecting broad-band decoupling (see BB) in the F_1 frequency domain.

The sequence proposed by R. Freeman and his group [2] uses a probe for adjacent nuclei with a gyration operator (see TANGO) and bilinear rotation decoupling (see BIRD) pulse in order to discriminate between directly bound protons and long-range coupled protons.

The sequence proposed by Reynolds et al. [3] applies two BIRD pulses, and is, in principle, superior to the COLOC [1] and Freeman's sequences [2], at least vor ^2J (CH) correlations. This pulse sequence also discriminate between $^2J(CH)$ and $^3J(CH)$ correlations via the involved vicinal proton-proton couplings.

Recently [4] a simple modification of Freeman's sequence [2] has been developed and designated by the authors as COLOC-S (COLOC-selective), which provides rapid long-range-correlation two-dimensional nuclear magnetic resonance spectra (see 2D-NMR).

Reference

1. Kessler H, Griesinger C, Zarbock J, Loosli HR (1984) J. Magn. Reson. 57: 331
2. Bauer C, Freeman R, Wimperis S (1984) J. Magn. Reson. 58: 526
3. Reynolds WF, Hughs DW, Perpick-Dumont M, Enriquez RG (1985) J. Magn. Reson. 63: 413
4. Krishnamurthy VV, Casid JE (1987) Magn. Reson. Chem. 25: 837

COMARO

Acronym for **C**omposite **M**agic-**A**ngle **Ro**tation

This new method belongs to broadband heteronuclear decoupling techniques in the presence of homonuclear dipolar and quadrupolar interactions and was introduced [1] for nuclear magnetic resonance (see NMR) purposes of spin-1 and spin-1/2 in solids and liquids crystals.

The pulse sequences are windowless and perform net rotations in the spin space about magic-angle axes.

COMARO-2 and COMARO-4 are modifications of the basic COMARO pulse sequence, where the numbers 2 and 4 refer to the number of net rotation axes combined in the sequence cycle [1].

Reference

1. Schenker KV, Suter D, Pines A (1987) J. Magn. Reson. 73: 99

composite pulses

is the abbreviation for special radio-frequency pulses used in Nuclear
Magnetic Resonance (see NMR) Spectroscopy and Imaging (see NMRI)

In recent years, methods have been developed which involve replacing conventional pulses by so-called "composite pulses", clusters of two or more square
pulses whose lengths (flip angle) and phases are chosen to correct for the well-
known pulse imperfections [1–11]. The design for such composite pulses was at
first intuitive, but assisted by computer simulation of the paths of the magnetization vectors during the sequence of pulses [1–5]. Then more sophisticated
techniques were developed, using recursive expansions [6,7], "reversed nutation
pulses" [8], average HAMILTONian theory [9] (see the HAMILTONian),
MAGNUS expansions [10], and fixed-point analyses [11], which went some way
toward designing composite pulse sequences according to specific performance
requirements. A design procedure, originally suggested by Levitt [2], is to allow a
computer program to find the combination of component pulse lengths and
phases which will best satisfy a given experimental requirement, for example, to
achieve greater than 99% inversion efficiency over a stated range of resonance
offset and radio-frequency field strength values. A very similar approach has been
used successfully in the design of selective radio-frequency pulses for NMR
imaging [12]. Recently [13] a numerical procedure has been presented for the
design of composite pulses according to specific experimental requirements.
Removing arbitrary restrictions on the pulse parameters allows the design of more
efficient composite pulses with fewer components and shorter overall length than
before. New dual-compensating composite inversion pulses with three to seven
components have been presented and their tolerance to setting-up errors discussed [13]. Composite radio-frequency pulses designed in this way should prove
useful in high-resolution NMR spectroscopy and especially in NMR imaging
where dual compensation is frequently required, but where the overall pulse
duration must be kept to a minimum in order to avoid excessive power deposition.

Recently [14] composite pulses have been described by a multipole theory:
From the results for a pure pulse from the multipole theory of NMR, it is possible
to get general analytical expressions for the decomposition of a single pulse into a
product of a number of constituent pulses. These pulses, which are represented as
WIGNER's rotation matrices, have the angles as functions of the off-resonance
frequency and the radio-frequency amplitude. By multiplying 3×3 matrices n
times it is possible to generate suitable analytical expressions for n composite
pulses describing the three components of the spin magnetization vector.

Recently [15] Shaka and Pines derived sequences of new composite pulses
which can provide rotations of arbitrary flip angles in the presence of large
resonance offset effects. These symmetric sequences use only 180° phase shifts, and
will have definitely the same symmetry properties as a single radio-frequency
pulse. In the case of two-level systems, these composite pulses behave like ideal
radio-frequency pulses, making them of potential use in a wide variety of
experimental situations in NMR spectroscopy according to the authors [15].

References

1. Levitt MH, Freeman R (1979) J. Magn. Reson. 33: 473
2. Freeman R, Kempsell SP, Levitt MH (1980) J. Magn. Reson. 38: 453

3. Levitt MH, Freeman R (1981) J. Magn. Reson. 43: 65
4. Levitt MH (1982) J. Magn. Reson. 48: 234
5. Levitt MH (1982) J. Magn. Reson. 50: 95
6. Levitt MH (1983) J. Magn. Reson. 55: 247
7. Shaka AJ, Freeman R (1984) J. Magn. Reson. 59: 169
8. Shaka AJ, Freeman R (1983) J. Magn. Reson. 55: 487
9. Tycko R (1983) Phys. Rev. Lett. 51: 775
10. Tycko R, Cho HM, Schneider E, Pines A (1985) J. Magn. Reson. 61: 90
11. Tycko R, Pines A, Guckenheimer J (1985) J. Chem. Phys. 83: 2775
12. Lurie DJ (1985) Magn. Reson. Imaging 3: 235
13. Lurie DJ (1986) J. Magn. Reson. 70: 11
14. Sanctuary BC, Cole HBR (1987) J. Magn. Reson. 71: 106
15. Shaka AJ, Pines A (1987) J. Magn. Reson. 71: 496

C

CONTIN
Abbreviation of a Computer Program for **Contin**uum Analysis in Spectroscopy

Continuum analysis in spectroscopy may be carried out with a FORTRAN (see FORTRAN) computer program, called CONTIN, developed by S.W. Provencher [1,2]. This is a generalized program capable of solving equations such as Eq. (1).

$$y_k = \int_{T_{min}}^{T_{max}} F_k(t_k, T)S(T)dT + \beta + \varepsilon_k \tag{1}$$

where y_k is the input data measured at times t_k, F_k is a known function (the so-called kernel), β corresponds to a constant baseline term, and S itself is the solution to be calculated. Recently [3] continuum analysis by using CONTIN was reported in the case of biological NMR (see NMR) relaxation data. Applied to biological relaxation data, the continuum analysis gives broad distributions of relaxation times with a structure that has never been described before [3]. Coming back to Eq. (1), ε_k, are the unknown error or noise components. In the biologists' application [3] of CONTIN the exponential kernel $F_k = \exp(-t_k/T_3)$ is used, where T stands for either T_1 or T_2 (see T_1 and T_2 as relaxation symbols). For most $F_k(t_k, T)$, estimating a unique S (T) in Eq. (1) is an ill-posed problem because there will exist a large set, Ω, of possible solutions all satisfying Eq. (1) to within the experimental error ε_k. Additionally, members of Ω may have large deviations from each other and therefore from the true solution, too. The main problem to be solved, then, is not the "solution" of Eq. (1), but the selection of a uniquely appropriate member in the total set Ω of solutions. The basic idea used by the CONTIN program to reduce the size of Ω is the *principle of parsimony*. This principle demands that, of all the members of the Ω ensemble that have been eliminated by logical constraints imposed by the inherent nature of the problem (e.g. S(T)>0, and S(T)=0 for $T<T_{min}$ and for $T>T_{max}$), the most appropriate solutions will be the simplest ones, i.e. the ones that reveal the fewest details or the least information that were not already known or expected. The same principle is the well-known basis of the maximum entropy method (see MEM). MEM has been used in many applications of data analysis [4]. Please note that this *principle of parsimony* has served science for nearly five centuries under the motto *Pluralitis non est ponenda sine necessitate* [5].

References

1. Provencher SW (1982) Comput. Phys. Commun. 27: 213
2. Provencher SW (1982) Comput. Phys. Commun. 27: 229
3. Kroeker RA, Henkelman RM (1986) J. Magn. Reson. 69: 218
4. Jaynes ET (1982) Proc. IEEE 70: 939
5. William of Occam (1488) Summa totius logicae, Paris

COSMIC

Abbreviation for **C**omputer **O**riented **S**tructure **M**odels from **I**nadequate derived **C**oupling Data

COSMIC is a typical PASCAL program (see PASCAL), written by Varian Associates, Palo Alto/USA [1] in order to automate analysis of INADEQUATE spectra (see INADEQUATE) to get carbon-carbon-connectivities in an one-dimensional procedure. In the meantime two-dimensional applications (see 2D-INADEQUATE) have become much more advantageous.

Reference

1. Ammann W, Richarz R, Wirthlin T (1982) Varian instruments at work, Z-12.

COSS

Acronym for **C**orrelation with **S**hift **S**caling

Like S. COSY (see p 242) COSS is a recently proposed pulse scheme for achieving chemical shift (CS) scaling along the f_1 axis of a two-dimensional correlated nuclear magnetic resonance (see 2D-NMR) spectrum [1]. Both the pulse sequences are simple modifications of the so-called *constant time experiments* [2,3] and have been designed for resolution enhancement purposes in correlated NMR spectroscopy.

Much more recently the combined use of COSS and S. COSY has been proposed for the recording of pure-phase shift-scaled NMR spectra [4] and can be best understood using the product operator formalism (see POF) developed by Sørensen et al. [5].

References

1. Hosur RV, Ravikumar M, Sheth A (1985) J. Magn. Reson. 65: 375
2. Bax A, Freeman R (1981) J. Magn. Reson. 44: 542
3. Rance M, Wagner G, Sørensen OW, Wüthrich K, Ernst RR (1984) J. Magn. Reson. 59: 250
4. Hosur RV, Sheth A, Majumdar A (1988) J. Magn. Reson. 76: 218
5. Sørensen OW, Eich GW, Levitt MH, Bodenhausen G, Ernst RR (1983) Prog. NMR Spectrosc. 16: 163

COSY

Acronym for **C**orrelation **S**pectroscopy

COSY is one of the most important two-dimensional techniques in nuclear magnetic resonance (see 2D-NMR). J. Jeener was the first, in 1971, to propose the idea of 2D FOURIER transformation (see FT and FFT) as a function of two time variables, yielding a 2D-NMR spectrum (see p 10) as a function of two frequency variables [1].

Because of the different kinds of COSY they will be explained under their individual names, acronyms, and abbreviations in this dictionary.

References

1. Jeener J (1971) Ampère International Summer School, Basko Polje, Yugoslavia

COSY-LR

Acronym for **C**orrelation **S**pectroscopy with **L**ong-**r**ange (coupling interactions)

COSY-LR is a special kind of two-dimensional δ–δ correlated nuclear magnetic resonance spectroscopy with the incorporation of fixed delays in the pulse sequence to optimize the detection of weak couplings. It has been shown that COSY and COSY-LR can lead to the discrimination between different pathways of couplings across single bonds: ^{4}J through coplanar of *gauche* arrangement of the bonds and ^{5}J through a *zig-zag* pathway or bond proximity [1]. Applications to configurational and conformational analysis have been presented [1].

Reference

1. Platzer N, Goasdoue N, Davoust D (1987) Magn. Reson. Chem. 25: 311

CP

Abbreviation of **C**ross-**P**olarization

CP is a very important technique in nuclear magnetic resonance (see NMR) of solids. The introduction of the cross-polarization method made it possible to overcome the low sensitivity of dilute spins, and provided a means of observing the NMR signals of those nuclei in solids [1].

CP is normally combined with high-power decoupling and magic angle spinning (see CP/MAS) in solid-state NMR.

However, CP cannot readily be applied to nuclei with low gyromagnetic ratios, such as, ^{57}Fe, ^{103}Rh, ^{187}Os, etc., because to achieve the HARTMANN-HAHN condition [2] (see HAHA) with those nuclei, very high values of the radio-frequency power are required.

A. Bax and co-workers [3] have proposed an interesting new technique utilizing off-resonance effects [4] which effectively reduce the radio-frequency power requirements for CP. But, in that method radio-frequency inhomogeneity can easily cause a mismatch of the HAHA condition, and the technique itself seems to be difficult to apply in practice.

Another method of CP using multiple-pulse irradiation of the I spins was proposed in 1984 by Quiroga and Virlett [5]. These authers have used a multipulse sequence, called HW-8 [6], to perform two conflicting demands simultaneously: (i) to suppress the homonuclear dipolar interaction, and (ii) to provide a spin-locking field for the I spins of the system. As we know, a perfectly tuned WAHUHA-4 [7] type homonuclear-decoupling pulse sequence (see WAHUHA and WAHUHA-4) cannot provide the I spin-locking field, one has to introduce the so-called pulse length "error" into the pulse sequence without destroying the homonuclear-decoupling effect. Therefore it is difficult to control the I spin-locking field strength provided by this pulse length "error".

Recently [8] an alternative method, called CP-TAPF, has been published (see p 57).

References

1. Pines A, Gibby MG, Waugh JS (1973) J. Chem. Phys. 59: 569
2. Hartmann SR, Hahn EL (1962) Phys. Rev. 128: 2042
3. Bax A, Hawkins BL, Maciel GE (1984) J. Magn. Reson. 59: 530
4. Bleich HE, Redfield AG (1977) J. Chem. Phys. 67: 5040
5. Quiroga L, Virlet J (1984) J. Chem. Phys. 81: 4774
6. Haeberlen U, Waugh JS (1968) Phys. Rev. 175: 453
7. Waugh JS, Huber LM, Haeberlen U (1968) Phys. Rev. Lett. 20: 180
8. Takegoshi K, McDowell CA (1986) J. Magn. Reson. 67: 356

CP-ENDOR
Acronym for **C**ircular **P**olarized **E**lectron **N**uclear **D**ouble **R**esonance

This new method in electron spin or paramagnetic resonance (see ESR or EPR) is in its application not limited to so-called ordered spin systems. The procedure may be used in simplification of electron spin resonance spectra of polycrystaline samples or radicals in solution, too [1]. It has been readily shown, that when using circular-polarized radio-frequency fields, transition probabilities of ENDOR lines (see ENDOR) will be dependent on the chirality of the effective radio-frequency field [2–4].

References

1. Schweiger A (1986) Chimia 40: 111
2. Schweiger A, Günthard HH (1981) Mol. Phys. 42: 283
3. Forrer J, Schweiger A, Berchten N, Günthard HH (1981) J. Phys. E. 14: 565
4. Schweiger A, Günthard HH (1982) Chem. Phys. 70: 1

CP/MAS
Abbreviation of **C**ross-**P**olarization/**M**agic **A**ngle **S**pinning

The combination of cross-polarization (see CP) and magic angle spinning (see MAS or MASS) became the most important method (when using high-power dipolar decoupling or gated high-power decoupling—GHPD) in obtaining high-resolution carbon-13 nuclear magnetic resonance spectra of solids.

Ultrahigh, stable speed spinning (i.e. $v_R > 5$ kHz) is a desirable technique in multinuclear CP/MAS experiments on high-field NMR spectrometers to average effectively the chemical-shift anisotropy (see CSA) and for several other reasons. For example, for carbon-13 the number of spinning sidebands increases with increasing magnetic field and thus reduces the sensitivity and complicates the interpretation of crowded spectra. Another important gain of ultrahigh spinning rates is observed for quadrupolar nuclei, as for example in aluminium-27 MAS studies of natural alumino-silicates and mixed-layered clay minerals [1]. Furthermore, for signal-to-noise ratio (see SNR) reasons, it is necessary to combine a large sample volume with high v_R frequencies in high-field multinuclear CP/MAS NMR studies.

Jakobsen and his group [1] have developed a special CP/MAS probe and a spinner assembly for a 7 mm (outer diameter) cylindrical rotor which routinely permits spinning speeds up to at least 10 kHz with a stability of a few Hz, using ordinary air for the bearing and drive. The air pressure in this assembly needs to be increased from only about 1 to 2 bar for all spinning rates. According to a paper published by Doty and Ellis [2] they calculated and tested a so-called maximum safespeed for the rotor machined from different materials such as Macor, Al_2O_3, and Si_3N_4 with samples of different densities. Today, this solid-state equipment developed by Jakobsen [1] is a commercially available product of Varian Assoc., Palo Alto/USA.

A somewhat different approach in high-speed MAS with v_R equal to 23 kHz has been achieved by use of a scaled-down version [3] of the MAS system developed at Delft University [4]. The only additional modifications besides the physical scaling down are (a) separation of the pressure for the drive and carrier gas and (b) use of helium or helium/air mixtures instead of normal air for driving purposes. As rotor materials have been tested Torlon (a polyamide-imide material manufactured by Amoco) and Kel-F (from 3M Corporation). $(CH_3)_3Si(CH_2)_3SO_3Na$, abbreviated DSS, has been used for referencing. Solid-state high-speed MAS spectra of fluorine-19 at 338.7 MHz, of hydrogen-1 at 360 MHz, and of cadmium-113 at 79.8 MHz have been reported [3].

Variable-temperature NMR spectroscopy with cross-polarization and magic-angle spinning (VT CP/MAS) is an important technique for the study of structure, dynamics, reactivity [5], and magnetic properties [6] in the solid state. The problem of temperature measurement and control is not trivial and has been considered in detail recently [7]. By use of the new developed technique of CP/MAS NMR chemical shift thermometry [8], the various factors contributing to the uncertainty in sample temperature have been identified and assessed [7]. The important factors demonstrated in this study [7] include

– radio-frequency heating of the sample,
– mismatches in the drive- and bearing-gas channel temperatures,
– sample temperature equilibration times, and
– JOULE/THOMPSON cooling and heating.

Techniques for the measurement of these contributions and the resulting temperature gradients have been reported. The authors [7] have demonstrated techniques that allow the uncertainty in sample temperature to be reduced to approximately ±1K, an improvement of nearly one order of magnitude.

Although general in scope, this study [7] emphasizes the temperature range from 77 K to ambient, a range encompassing most of the VT CP/MAS studies in the chemical literature so far.

As a facet of this detail study, the design of an improved VT accessory, which provides greater reliability, convenience, and performance, has been reported, too [7].

References

1. Langer V, Daugaard P, Jakobsen HJ (1986) J. Magn. Reson. 70: 472
2. Doty FD, Ellis PD (1981) Rev. Sci. Instrum. 62: 1868
3. Dec SF, Wind RA, Maciel GE, Anthonio FE (1986) J. Magn. Reson. 70: 355
4. Wind RA, Anthonio FE, Duijvestijn MJ, Schmidt J, Trommel J, de Vette GMC (1983) J. Magn. Reson. 52: 424
5. Lyerla JR, Yannoni CS, Fyfe CA (1982) Acc. Chem. Res. 15: 208
6. Haw JF, Campbell GC (1986) J. Magn. Reson. 66: 558
7. Haw JF, Campbell GC, Crosby RC (1986) Anal. Chem. 58: 3172
8. Campbell GC, Crosby RC, Haw JF (1986) J. Magn. Reson. 69: 191

CPMG

Abbreviation of an important pulse sequence used in nuclear magnetic resonance (see NMR) with the following author's names: C = Carr, P = Purcell, M = Meiboom, and G = Gill

In 1954, H.Y. Carr and E.M. Purcell [1] published a pulse sequence which was modified in 1958 by S. Meiboom and D. Gill [2]. Later on the combination of both sequences became very important in NMR spectroscopy [3], especially as a spin-echo experiment for measurements of spin-spin relaxation times (T_2). The CPMG spin-echo sequence may be written schematically with Eq. (1).

$$D1 - P1 - (D2 - P2_{PH2} - D2)_n - FID \tag{1}$$

Abbreviations in Eq. (1) have the following meanings:

D1   a relaxation delay,
P1 $90°$ radio-frequency pulse,
D2   a fixed delay,
P2 $180°$ radio-frequency pulse,
PH2   = 90, 90, 90, 90, 180, 180, 180, 180,

(n must be even; D2 = approx. 1 msec to eliminate J-modulation and diffusion effects) phase cycle, FID = free induction decay.

References

1. Carr HY, Purcell EM (1954) Phys. Rev. 94: 630
2. Meiboom S, Gill D (1958) Rev. Sci. Instrum. 29: 688
3. Freeman R, Hill HDW (1975) In: Jackman LM, Cotton FA (eds) Dynamic nuclear magnetic resonance spectroscopy, Academic, New York, p 136

CP-T

Abbreviation of the Carr and Purcell-Technique

Carr and Purcell showed in 1954, that a simple modification of the famous *Hahn spin-echo* experiment [1] (see SE and FID) efficiently reduces the effect of diffusion on the determination of T_2, the spin-spin relaxation time, in nuclear magnetic resonance (see NMR) spectroscopy [2]. CP-T may be characterized as a

$90°, \tau, 180°, 2\tau, 180°, 2\tau, 180°, 2\tau, \ldots$ sequence
(or, more commonly, a *Carr-Purcell sequence*).

For further details the interested reader is referred to a critically written monograph [3].

References

1. Hahn EL (1950) Phys. Rev. 80: 580
2. Carr HY, Purcell EM (1954) Phys. Rev. 94: 630
3. Farrar C, Becker ED (1971) Pulse and Fourier transform NMR: Introduction to theory and methods, Academic, New York, Sect. 2.4

C

CP-TAPF

Acronym standing for Cross-Polarization using the Time-Averaged Precession Frequency

CP-TAPF is a simple technique [1] to reduce radio-frequency power requirements for magnetization transfer experiments (see MT) in solid-state nuclear magnetic resonance (see NMR) with cross-polarization (see CP).

Reference

1. Takegoshi K, McDowell CA (1986) J. Magn. Reson. 67: 356

CRAMPS

Acronym for Combined Rotation and Multiple-Pulse Spectroscopy

CRAMPS [1] became an interesting technique in nuclear magnetic resonance (see NMR) of solid polymers [2]. CRAMPS has been used in 8- or 24-pulse cycles for proton magnetic resonance especially at higher magnetic fields. Because chemical shifts scale linearly with the static magnetic field, the resolution at higher fields can be accordingly greater. But, the use of very high frequencies in applying CRAMPS requires a relatively uniform B_1 field, which implies a ratio much greater than one for the length to width of the inductor. As the frequency increases, the inductance used for tuning purposes decreases for reasonable values of tuning capacitors, which means that the length to width ratio of the inductor becomes less favorable for a fixed coil diameter. One can solve this problem by using doubly wound inductances [3].

Proton NMR spectroscopy studies using CRAMPS have been reported for poly(methyl methacrylate) and isotactic poly(styrene) in comparison to 4,4'-dimethylbenzophene and 2,6-dimethylbenzoic acid [2], for example.

References

1. Gerstein BC, Pembleton RG, Wilson RC, Ryan LM (1977) J. Chem. Phys. 66: 361
2. Gerstein BC (1986) High-resolution solid-state NMR of protons in polymers. In: Komoroski RA (ed) High resolution NMR spectroscopy of synthetic polymers in bulk, VCH Publishers, Deerfield/USA
3. Fry CG, Iwamiya JT, Apple TM, Gerstein BC (1985) J. Magn. Reson. 63: 214

CS

Acronym for **C**hemical **S**hift used in Nuclear Magnetic Resonance (see NMR) and in Nuclear Quadrupole Resonance (see NQR), too

The significance of NMR spectroscopy is based on its ability to distinguish a particular nucleus with respect to its environment in a given molecule. That is, one observes that the resonance frequency of an individual nucleus is influenced by the distribution of electrons in the chemical bonds of the molecule under investigation. The distinct value of the resonance frequency of a particular nucleus such as a proton is therefore dependent upon molecular structure. This effect is known as the chemical shift of the resonance frequency or more simply as the *chemical shift*, or CS.

C

CS is related to the so-called *shielding* or *screening constant*. σ, given with Eq. (1).

$$B_{local} = B_0(1 - \sigma) \tag{1}$$

Equation (1) describes the following situations: The applied magnetic field, B_0, induces circulations in the "cloud" of electrons surrounding the nucleus such that, following LENZ's rule, a magnetic moment μ, opposed to B_0, is produced. Thus, the local at the given nucleus is smaller than the applied field.

Any discussion or interpretation of the differential shielding of individual nuclei assumes that a system of measurement for the chemical shift has been established.

In order to define the position of a resonance signal in the NMR spectrum it would principally possible to measure the strength of the external magnetic field, B_0, or the corresponding resonance frequency, v, at which the signal of interest appears. These parameters are, however, not suited for the characterization of CS because NMR spectrometers operate at different B_0 fields (1.4, 2.3, 5.2 T, etc.) and the resonance frequency differs with the field strength. In addition, an absolute determination of the field strength and/or resonance frequency is technically difficult to realize.

Hence, one is obliged to determine the resonance position relative to that of a reference compound or standard. For proton, carbon-13, and silicon-29 NMR spectroscopy tetramethyl silane (see TMS) is the commonly used reference compound.

With Eq. (2) we can define a *dimensionless quantity* for CS:

$$\delta = \frac{v_{substance} - v_{reference}}{v_0} \tag{2}$$

In Eq. (2), v_0 stands for the spectrometer frequency (for example 60 MHz) and as unit for the δ-*scale* of CS one uses therefore parts per million, abbreviated with *ppm*.

Pars pro toto we will summarize here the origins of *proton chemical shifts*:

a) the influence of the electron density at the proton,
b) the influence of the electron density at the neighbouring carbons atoms,
c) the influence of induced magnetic moments of neighbouring atoms and bonds,
d) ring current effects of cyclic conjugated π-systems,
e) electric field effects of polar groups,

f) dia- and paramagnetic anisotropy effects,
g) the influence of van der WAAL's effects,
h) the influence of hydrogen bonding,
i) isotopic effects,
j) medial effects.

Multi-nuclear referencing of NMR spectra seems to be a problem so far.

For 2 years the IUPAC I-5 (Spectroscopy) Commission has been considering the need for updating the recommendations (1975) [1] on reference standards for NMR purposes.

Apart from the few suggestions which have been received by P. Granger, the Commission has had to rely on scarce papers related to certain nuclei and the common literature with the more or less accepted references, which are not necessarily the best ones [2] as shown for instance in the case of oxygen-17 [3].

As referencing also depends on several other parameters [4] and on the methods applied, we have to discuss the referencing methods themselves and to propose more precise procedures [2].

This early stage of very high resolution NMR (see UHRNMR) [5] will lose part of its interest if no precise measurements of the chemical shifts are available [2].

References

1. a) (1976) Pure Appl. Chem. 45: 219;
   b) (1978) Org. Magn. Reson. 11: 267
2. Granger P (1988) Magn. Reson. Chem. 26: 179
3. Gerothanassis IP, Lauterwein J (1986) Magn. Reson. Chem. 24: 1034
4. Granger P (1987) Eur. Spectrosc. News No. 72: 22
5. Allerhand A, Maple SR (1987) Anal. Chem. 59: 441A

CSA
Acronym for Chemical Shift Anisotropy

CSA is one of the four different mechanisms dominating carbon-13 spin-lattice relaxation in nuclear magnetic resonance (see NMR). This kind of relaxation is caused by the modulation of an anisotropic nuclear screening tensor by the BROWNian motion. Equation (1) has been formulated [1] for the chemical shift anisotropy relaxation rate in the case of an axially symmetric system:

$$1/T_1^{CSA} = 2/15\gamma_c^2\, H_0^2(\sigma_{||} - \sigma_\perp)^2\, \tau_c \tag{1}$$

In Eq. (1), $\sigma_{||}$ and $\sigma_\perp$ are the components of the chemical shift tensor parallel and perpendicular to the external magnetic axis respectively. A noteworthy property of the CSA relaxation rate is its dependence upon the square of the external magnetic field. Detection of this mechanism is thus accomplished by measurements at two different magnetic field strengths. Furthermore, T_1^{CSA} is governed by the same correlation time (τ_c) as T_1^{DD} (see DD for the dipole-dipole relaxation mechanism). CSA seems to be a relatively ineffective mechanism and it has been shown to become a substantial contributor only if, for some reasons, both DD and spin-rotation (see SR) are insignificant [2]: non-protonated carbons at low temperatures and high magnetic fields. In the so-called extreme narrowing limit $(\omega_0 \cdot \tau_c \ll 1)$ $T_1^{CSA}/T_2^{CSA} = 7/6$ is valid.

References

1. Carrington A, McLachlan AD (1967) Introduction to magnetic resonance, Harper and Row, New York, chap 11
2. An illustrative example of anisotropic overall motion has been described by Levy GC, White DM, Anet FAL (1972) J. Magn. Reson. 6: 453

CSA
Abbreviation of Chiral Solvating Agent

Mislow and Raban [1] proposed in 1967 and Pirkle [2] experimentally demonstrated that chiral solute enantiomers exhibit different nuclear magnetic resonance (see NMR) spectra when dissolved in non-racemic chiral solvent. Since that time, many examples of the application of *chiral solvents* and other diamagnetic *chiral solvating agents* (CSA's) for NMR studies of chiral solutes (Pirkle and Hoover, 1982) have appeared in the Reference [3].

The principal applications of CSA's are:

- determinations of enantiomeric purity,
- simple demonstrations of chirality,
- determinations of absolute configuration, and
- determinations of enantiomerization kinetics.

An excellent comprehensive review of NMR applications of CSA's appeared in 1982 [4]. This review is especially authoritative in light of the fact that Pirkle and co-workers carried out a substantial portion of the experimental work done in this field.

In 1983, G.R. Weisman [5] published a paper dealing with NMR analysis using CSA's. The purpose of this paper [5] was to acquaint organic chemists with the principles and practical use of CSA's for NMR determination of enantiomeric purity and for demonstration of chirality. With an aim toward providing a concise and useful reference source and complementing the review of Pirkle and Hoover [4], a compilation of CSA/NMR applications which had been reported, categorized according to chiral solute functional groups, was included in this paper [5]. The most common CSA's are of the structural types **1** and **2** (see below) and they have been used successfully in $CDCl_3$, CD_2Cl_2, CCl_4, CS_2, and C_6D_6 as solvents. C_6D_6 has been recommended as the solvent of choice for **1c**, and pyridine has been useful for salts formed by the reaction of phosphorus acids with CSA's of type **2**.

Normally, lower sample temperatures will favor association, and several examples for increasing the relative difference of chemical shifts (see CS) with lowering the temperature have been reported [6,7].

Proton Nuclear Magnetic Resonance (PNMR) was, in the past, the method of choice for CSA application because of its ubiquity and high sensitivity. PNMR became more and more popular in this area after introduction of the *pulse FOURIER transform mode* (see PFT) and the use of superconducting solenoids in high-field NMR. For example, recently [8] we learned, that the determination of the enantiomeric purity of tertiary amines may be possible under using α-methoxy-α-(trifluoromethyl)phenylacetic acid as CSA in PNMR [8].

In the meantime PNMR has to compete with different kinds of *heteronuclear* magnetic resonance. The following hetero-nuclei have been used in the field of

$$\begin{array}{c} CH_3 \\ | \\ Ar-CH-N \diagdown^{R}_{R'} \end{array}$$

2

$$\begin{array}{c} CF_3 \\ | \\ Ar-CH-OH \end{array}$$

1

	Ar	R	R'
a	Ph	H	H
b	!-Naphthyl	H	H
c	Ph	Me	Me
d	1-Naphthyl	Me	Me
e	Ph	H	Me
f	p-NO$_2$-Ph	H	H
g	Ph	H	CH$_2$Ph
h	2-Naphthyl	H	H

	Ar
a	Ph
b	1-Naphthyl
c	9-Anthryl
d	9-(10-Methylanthryl)
e	9-(10-Bromoanthryl)

CSA/NMR applications:

– carbon-13, fluorine-19, nitrogen-15,
– and phosphorus-31.

Modern PFT measurements of these nuclei under broad-band decoupling (see BB) and different manipulations of the free induction decay (see FID) are today used in addition to proton resonance.

References

1. Mislow K, Raban M (1967) Top. Stereochem. 1: 1
2. Pirkle WH (1966) J. Am. Chem. Soc. 88: 1837
3. see the literature compilation in ref. [5].
4. Pirkle WH, Hoover DJ (1982) Top. Stereochem. 13: 263
5. Weisman GR (1983) Nuclear magnetic resonance analysis using chiral solvating agents. In: Morrison JA (ed) Asymmetric synthesis, Academic, New York, vol 1 (Analytical Methods) chap 8
6. Pirkle WH, Burlingame TG, Tetrahedron Lett. 1967: 4039.
7. Pirkle WH, Hoekstra MS (1975) J. Magn. Reson. 18: 396
8. Villani FJ Jr, Costanzo MJ, Inners RR, Sutter MS, McClure DE (1986) J. Org. Chem. 51: 3715

CT
Abbreviation of **C**oherence **T**ransfer used in Nuclear Magnetic Resonance (see NMR)

For the correlation of NMR resonances within spin systems—homo- and heteronuclear—most two-dimensional experiments (see COSY, MQF-COSY, and RELAY) rely upon the process of CT. Single-quantum coherences (magnetizations) can be transferred between coupled nuclei or into multi-quantum coherences. These experiments, which include polarization transfer (see PT) experiments (see INEPT and DEPT), can be analyzed as combinations of CT processes and periods of free precession. In particular, two-dimensional NMR (see 2D-NMR) provides a direct demonstration [1–3] of the different pathways of CT [4] induced by a mixing process.

CT has been rationalized in terms of so-called *coherence transfer selection rules* [5,6] which enable one to predict "*allowed*" and "*forbidden*" transfer processes in various NMR situations such as 2D correlation spectroscopy (see COSY) [2,3], multiple-quantum filtering [6–8], and multiple-quantum spectroscopy (see MQS) [2,5].

Recently [9] relaxation-induced violations of the above mentioned rules in NMR have been discussed and analyzed in detail.

A theoretical analysis of homonuclear rotating-frame CT in liquids has been presented recently [1]. A general procedure has been developed to describe CT under the influence of a phase-modulated radio-frequency field. The mechanism of CT has been discussed in detail and compared with other well-known homonuclear correlation techniques in NMR [10].

References

1. Ernst RR, Bodenhausen G, Wokaun A (1987) Principles of NMR in one and two dimensions, Clarendon, Oxford
2. Aue WP, Bartholdi E, Ernst RR (1976) J. Chem. Phys. 64: 2229
3. Bax A (1982) Two-dimensional nuclear magnetic resonance in liquids, Delft University Press, Dordrecht
4. Bodenhausen G, Kogler H, Ernst RR (1984) J. Magn. Reson. 58: 370
5. Braunschweiler L, Bodenhausen G, Ernst RR (1983) Mol. Phys. 48: 535
6. Piantini V, Sørensen OW, Ernst RR (1982) J. Am. Chem. Soc. 104: 6800
7. Shaka AJ, Freeman R (1983) J. Magn. Reson. 51: 169
8. Müller N, Ernst RR, Wüthrich K (1986) J. Am. Chem. Soc. 108: 6482
9. Müller N, Bodenhausen G, Ernst RR (1987) J. Magn. Reson. 75: 297
10. Bazzo R, Boyd J (1987) J. Magn. Reson. 75: 452

CVA

Abbreviation of Characteristic Vector Analysis

Characteristic Vector Analysis may be a useful approach for understanding carbon-13 nuclear magnetic resonance spectra. The well-known set of carbon-13 NMR substituent chemical shifts (see SCS) of 43 monosubstituted cyclohexanes has been used recently in CVA [1]. All variability in the carbon-13 spectra generated by different substituents and their configuration could be explained by three orthogonal characteristic vectors [1]. Experimental SCS values are thus linear combinations of three independent phenomena or effects acting at each carbon atom simultaneously with varying degrees of intensity. The largest contribution, described by V_1, n is associated with the polarizing power of the substituent through the involved σ bonds. The contribution to SCS by V_2, n is due to a field-inductive effect of the substituent. A contribution associated with V_3, n is significant only for C–γ and C–δ and results in an upfield or downfield shift effect in relation to the conformational energy of the substituent [1]. The mathematical foundation of CVA is well-known and established [2–8].

References

1. Zalewski RI (1986) J. Chem. Soc., Perkin Trans. II: 495
2. Anderson TW (1958) Introduction to multivariate analysis, Wiley, New York
3. Kendall MG (1965) A course in multivariate analysis, Griffin, London
4. Press SJ (1973) Applied multivariate analysis, Holt, Rinehard, and Winston, New York
5. Mather PM (1976) Computational methods of multivariate analysis in physical geography, Wiley, New York
6. Simonds JL (1963) J. Opt. Soc. Am. 53: 968
7. Woud H (1974) Ph.D. Thesis, McGill University of Montreal
8. Malinowski ER, Hovery DG (1980) Factor analysis in chemistry, Wiley-Interscience, New York

CW
Abbreviation of Continuous Wave

CW is typical for sequential excitation in nuclear magnetic resonance (see NMR) and electron spin resonance (see ESR), too. CW-NMR has been defined [1] as a form of high-resolution NMR in which nuclei of different field/frequency ratio at resonance are successively excited by sweeping (see SWEEP) the magnetic field or the radio-frequency. Today, most of the modern NMR spectrometers are in operation with pulse FOURIER transform (see PFT) instead of CW.

C

Reference
1. This definition was specified under the jurisdiction of the ASTM Committee E-13 on Molecular Spectroscopy.

CW-ENDOR
Acronym for Continuous Wave-Electron Nuclear Double Resonance

The CW method of ENDOR (see p 87) is the conventional procedure used in electron spin or electron paramagnetic resonance (see ESR or EPR) spectroscopy for simplification or assignment purposes.

But, in some cases, one has to deal with low-frequency nuclear spin transitions which are not easily accessible with CW-ENDOR techniques. Although spectacular improvements have been accomplished in this area [1–4], the electron spin-echo envelope modulation (see ESEEM) has became more and more favored [5].

References
1. Kirmse R, Abram U, Böttcher R (1982) Chem. Phys. Lett. 90: 9
2. Kirmse R, Stach J, Abram U, Dietzsch W, Böttcher R, Gribnau MCM, Keijzers CP (1985) Inorg. Chem. 23: 3333
3. Böttcher R, Kirmse R, Stach J, Keijzers CP (1985) Mol. Phys. 55: 1431
4. Böttcher R, Kirmse R, Stach J, Reijerse EJ, Keijzers CP (1986) Chem. Phys. 107: 145
5. Reijerse EJ, Keijzers CP (1987) J. Magn. Reson. 71: 83

CYCLOPS
Acronym for Cyclically Ordered Phase Sequence

CYCLOPS is a special sequence which was introduced in the field of nuclear magnetic resonance (see NMR) in 1975 [1] as a result of an analysis of the critical factors in the design of sensitive high-resolution NMR spectrometers. The analysis dealt with factors which influence the performance of FOURIER transform (see FT and FFT) NMR spectrometers including field inhomogeneity, probe design, transient circuit behaviour, so-called JOHNSON's noise, non-linear analysis, phase-sensitive detection in quadrature, and signal processing, too. For more details, the interested reader is referred to ref. [1].

Reference
1. Hoult DI, Richards RE (1975) Proc. Royal. Soc. (London), A. 344: 311

DAISY
Acronym for **D**üsseldorfer **A**nalyse- und **I**terationssystem

DAISY was developed at the University of Düsseldorf by G. Hägele and co-workers and has been documentated recently [1] in a monograph like a manual which describes different methods for an automated analysis and simulation of nuclear magnetic resonance (see NMR) spectra. DAISY consists of a FORTRAN (see p 114) package of seven main programs for the automated analysis of high-resolution NMR spectra. DAISY permits not only the analysis of spectra arising from nuclei with spin 1/2 but also analysis of spectra which involve nuclei with higher spin quantum numbers and spin systems with dipolar coupling such as complex spectra of organic molecules oriented in liquid-crystalline phases.

D

The authors [1] have reviewed the history of computer-assisted analysis of NMR spectra (see LAOCOON, NMRIT/NMREN, DAVINS, etc.) and have discussed the advantages of DAISY, especially from the view of computational efficiency.

However, H.H. Limbach [2] recently pointed out that a book or manual in this field written in English would be more advantageous.

References
1. Hägele G, Engelhardt M, Boenigk W (1987) Simulation und automatisierte Analyse von Kernresonanzspektren, VCH-Verlagsgesellschaft, Weinheim
2. Limbach HH (1988) Nachr. Chem. Techn. Lab. 36: 288

DANTE
Acronym for **D**elays **A**lternating with **N**utations for **T**ailored **E**xcitation

Selective excitation of individual carbon-13 resonance lines by using DANTE pulse sequences permits a "deconvolution" of overlapping multiplets in carbon-13 spectra under gated decoupling (see GD). Using DANTE pulse sequence it is possible to saturate unwanted signals (such as solvent signals; for solvent suppression techniques see SST), too. The method is useful in the case of selective irradiation of a proton signal under direct observation of a given different nucleus, i.e. not a proton.

References
1. Morris GA, Freemann R (1978) J. Magn. Res. 29: 433
2. for its application see the Journal 'Material und Strukturanalyse' (April 1982) in connection with the JEOL PG-200 pulse programmer (see JEOL)

DAVINS
Acronym for **D**irect **A**nalysis of **V**ery **I**ntricate **N**MR **S**pectra

DAVINS is the name of a program for automated analysis of high-resolution nuclear magnetic resonance specta [1,2]. Principles and computational strategy

[1] on one hand, illustrative applications of the computer program on the other [2], have been published. After working out the quantum statics of nuclear spin systems in its final form by the mid-1950's, a plethora of programs for generating theoretical NMR spectra from an ambient set of chemical shifts and coupling constants was developed (see LAOCOON) culminating in Hägele's CYMTRY (see ref [3]), which takes maximum advantage of simplifications arising from all operators commuting with the nuclear spin HAMILTONian (see p 122) and which has been applied to a system of spins consisting of 22 spins-1/2 (for a precursor of this program, see ref. [4]). The problem of NMR spectral analysis or synthesis (sometimes somewhat inappropriately called simulation) may hence be regarded as solved, not only in principle but also in practice [1], but it does not lend itself to such a final solution, because the crucial obstacles are not of a formal mathematical nature [1]. The currently most widely used analysis programs [5] are all based on the assumption that, from judiciously guessed parameters, one can synthesize a theoretical trial spectrum, corresponding to the experimental spectrum closely enough so that a sufficient number of energy levels or transitions can be assigned for a subsequent iterative refinement of the NMR parameters. In the author's opinion this a tribute to the ingenuity, intuition, experience, diligence, and not infrequently also the luck of the practitioners of this "art", that a large number of quite complicated NMR spectra could successfully be analyzed with this kind of approach [1]. The authors [1]: "A general iterative method for analyzing high-resolution NMR spectra for chemical shifts and isotropic coupling constants is presented that differs from conventional procedures in that it does not depend on the assignment of energy levels or transitions; makes use of the full information content of a spectrum; and is in principle, and in the great majority of cases also in practice, capable of arriving at the correct solution from starting parameters chosen randomly within liberally specified boundaries. The basic equations are derived from a generalized formulation of the least-squares problem. One is led to an ordered sequence of correlation matrices W_k in data space, whose off-diagonal elements provide a general mechanism for smoothing out the multitude of false local minima on the error hypersurface of the ordinary least-squares method. The conditions to be satisfied by the W_k matrices to establish a unique pathway to the global minimum are stated in a formally rigorous way. It is demonstrated that although specific information about the W_k matrices is not accessible in practice, and paradoxically only because of this, one may conclude that they must be symmetric band matrices with diagnol elements $w_{ii} = 1$ and off-diagnol elements w_{ii} decaying monotonically with absolute distance from the principal diagonal and satisfying the condition $0 \leq w_{ii} < 1$. The characteristics of an efficient computational strategy based on this formalism are outlined."

References

1. Stephenson DS, Binsch G (1980) J. Magn. Reson. 37: 395
2. Stephenson DS, Binsch G (1980) J. Magn. Reson. 37: 409
3. Hägele G (1978) paper presented at the 5th Meeting of the German NMR Discussion Group, Ettal, September 1978; Lueg V (1977) Doctoral dissertation, University of Düsseldorf
4. Lueg V, Hägele G (1977) J. Magn. Res. 26: 505
5. Diehl P, Kellerhals H, Lustig E (1972) In: Diehl P, Fluck E, Kosfeld R (eds) NMR: Basic principles and progress. Springer, Berlin Heidelberg New York, vol 6

DD
Acronym for **D**ipole-**D**ipole Interaction

Relaxation by magnetic dipole-dipole interactions is the most important of the four different mechanisms dominating carbon-13 spin-lattice relaxation in nuclear magnetic resonance (see NMR).

Spin-lattice relaxation involves energy exchange between the spin system and its environment, the lattice. Eventually a nuclear spin in its excited state will return to the ground state by dissipating its excess energy to the lattice. Since spontaneous emission is insignificant in NMR, such transitions occur through interaction with the lattice. This stipulates the presence of an oscillating magnetic field at the site of the nucleus. A further requirement is that the frequency of the local magnetic field, H_{loc}, be equal to the LARMOR frequency of the nucleus in question. Most frequently, these fields are generated by magnetic dipoles of nearby magnetic nuclei such as a proton attached to the carbon-13 in question. The magnetic field produced at the site of a nucleus j by a nucleus i is given by Eq. (1) [1].

$$H_{loc} = \pm \mu_i \frac{3\cos^2\theta_{ij} - 1}{r_{ij}^3} \tag{1}$$

In Eq. (1), μ_i represents the magnetic dipole moment of nucleus i, r_{ij} is the internuclear distance, and θ_{ij} is the angle between r_{ij} and the external field H_o. From Eq. (1), it is seen that H_{loc} rapidly decreases with increasing separation of the two interacting nuclei. In order to derive the general equation for dipolar relaxation in a simple two-spin system defined by the two nuclei carbon-13 and proton, let us have a look on their energy level diagram, given in Fig. 6 [2].

In Fig. 6 the transition probabilities W_{1H} and W_{1C} refer to transitions of a proton and a carbon-13 spin, respectively, while W_2 and W_0 denote the transition probabilities for a double and zero quantum jump. Quantum-mechanical treatment of H_{loc} leads to a close relationship between the spectral density function, $J(\omega_{ij})$, and the transition probability, W_{ij} [2]:

$$W_{ij} = F_{ij}^2 J_m(\omega_{ij}) \tag{2}$$

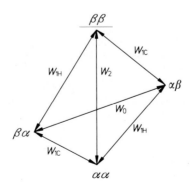

Fig. 6

Equations (3–6) are the solutions from a calculation of the matrix elements F_{ij} accounting for the dipolar interactions of the spin states involved; they are corresponding to the four transition probabilities [2]:

$$W_0 = \left(\frac{1}{20}\right) K^2 J_0(\omega_H - \omega_C) = \left(\frac{1}{10}\right) K^2 \frac{\tau_c}{1 + (\omega_H - \omega_C)^2 \tau_c^2} \tag{3}$$

$$W_{1C} = \left(\frac{3}{40}\right) K^2 J_1(\omega_C) = \left(\frac{3}{20}\right) K^2 \frac{\tau_c}{1 + \omega_C^2 \tau_c^2} \tag{4}$$

$$W_{1H} = \left(\frac{3}{40}\right) K^2 J_1(\omega_H) = \left(\frac{3}{20}\right) K^2 \frac{\tau_c}{1 + \omega_H^2 \tau_c^2} \tag{5}$$

$$W_2 = \left(\frac{3}{10}\right) K^2 J_2(\omega_H + \omega_C) = \left(\frac{3}{5}\right) K^2 \frac{\tau_c}{1 + (\omega_H + \omega_C)^2 \tau_c^2} \tag{6}$$

The subscript in J of these equations above indicates the change in the total magnetic quantum number upon transition, K is equal to $\gamma_H \gamma_C \hbar r_{CH}^{-3}$.

It has been shown [1,3] that in the absence of an irradiating field at the proton resonance frequency, the rate of change of the longitudinal magnetization, $M_z(t)$, is governed by the two different time constants, given with Eqs. (7) and (8):

$$T_{1C} = (W_0 + 2W_{1C} + W_2)^{-1} \tag{7}$$

$$T_{1CH} = (W_2 - W_0)^{-1} \tag{8}$$

Spin-lattice relaxation becomes now independent of the coupling term, T_{1CH}, only at saturation of the proton transitions and can therefore be described by a single time constant $T_1^{DD} = T_{1C}$. The theoretical prediction of different recovery rates in the proton-coupled and decoupled case has been verified experimentally [4]. Equation (9), a combination of equations (3–7), describes the rate for

$$1/T_1^{DD} = \frac{\hbar^2 \gamma_H^2 \gamma_C^2}{20 r_{CH}^6} \{J_0(\omega_H - \omega_C) + 3J_1(\omega_C) + 6J_2(\omega_H + \omega_C)\} \tag{9}$$

dipolar relaxation of a carbon-13 nucleus by a single proton.

For further details and for comparing the main characteristics of the four different mechanisms relevant the carbon-13 relaxation see ref. [5].

References

1. Lyerla JR Jr, Grant DM (1972) In: McDowell CA (ed) Int. Rev. Science Phys. Chem. Series, Medical and Technical Publishing, Chicago, vol 4 chap 5
2. Kuhlmann KF, Grant DM, Harris RK (1970) J. Chem. Phys. 52: 3439
3. Lyerla JR Jr, Levy GC (1974) In: Levy GC (ed) Topics in carbon-13 nuclear magnetic resonance spectroscopy. Wiley-Interscience, New York, vol 1 chap 3
4. Campbell ID, Freeman R (1973) J. Chem. Phys. 58: 2666
5. Wehrli FW, Wirthlin T (1976) Interpretation of carbon-13 NMR spectra, chap 4 (Nuclear Spin Relaxation; F.W.W.), Heyden, London

DEFT

Acronym for **D**riven **E**quilibrium **FOURIER** **T**ransform

DEFT is used in nuclear magnetic resonance (see NMR) spectroscopy for a sensitivity enhancement by recycling of magnetization [1,2] and using a pulse sequence given by Eq. (1), where the signal is observed in τ-intervals, and where the last pulse restores part of the magnetization along the z-axis. DEFT is an alternative approach to the so-called spin-echo FOURIER transform (see SEFT).

$$\pi/2 - \tau - \pi - \tau - \pi/2 \tag{1}$$

Both DEFT and SEFT methods fail in the presence of homonuclear couplings and can only be used for dilute nuclei such as carbon-13.

References

1. Waugh JS (1970) J. Mol. Spectrosc. 35: 298
2. Becker ED, Ferretti JA, Farrar TC (1969) J. Am. Soc. 91: 7784

D

DEPT
Acronym standing for **D**istortionless **E**nhancement by **P**olarization **T**ransfer

A typical procedure used in NMR spectroscopy as an interpretation aid for carbon-13 NMR spectra in order to get a special editing corresponding to different carbon species with different numbers of attached protons (see APT). The method can be seen as an alternative to experiments such as INEPT (see p 143) or APT (see p 23).

Reference

1. Doddrell DM, Pegg DT, Bendall MR (1982) J. Magn. Reson. 48: 323

DEPT-INADEQUATE
Acronym standing for **D**istortionless **E**nhancement by **P**olarization **T**ransfer-**I**ncredible **N**atural **A**bundance **D**ouble **Qua**ntum **T**ransfer **E**xperiment

S.W. Sparks and P.D. Ellis [1] have presented a nuclear magnetic resonance (see NMR) hybrid experiment, named DEPT-INADEQUATE. In this experiment the polarization-transfer technique DEPT [2] has been employed to the INADEQUATE procedure [3] to enhance signal intensity. Similar in effectiveness to INEPT-INADEQUATE (see p 144), the DEPT-INADEQUATE experiment has been analyzed [1] to

a) quantitatively estimate transfer efficiencies,
b) assess the desirability of extensive phase-cycling schemes, and
c) determine methods of furthering the utility of polarization-transfer INADEQUATE experiments.

The authors [1] have calculated results for several spin systems and compared those with experimental observations on simple compounds. Effective enhancement relative to proton-decoupled INADEQUATE will be of a factor of 2 or greater, similar to the INEPT-INADEQUATE (see p 144) experiment. Polarization-transfer to non-protonated carbons using DEPT-INADEQUATE has been demonstrated. The relative peak intensities in INADEQUATE NMR spectra have been shown to give additional information for assignment purposes.

On the other hand, several experimental difficulties have been carefully discussed and solutions proposed which will help to suppress spectral artifacts and to maximize the efficiency of polarization-transfer [1].

References

1. Sparks SW, Ellis PD (1985) J. Magn. Reson. 62: 1
2. Doddrell DM, Pegg DT, Bendall MR (1982) J. Magn. Reson. 48: 323
3. Bax A, Freeman R, Frenkiel TA, Levitt MH (1981) J. Magn. Reson. 43: 478

DESPOT

D

Acronym for **D**riven-**E**quilibrium **S**ingle-**P**ulse **O**bservation of T_1

This is a new method among others in the field of nuclear magnetic resonance (see NMR) for the measurement of the spin-lattice relaxation time (T_1) which permits a very rapid determination of this interesting NMR parameter [1–3].

Until recently, there have been two principal difficulties with the convenient use of T_1 in studies of chemically related problems: The first relates to the devolution of measured T_1's into their intra- and intermolecular components [4–6]. The second concerns the excessive time requirements, especially on samples yielding low spectral signal-to-noise ratios (see SNR).

In fact, however, a procedure has been proposed [1] that does enable a tremendous reduction in experimental time. Nevertheless, despite attention having been refocused [2] on this method it does not appear to have found significant use in NMR. Recently [3] a reevaluation of the accuracy of this originally called "variable nutation angle method" [1] has been presented. Specifications of the criteria for its optimal application and present developments in a decrease of the necessary experimental time have been carefully discussed [3].

References

1. Christensen KA, Grant DM, Schulman EM, Walling C (1974) J. Phys. Chem. 78: 1971
2. Gupta RK (1977) J. Magn. Reson. 25: 231
3. Homer J, Beevers MS (1985) J. Magn. Reson. 63: 287
4. Homer J, Valdivieso Cedeno ER (1983) J. Chem. Soc., Transactions II, 79: 1021
5. Homer J, Valdivieso Cedeno ER (1984) J. Chem. Soc., Transactions II, 80, 375
6. Homer J, Valdivieso Cedeno ER (1983) Tetrahedron 39: 2847

Addendum

Recently, the conditions governing the most satisfactory implementation of the DESPOT method of measuring NMR spin-lattice relaxation times (T_1) have been discussed in detail [1]. The determination of the number of non-acquisitions, "dummy"-pulses, etc. required to drive a spin system to the DESPOT equilibrium conditions has been re-examined. Spectrometer pulse width-nutation angle calibration (see NNMR) and sample diffusion have been discussed, too.

Reference

1. Homer J, Roberts JK (1987) J. Magn. Reson. 74: 424

DFFT
Acronym for **D**ouble **F**orward **FOURIER T**ransformation

DFFT has been proposed for data reduction [1] in nuclear magnetic resonance operating in the pulse FOURIER transform (see PFT) mode. The procedure required for this data reduction is as follows:

The $N/2$ data points of the *absorption part*, supplemented by an equal number of zeros, are FOURIER transformed into the time domain by a second forward FOURIER transformation. This results in a time domain signal which consists of two "pseudo"-free induction decays (see FID), both with $N/2$ data points. Owing to the addition of $N/2$ zeros to the absorption part, both pseudo-FID's are not independent but are, except for the sign, related by a HILBERT transformation (see HT) as shown by Bartholdi and Ernst [2]. Hence, each pseudo-FID contains the same information about the absorption and one of them can now be stored for documentation, or manipulated by all common weighting functions. Simply by addition of $N/2$ zeros to the pseudo-FID and FT one regains the corresponding frequency spectrum. In spite of the two additional forward FT's the genuine absorption part of the phased spectrum can be reproduced almost perfectly.

Roth and Kimber [1] pointed out that a very interesting aspect is that, after discarding the original *dispersion part*, the whole approach yields a *new dispersion part*. In contrast to the genuine dispersion part calculated in the last step of the procedure is, except for sign, a HILBERT transform of the absorption part.

The described approach and its results show that, simply by two successive forward FT's, the amount of data can be reduced by a factor of two, without losing any relevant information or flexibility. In addition, a saving in computational time can also be achieved, because all calculations within parameter optimization for data manipulation in the time domain include only half of the original number of data points. This halving of the memory space and computing time could be especially advantageous if the DFFT procedure can be implemented into two-dimensional NMR spectroscopy (see 2D-NMR).

References
1. Roth K, Kimber BJ (1985) Magn. Reson. Chem. 23: 832
2. Bartholdi E, Ernst RR (1973) J. Magn. Reson. 11: 9

DFT
Acronym for **D**igital **FOURIER T**ransformation

The DFT method is one of the modelling methods used in non-linear spectral analysis for nuclear magnetic resonance (see NMR) spectra. Following R. Kaiser [1], three steps can be identified for these modelling methods:

a) Prior knowledge is used to develop a parametric control for the process producing a record in the time domain.
b) The model parameters are fitted to match data obtained from an observed time record.
c) The spectrum which the model would produce with the parameter values so determined is computed.

DFT can be interpreted in the same way [1]:

a) The model is a finite sum of non-interacting, undamped simple harmonic oscillators (see SHO) at equally spaced frequencies. The frequencies themselves are the harmonics of the duration of the time record. Parameters are the complex amplitudes of the oscillators.
b) The parameters are determined by a precise fitting procedure to the time record.
c) The spectrum itself corresponds to the sum of delta-functions of the model SHO's.

Reference

1. Kaiser R: paper presented at the 8th European ENC, June 3–6, 1986, Spa/Belgium.

D

DFTS
Acronym for **D**ispersive **FOURIER T**ransform **S**pectrometry

DFTS is a special technique used in infrared spectroscopy [1]. Normally the sample is held between a MICHELSON interferometer and the detector. But, in DFTS the sample is placed in one arm of the interferometer. Thus the optical pathlength is increased and the phase can shift dramatically. DFTS has been used for transmission and reflection measurements of solid samples, liquids, gases and vapors. A very large number of far-infrared DFTS measurements has been described in the research literature. A bibliography with more than 130 references in this field was published in 1979 [2]. It should be noted that very little work in the mid or near infrared has been described. But, here we will give some application notes typical for DFTS.

The precise knowledge of the refractive index spectrum of relevant atmospheric species in the far-infrared region is starting to become necessary in view of the interest in telecommunications, radar, and remote sensing using submillimeter and near-millimeter wavelengths, especially for the calculations of electromagnetic propagation through turbulent atmospheres. It may be easier to calculate the integrated absorption of a line from the refractive index dispersion across it for this purpose because of the need to measure the contributions from the wings of the line if the conventional absorption mode were to be used (3).

One of the most strongly absorbing liquids in the far-infrared region is water, and the refractive index and the absorption spectra of water have been measured via DFTS in the reflection made [4–7].

A very large number of studies of dispersive FOURIER transform spectra of solid samples, including alkali halides [8,9], cadmium telluride [10], polymers [11], quartz [12], soda-lime glass [13], silicon [14], wire grids [15,16], and even hydrated lysozyme [17], have been reported so far.

References

1. Bell EE (1971) Aspen Int. Conf. on Fourier Spectroscopy, 1970 (papers edited by G.A. Vanasse, A.T. Stair, and D.J. Baker), AFCRL-71–0019, p 71
2. Birch JR, Parker TJ (1979) Infrared Phys. 19: 201
3. Birch JR (1981) Proc. Soc. Photo-Opt. Instrum. Eng. 289: 363
4. Afsar MN, Hasted JB (1977) J. Opt. Soc. Am. 67: 902

5. Afsar MN, Hasted JB (1978) Infrared Phys. 18: 843
6. Birch JR, Bennouna M (1981) Infrared Phys. 21: 224
7. Bennouna M, Cachet H, Lestrade JC, Birch JR (1981) Proc. Soc. Photo-Opt. Instrum. Eng. 289: 266
8. Staal PR, Eldrige JE (1979) Infrared Phys. 19: 625
9. Memon A, Parker TJ (1981) Intl. J. Infrared and Millimeter Waves 2: 839
10. Parker TJ, Birch JR, Mork CL (1980) Sol. St. Comm. 36: 581
11. Birch JR, Dromey JD, Lesurf J (1981) Infrared Phys. 21: 225
12. Parker TJ, Ford JE, Chambers WG (1978) Infrared Phys. 18: 215
13. Birch JR, Cook RJ, Pardoe GWF (1979) Sol. St. Comm. 30: 693
14. Birch JR (1978) Infrared Phys. 18: 613
15. Mok CL, Chambers WG, Parker TJ, Costley AE (1979) Infrared Phys. 19: 437
16. Beunen JA, Costley AE, Neill GF, Mok CL, Parker TJ, Tait G (1981) J. Opt. Soc. Am. 71: 184
17. Golton I (1980) Ph.D. thesis, University of London

D

DICD
Acronym for **D**ispersion-**I**nduced **C**ircular **D**ichroism

The DICD effect has been established experimentally for some organic and inorganic chromophores [1–3]. Theoretical studies [4,5] of the DICD effect led to the suggestion that it could serve as a fundamental basis of a spectroscopic method complementary to the normal absorption for the spectral assignment of *magnetic dipole allowed transitions* of achiral species in solution. These expectations for both organic chromophores [6] and inorganic systems [7] have been consolidated in the meantime.

Recently a DICD and normal absorption study of the n-Π^* carbonyl transition for the isoelectronic substituent series $-CH_3$, $-NH_2$, and $-OH$ has been reported [8].

References
1. Bosnich B (1967) J. Am. Chem. Soc. 89: 6143
2. Hayward LD, Totty RN (1971) Can. J. Chem. 49: 624
3. Axelrod E, Bartl G, Bunnenberg E: Tetrahedron Lett. 1969: 5031
4. Schipper PE (1975) Mol. Phys. 29: 1705
5. Schipper PE (1981) Chem. Phys. 57: 105
6. Schipper PE, O'Brien JM, Ridley DD (1985) J. Phys. Chem. 89: 5805
7. Schipper PE, Rodger A (1985) J. Am. Chem. Soc. 107: 3459
8. Schipper PE, Rodger A (1988) Spectrochimica Acta 44a: 575

DIGGER
Acronym for **D**iscrete **I**solation from **G**radient-**G**overned **E**limination of **R**esonances

DIGGER is a new technique for *in vivo* volume-selected nuclear magnetic resonance (see NMR) spectroscopy [1]. The heart of this method is an extension of a somewhat older procedure [2] for performing water suppression *in vivo* which in turn is related to techniques developed for high-resolution NMR spectroscopy [3]. The method is, in principle, similar to the VSE procedure (see VSE). However, it does not rely upon phase cycling since a well defined z-slice is generated in a single gradient/pulse episode.

References

1. Doddrell DM, Bulsing JM, Galloway GJ, Brooks WM, Field J, Irving M, Baddeley H (1986) J. Magn. Reson. 70: 319
2. Doddrell DM, Galloway GJ, Brooks WM, Field J, Bulsing JM, Irving M, Baddeley H (1986) J. Magn. Reson. 70: 176
3. Bulsing JM, Doddrell DM (1986) J. Magn. Reson. 68: 52

DISCO

Acronym for **D**ifferences and **S**ums of Traces within **COSY**

This new technique in nuclear magnetic resonance (see NMR) has been suggested for the purpose of extracting accurate coupling constants from two-dimensional phase-sensitive homonuclear correlated spectra [1]. In comparison with conventional procedures for obtaining coupling constants, the advantages are as follows:

The "decoupling effect" achieved by DISCO allows the extraction of J values from multiplets with broad lines. In addition, the assignments of the coupling partners is given directly.

DISCO secures good results even in cases where double resonance methods (see DR or NMDR) are no longer applicable: effective conventional homonuclear decoupling is impossible if a broad multiplet has to be irradiated. The well-known criterion $\gamma B_2 \gg J$ is very critical [2,3].

All coupling constants can be extracted from a single COSY spectrum (see COSY). The decision as to which nuclei are to be "decoupled" can be taken after the measurement.

When difference decoupling is applied to multiplets with broad lines, the observed splittings are not identical with the coupling constants. Further, when DISCO is applied, the number of lines is reduced by a factor of two, but when applying difference decoupling it is increased by half of the number of lines of the corresponding multiplet. In the DISCO technique, firstly pseudo-one-dimensional spectra are created by co-addition of the individual traces of cross-peaks, and by a similar addition of the diagonal peaks in a spectrum which is corrected in phase to pure-absorption line shapes. The sum and the difference of the pseudo-one-dimensional spectra of a cross-peak and its diagonal peak will give two spectra, with a pattern in which the active coupling ceases to exist ("spin decoupling") but is shifted by half the value of the same. If diagonal peaks are not available owing to overlapping signals, sums and differences of the pseudo-one-dimensional spectra of two cross-peaks can be applied to simplify spectral patterns for the evaluation of values.

The DISCO procedure has been demonstrated with the example of the AMXY spin system of a tyrosine derivative, and the complex pattern of the Pro 8 spin system of the cyclic decapeptide antamanide [1].

References

1. Kessler H, Müller A, Oschkinat H (1985) Magn. Reson. Chem. 23: 844
2. Hoffman RA, Forsén S (1966) In: Emsley JW, Feeney J, Sutcliffe LH (eds) Prog. Nucl. Magn. Reson. Spectrosc. Pergamon Oxford, vol 1 p 15
3. Abragam A (1961) The principles of nuclear magnetism, Oxford University Press, London

DISPA
This acronym stands for a Plot of **Disp**ersion versus **A**bsorption in Nuclear Magnetic Resonance (see NMR)

DISPA [1] is a new method for evaluating static magnetic field inhomogeneity in an NMR sample. Deviation from circularity of a DISPA plot provides a measure of the deviation of an experimental NMR lineshape from an ideal LORENTZian lineshape. From DISPA plots constructed from spectra simulated via specified z-gradient (so-called "spinning shim"; see SHIM) inhomogeneity terms, it is clear that the magnitude and direction of displacement of an experimental DISPA curve from its reference circle can be used to discriminate and quantitate the inhomogeneity in each orders of z (z, z^2, z^3, z^4, and z^5). Like an index for monitoring field inhomogeneity, DISPA is much more sensitive than monitoring the absolute area under a free induction decay (see FID) (especially for higher order spinning shims), and much faster and more discriminating (between different z^n terms) than any iterative fitting procedure of the frequency-domain absorption spectrum. DISPA has already been proved for automated phasing of FOURIER transform NMR spectra.

It has been suggested that DISPA may offer a simple, rapid, sensitive method for testing shim coil designs and ultimately for manual or automated shimming of the static magnetic field of a superconducting solenoid as well.

Reference
1. Craig EC, Marshall AG (1986) J. Magn. Reson. 68: 283

DMRGE
a possible acronym for an approach recently (1) proposed and titled "**D**esign of **M**agnetic **R**esonance Experiments by **G**enetic **E**volution"

Ray Freeman and Wu Xili [1] have pointed out that technology has often benefitted from a careful study of natural processes. For the authors it was fascinating to see the adaptation of the principles of genetic evolution [2] to rather complex problems arising from aerodynamics and other engineering disciplines [3]. According to the famous DARWINian evolution theory, small variations caused by random mutation of one of the involved genes are favored if the progeny is thereby much better adapted to its environment, which is normally hostile.

A cumulative effect of a long sequence of such "desirable changes may be the driving force behind the evolution of all living organisms in nature. Random change is thereby converted into a direct variation by the process of natural selection.

The authors [1] have considered the application of these principles to the problem of designing new pulse sequences for nuclear magnetic resonance (see NMR) spectroscopy purposes.

In their opinion, a considerable amount of thought has been directed to the fundamental problem of exciting a spin system in NMR in a manner that is definitely not uniform over the frequency band under investigation.

The discussed different methods range from the simple selective pulse [4,5] through the so-called tailored excitation [6,7] and solvent suppression sequences [8–11] (see SST) to GAUSSian [12] pulses (see p 119) and sinc-function-shaped pulses [13]. The rational background for designing such pulse sequences is commonly based on the approximate FOURIER transform (see FT) relation between the time-domain pulse envelope (see FID) and the frequency-domain excitation pattern.

Intuition, trial-and-error methods, recursive expansion and computer approaches for optimization have also been employed so far. But, all these methods have one factor in common, that the final goal for the frequency-domain excitation is known a priori. In one way or another all these approaches are looking for a minimum in a multi dimensional hyperspace [1].

The interested reader is referred for further details, ideas, and application possibilities to the original paper [1].

References

1. Freeman R, Wu Xili (1987) J. Magn. Reson. 75: 184
2. Dawkins R (1986) The blind watchmaker, Longmans, New York
3. Rechenberg I (1973) Evolutionsstrategie: Optimierung Technischer Systeme nach Prinzipien der Biologischen Evolution, Frommann-Holzboog
4. Freeman R, Wittekock S (1969) J. Magn. Reson. 1: 238
5. Morris GA, Freeman R (1978) J. Magn. Reson. 29: 433
6. Tomlinson BL, Hill HDW (1973) J. Chem. Phys. 59: 1775
7. Freeman R, Hill HDW, Tomlinson BL, Hall LD (1974) J. Chem. Phys. 61: 4466
8. Redfield AG, Kunz SD, Ralph EK (1975) J. Magn. Reson. 19: 114
9. Turner DL (1983) J. Magn. Reson. 54: 146
10. Hore PJ (1983) J. Magn. Reson. 54: 539
11. Hore PJ (1983) J. Magn. Reson. 55: 283
12. Bauer CJ, Freeman R, Frenkiel T, Keeler J, Shaka AJ (1984) J. Magn. Reson. 58: 442
13. Mansfield PJ, Maudsley AA, Baines T (1976) J. Phys. Soc. E. Sci. Instrum. 9: 67

DMSO or DMSO-d$_6$

Abbreviation of **Dim**ethylsulf**o**xide, a commonly used solvent in its protonated or deuterated (DMSO-d$_6$) form in nuclear magnetic resonance (see NMR).

DNCP

Acronym for **D**ipolar-**N**arrowed **CARR-PURCELL** Sequence used in Solid-State Nuclear Magnetic Resonance (see NMR)

A pulse sequence in NMR that involves both the spin-echo (see SE) sequences and 180° refocusing pulses, incorporated into appropriate combinations of dipolar echo-sequences. It is capable of removing both dipolar and shielding interactions and leaving relaxation under multiple-pulse sequences as the major broadening interaction and has been named DNCP [1]. DNCP can be used to measure lifetimes under homonuclear dipolar decoupling in the field of solid polymers [1,2].

References

1. Dybowski CR, Pembleton FG (1979) J. Chem. Phys. 70: 1962
2. Gerstein BC (1986) High-resolution solid-state of protons in polymers. In: Komoroski R (ed) High resolution NMR spectroscopy of synthetic polymers in bulk, VCH Publishers, Deerfield/USA

DNMR
Acronym for **D**ynamic **N**uclear **M**agnetic **R**esonance

Nuclear magnetic resonance can be applied to study fast reversible reactions because the line shape of NMR signals is sensitive to chemical exchange processes if these affect the NMR parameters of the nucleus (proton, carbon-13, etc.) in question. The NMR spectra of many compounds are therefore temperature dependent. In the following the physical basis of this phenomenon, known today as DNMR, will be discussed and applications will be illustrated [1]. DNMR is a title of some well-established monographs [2,4,5] in the field of nuclear magnetic resonance.

For a quantitative description concerning the correlation between line shape on one hand and mechanism and kinetics of dynamic processes on the other, we must find a relationship of the life time of the protons (or other nuclei) in positions A and B and the line shape of the nuclear magnetic resonance signal. Following BLOCH's equations (see p 29), the shape of a resonance signal is a function of frequency, v, and transverse relaxation time, T_2. In the case of a dynamic process that causes a change of LARMOR frequencies $v_A \rightleftharpoons v_B$, the BLOCH's equations must be modified. In addition to the normal relaxation effects, time-dependent changes of magnetization at each site caused by the chemical equilibrium (1).

$$A \underset{k_B}{\overset{k_A}{\rightleftharpoons}} B \tag{1}$$

Therefore, the x, y-magnetization at site A, M_A is increased through arriving nuclei by an amount proportional to $k_B M_B$. On the other hand, because of the departing nuclei, it suffers a loss proportional to $k_A M_A$. An analogous situation is true for site B. The explicit consideration of these terms finally leads to Eq. (2) for the line shape $g(v)$ of the nuclear magnetic resonance signal

$$g(v) = [(1 + \tau\pi\Delta)P + QR]/(4\pi^2 P^2 + R^2) \tag{2}$$

where

$$P = (0.25\Delta^2 - v^2 + 0.25\delta v^2)\tau + \Delta/4\pi$$

$$Q = [-v - 0.5(\rho_A - \rho_B)\delta v]\tau$$

$$R = 0.5(\rho_A - \rho_B)\delta v - v(1 + 2\pi\tau\Delta)$$

in which τ represents $\tau_A \tau_B/(\tau_A + \tau_B)$. The other symbols above have the following meanings:

τ_A resp. τ_B — average life times [s] of nuclei in position A and B, respectively;
ρ_A resp. ρ_B — molar fractions of components A and B, respectively;
$\delta_v = v_A - v_B$ — difference of the resonance frequencies in position A and B [Hz];

Δ width at half-height [Hz] of the signal in the absence of exchange ($\tau \to \infty$), in which case Δ_A would be set equal to Δ_B of simplicity reasons;

v variable frequency [Hz].

For the average life times τ_A and τ_B the relationships in Eq. (3) also hold:

$$\tau_A = \tau/\rho_B = 1/k_A \quad \text{and} \quad \tau_B = \tau/\rho_A = 1/k_B \tag{3}$$

The significance of Eq. (2) must be seen as the consequence that it opens up the possibility of determining rate constants if a reversible chemical reaction is accompanied by a change in LARMOR frequency of one or more nuclei. The order of magnitude of rate constants that can be determined by nuclear magnetic resonance spectroscopy lies between 10^{-1} and 10^5 s^{-1} (so-called NMR time-scale). The dynamic processes of interest are first-order reactions and are characterized by energy barriers of between 20 and 100 kJ mol^{-1} (i.e. 5 to 25 kcal/mol).

In the case of a kinetic investigation the NMR spectrum of the spin system of interest is measured over as wide a range of temperature as possible. Rate constants determined by the comparison of theoretical with experimental spectra for a series of different temperatures are then substituted in the ARRHENIUS Eq. (4). An ARRHENIUS plot yields the ARRHENIUS activation energy, E_a, and the frequency factor, A, for the reaction under investigation.

Thereby it is assumed that both quantities are temperature-independent.

$$\ln k = -E_a/RT + \ln A \tag{4}$$

Also of general interest are the activation parameters of the transition state theory, the enthalpy of activation, $\Delta H^{\ddagger}$, and the entropy of activation, $\Delta S^{\ddagger}$, which can be calculated by using Eqs. (5) and (6), respectively:

$$\Delta H^{\ddagger} = E_a - RT \tag{5}$$

$$\Delta S^{\ddagger} = R[\ln(hA/\kappa k_B T) - 1] \tag{6}$$

where T is the absolute temperature [K], R is the universal gas constant (8.31 JK^{-1}), k_B is BOLTZMANN's constant (1.3805×10^{-23} JK^{-1}), h is PLANCK's constant (6.6256×10^{-34} J s) and κ is the so-called transmission coefficient, which is usually set equal to 1.

From the relation

$$\Delta G^{\ddagger} = \Delta H^{\ddagger} - T\Delta S \tag{7}$$

the free energy of activation at a specific temperature can also be obtained.

Another possibility for the determination of the free energy of activation is based on the well-known EYRING Eq. (8):

$$k = \kappa \frac{k_B T}{h} \exp(-\Delta G^{\ddagger}/RT) \tag{8}$$

DNMR can be used to study the internal dynamics of organic molecules, e.g.

- hindrance to internal rotation,
- inversion of configuration,
- ring inversion,

– valence tautomerism,
– dynamic processes in organometallic compounds and carbocations.

Intermolecular exchange processes can be studied by DNMR, too.
Today, we have various types of computer programs for DNMR [3].

References

1. Günther H (1980) NMR spectroscopy: An introduction, J. Wiley, Chichester, chap 8
2. Binsch G (1975) In: Jackman LM, Cotton FA (eds) Dynamic nuclear magnetic resonance spectroscopy, Academic, New York, p 45
3. Stephenson DS, Binsch G (1978) Quantum Chemistry Program Exchange 11: 365
4. Sandström J (1982) Dynamic NMR spectroscopy, Academic, New York
5. Kaplan JI, Fraenkel G (1980) NMR of chemically exchanging systems, Academic, New York

D

DNP
Abbreviation of **D**ynamic **N**uclear **P**olarization

In nuclear magnetic resonance (see NMR) one can find a lot of different situations leading to the apparent effect of DNP, such as CIDNP (see p 48 and related acronyms in this dictionary) or PENIS (see p 211).

DOUBTFUL
Acronym or abbreviation for **Doub**le **Q**uantum **T**ransitions for **F**inding **U**nresolved **L**ines

An experiment given by Kaptein [1] permits the selection of multiplets in strong overlapping proton spectra under using double-quantum coherence (see DQC or INADEQUATE).
R. Kaptein [1] looked for the following kinds of application:

a) analysis of complex mixtures (typical in chemistry) of liquid fuels and coal;
b) application in metabolic studies of living cell suspensions;
c) detection of resonance lines in the immediate neighbourhood of strong solvent absorptions;
d) detection of small molecules in the direct presence of macromolecules.

Reference

1. Kaptein R et al. (1982) J. Am. Chem. Soc. 104: 4286

DQENMR
Acronym for **D**euterium **Q**uadrupole **E**cho **N**uclear **M**agnetic **R**esonance used in solid-state

Solid-state deuterium spectroscopy is becoming an increasingly popular method for studying molecular motion in a variety of solid [1–4] and semisolid materials [5–8]. The method is especially useful because deuterium quadrupole couplings, typically in the order of 200 kHz, become motionally averaged by processes with rates in the order of 10^5 to $10^7\,s^{-1}$, a time-scale which is not easily studied by other techniques.

If accurate rate information is to be obtained from an analysis of partially collapsed deuterium powder patterns it is necessary to account for lineshape distortions which arise from numerous instrumental limitations, some of which have been reported in the past [6,9–11]. Most of the important limitations are imposed by the finite recovery time of the receiver electronics and by the finite power available from the radio-frequency transmitter. Davis et al. [12] effectively solved the former problem by using a solid echo [13]. In fact, the dependence of the partially collapsed deuterium powder pattern upon echo pulse spacing [10] is a most important source of kinetic information which would not be available by continuous wave (see CW) methods.

Recently [14] the effects of chemical exchange during radio-frequency pulses on DQE spectra have been evaluated numerically for single pulses as well as for LSE composite pulses (see LSE). In principle, the inclusion of exchange effects during pulses of finite length requires lengthwise calculations of the evolution of the full density matrix (see DM). Fortunately, exchanging lineshapes (see DNMR) can be simulated accurately using only the smaller single-quantum transition (see SQT) subset of spin-density-matrix elements provided that the lengths of the pulse-echo intervals are properly adjusted [14].

References

1. Spiess HW (1976) Chem. Phys. 6: 217
2. Hentschel D, Sillescu H, Spiess HW (1984) Polymer 25: 1078
3. Meirovitch E, Belsky I, Vega S (1985) Mol. Phys. 56: 1129
4. Meirovitch E, Freed J H (1979) Chem. Phys. Lett. 64: 311
5. Davis JH (1983) Biochim. Biophys. Acta 737: 117
6. Griffin RG (1981) Methods Enzymol. 72: 108
7. Jacobs RE, Oldfield E (1981) Progr. NMR Spectrosc. 4: 113
8. Rice DM, Blume A, Herzfeld J, Witterbort RJ, Huang TH, Das Gupta SK, Griffin RG (1981) In: Sarma RH (ed) Proceedings of the Second SUNYA Conversations in the Discipline: Biomolecular Stereodynamics II, Academic, New York
9. Bloom M, Davis JH, Valic MI (1980) Can. J. Phys. 58: 1510
10. Spiess HW, Sillescu H (1980) J. Magn. Res. 42: 381
11. Heinrichs PM, Levitt JM, Linder M (1984) J. Magn. Reson. 60: 280
12. Davis JH, Jeffrey KR, Bloom M, Valic MI, Higgs TP (1976) Chem. Phys. Lett. 42: 390
13. Solomon I (1958) Phys. Rev. 110: 61
14. Barbara TM, Greenfield MS, Vold RL, Vold RR (1986) J. Magn. Reson. 69: 311

DQF-COSY

Acronym for **D**ouble **Q**uantum **F**iltered-**C**orrelation Spectroscopy in Nuclear Magnetic Resonance (see NMR)

DQF has been used in a typical COSY experiment (see COSY) to reduce the intensity of the diagonal [1]. This reduction coming into use because, in contrast to the common COSY spectrum, diagonal peak multiplets have anti-phase fine structure in the DQF-COSY experiment.

In general, a p-quantum filter can be applied to remove peaks arising from spin systems with less than p spins [2,3].

The analysis of the fine structure of cross-peaks in multiple-quantum filtered COSY spectra—abbreviated to MQF-COSY—has been proposed recently [4].

References

1. Rance M, Sørensen O W, Bodenhausen G, Wagner G, Ernst RR, Wüthrich K (1983) Biochem. Biophys. Res. Commun. 117: 479
2. Piantini U, Sørensen OW, Ernst RR (1982) J. Am. Chem. Soc. 104: 6800
3. Shaka AJ, Freeman R (1983) J. Magn. Reson. 51: 169
4. Boyd J, Redfield C (1986) J. Magn. Reson. 68: 67

DRIFTS

Acronym for **D**iffuse **R**eflectance **I**nfrared **FOURIER** **T**ransform **S**pectroscopy

DRIFTS is a new technique used in FOURIER transform-infrared (see FT-IR) spectroscopy [1]. The investigation of carbon-supported metal carbonyl clusters has been successfully using this technique and incorporating a modified controlled environment cell [2]. However, substantial modifications must be made to the commercially available equipment to gather the data [1].

References

1. Venter JJ, Vannice MA (1987) J. Am. Chem. Soc. 109: 6204
2. Venter JJ, Vannice MA (1989) J. Am. Chem. Soc. 111: 2377

DRS

Acronym for **D**iffuse **R**eflectance **S**pectroscopy

DRS is a method used in infrared spectroscopy (see IR and FT-IR).

If light is directed onto a sample it may either be transmitted or reflected. Since some of the light is absorbed and the remainder is reflected, study of the diffuse reflected light can be used to measure the amount absorbed. But, the low efficiency of this DR process makes it extremely difficult to measure [1] and it was therefore speculated that IR-DR measurements would be futile [1]. Initially, an integrating sphere was used to capture all of the reflected light [2]. But, more recently improved DR cells have been designed which permit the measurement of DR spectra utilizing modern FT-IR instrumentation [3].

The requirement, by definition, for reflectance to be diffuse is that the intensity of reflected light is isotropic. With a sufficiently large number of particles—as found in a powder—an isotropic scattering distribution can be achieved, so that the emerging light will still be diffuse [4].

Following the KUBELKA-MUNK theory one can define the function $f(R_\infty)$ [5–7] with Eq. (1) where R_∞ is the absolute reflectance of an infinitely thick layer, k the absorption and s the scattering coefficients.

$$f(R_\infty) = \frac{(1 - R_\infty)^2}{2R_\infty} = k/s \tag{1}$$

In practice, a standard is used and the following ratio is calculated

$$R' = R'\,(\text{sample})/R'\,(\text{standard})$$

where finely ground potassium bromide has been recommended as a DR standard [3].

References

1. Kortüm G, Delfts H (1964) Spectrochim. Acta 20: 504
2. Willey RR (1976) Appl. Spectrosc. 30: 593
3. Fuller MP, Griffiths PR (1978) Anal. Chem. 50: 1906
4. Kubelka P, Munk F (1931) Z. Techn. Phys. 12: 593
5. Kubelka P (1948) J. Opt. Soc. Amer. 38: 448
6. Hecht HG (1976) J. Res. Natl. Bur. Stand., Sect. A, 80: 567

DSP-NLR
Abbreviation of **D**ual **S**ubstituent **P**arameter-**N**on-**L**inear **R**esonance
Approach in Nuclear Magnetic Resonance Spectroscopy

D

This approach [1,2] is an extension of the earlier DSP procedure [3] which allows an observed substituent effect to be separated into both polar and resonance components. The DSP and DSP-NLR equations, as applied to experimentally evaluated carbon-13 substituent chemical shift (see SCS) data, are given in Eqs. (1) and (2).

$$\delta_{SCS} = \rho_I \sigma_I + \rho_R \sigma_R^0 \tag{1}$$

$$\delta_{SCS} = \rho_I + \sigma_I + \rho_R \sigma_R^0/(1 - \varepsilon\sigma_R^0) \tag{2}$$

δ_{SCS} represents the measured SCS (see p 243), σ_I and σ_R^0 are the polar and resonance substituent constants and ρ_I and ρ_R are the so-called transmission coefficients. In Eq. (2), the effective resonance substituent $\bar{\sigma}_R = \sigma_R^0/(1 - \varepsilon\sigma_R^0)$ varies in a non-linear manner with a parameter, called ε, representing the electron demand placed on the substituent. Early studies in the beginning of carbon-13 NMR [4,5] suggested that SCS values can be used additively, i.e. in a poly-substituted benzene carbon-13 chemical shift for a given site may be calculated by adding appropriate SCS values of all substituents to the chemical shift of benzene. More recent studies [6–9], in which chemical shifts have been more accurately measured at very high dilution, suggest that strict additivity does not always apply.

References

1. Bromilow J, Brownlee RTC, Craik DJ, Sadek M, Taft RW (1980) J. Org. Chem. 45: 2429
2. Bromilow J, Brownlee RTC, Craik DJ, Sadek M (1986) Magn. Reson. Chem. 24: 862
3. a) Wells PR, Ehrenson S, Taft RW (1968) Prog. Phys. Org. Chem. 6: 147;
   b) Ehrenson S, Brownlee RTC, Taft RW (1973) Prog. Phys. Org. Chem. 10: 1
4. Lauterbur PC (1961) J. Am. Chem. Soc. 83: 1846
5. Nelson GL, Levy GC, Cargioli JD (1972) J. Am. Chem. Soc. 94: 3089
6. Lynch BM (1977) Can. J. Chem. 55: 541
7. Bromilow J, Brownlee RTC, Topsom RD, Taft RW (1976) J. Am. Chem. Soc. 98: 2020
8. Ref. [1].
9. Brownlee RTC, Sadek M (1981) Aust. J. Chem. 34: 1593

DTGS
Abbreviation of **D**euterated **T**riglycine **S**ulfate

DTGS is the most commonly used material for *pyroelectric detectors* in the field of infrared spectroscopy (see IR and FT-IR). The DTGS element is usually mounted so that the thermal resistance between the element and its surroundings is large and the thermal time constant is therefore consequently long. The thermal circuit can be represented by an electric analog comprising a capacitance shunted by a large resistance. The voltage resonansivity of such a pyroelectric detector at frequency f is given by Eq. (1).

$$R_v = \frac{p(T)}{\rho C_p d} \frac{R_E}{[1 + (2\pi f R_E C_E)^2]^{1/2}} \tag{1}$$

In Eq. (1), ρ stands for the density of the crystal, C_p represents the specific heat, d is the spacing between the electrode surface, $p(T)$ is the pyroelectric coefficient at temperature T, R_E is the symbol for the feedback or load resistance, and C_E is the effective capacitance.

To a fairly good approximation, R_v is proportional to $1/f$. For a rapid-scanning interferometer, therefore, the scan speed v should be as low as possible. The capacitive element is principally noiseless, and the long thermal time constant leads consequently to low thermal noise values. In fact, the noise performance attainable with DTGS *pyroelectric bolometers* is dominated by the noise performance of its associated amplifiers.

For wide-frequency bandwidths such as those encountered in modern rapid-scanning FT-IR spectrometers, the so-called JOHNSON's noise in the load resistor is the dominating factor.

For further details the interested reader is referred to the literature [1].

Reference

1. Griffiths PR, de Haseth JA (1986) In: Elving PJ, Winefordner JD (eds) Fourier transform infrared spectrometry, J. Wiley, New York

E. COSY
Acronym for **E**xclusive **Co**rrelated **S**pectroscopy

The E. COSY pulse sequence is identical to the pulse sequence for multiple-quantum-filtered COSY (see MQF-COSY) and was introduced by R.R. Ernst and coworkers [1] for two-dimensional NMR spectroscopy. For weak coupling, coherence transfer in the E. COSY-experiment is restricted to taking place only between connected transitions with a common energy level. This restriction results in simplified cross-peak multiplet patterns which exhibit nearly ideal features for measurement and assignment of J coupling constants. The E. COSY-experiment can be understood as a linear combination of MQF-COSY spectra (see p 180) of different orders.

Recently practical aspects of the E. COSY technique for the measurement of scalar spin-spin coupling constants in peptides have been discussed [2]. Guidelines have been presented for experimental setup and spectral assignments. The features of the cross-peak multiplet patterns in E. COSY spectra have been illustrated by the spectra of valine, phenylalanine, and proline residues in the decapeptide antamanide [2].

References
1. Griesinger C, Sørensen OW, Ernst RR (1985) J. Am. Chem. Soc. 107: 6394; (1985) J. Am. Chem. Soc. 107: 7778
2. Griesinger C, Sørensen OW, Ernst RR (1987) J. Magn. Reson. 75: 474

ECM
Abbreviation of **E**xciton **C**hirality **M**ethod

The application of the *coupled oscillator theory* developed in 1930 by W. Kuhn [1] to various natural products has been termed the ECM [2] in the field of chiroptical methods.

Numerical calculations of *exciton CD COTTON* (see CD) effects by ECM have been presented [2]. Quite similar, theoretical calculations of CD spectra (see CD) by a dipole velocity molecular orbital method have been documented, too [2].

The exciton mechanism has been successfully employed in studies of chiroptical properties of biopolymers such as proteins, polypeptides, nucleic acids, oligonucleotides, dyes absorbed on those biopolymers, and inorganic metal complexes. Similarly, in the field of organic chemistry, CD due to the exciton coupling mechanism provides very useful and unambiguous information on absolute configuration and conformation. Because ECM is based on sound theoretical calculations, the absolute stereochemistry of organic compounds exhibiting typical split *CD COTTON effects* due to chiral exciton coupling is therefore assignable in a non-empirical manner. The method has thus been extensively used for determining the absolute configuration of natural and synthetic organic compounds [2].

References
1. Kuhn W (1930) Trans. Faraday Soc. 26: 293
2. Harada N, Nakanishi K (1983) Circular dichroic spectroscopy-exciton coupling in organic chemistry, University Science Books, Mill Valley/USA

EFG
Abbreviation of Electric Field Gradient

EFG is important in the field of nuclear quadrupole resonance (see NQR) spectroscopy, especially in the study of hydrogen bonding by nitrogen-14 and deuterium nuclear quadrupole spectroscopy. Because the lone pair of nitrogen in molecules such as pyridine, pyrimidine, and other nitrogen-containing bases makes a major contribution to the EFG of nitrogen [1], hydrogen bonding to such a nitrogen is manifested by changes in the EFG parameters. The effects of hydrogen bonding on the EFG parameters of deuterium have been studied in detail, and correlations between the deuterium quadrupole coupling constant (see QCC) and X . . . Y distance in X–H . . . Y hydrogen bonds have been put forward for X, Y = O, O [2] and N, O [3]. These results show that the D QCC undergoes very substantial reduction as the strength of the hydrogen-bond increases, as reflected in a shortened distance. Furthermore, the ^{2}H NQR data demonstrate substantial departures from linearity in some O–H . . . O hydrogen-bonding systems [4]. ^{14}N and ^{2}H NQR of a hydrogen-bonded pyrrole-pyridine (1:1) complex were observed at 77 k by field cycling double resonance [5]. In this study the important internuclear nitrogen-nitrogen distance was first estimated at 2.92 Å. The effect of hydrogen bonding on the ^{14}N electric field gradient tensors was also evaluated, and substantial charge transfer through the hydrogen bond was observed [5].

References
1. Lucken EAC (1969) Nuclear quadrupole coupling constants, Academic, London
2. Soda G, Chiba T (1969) J. Chem. Phys. 50: 439
3. Hunt MJ, Mackay AL (1974) J. Magn. Reson. 15: 402
4. Brown TL, Butler LG, Curtin D, Hiyama Y, Paul IC, Wilson RB (1982) J. Am. Chem. Soc. 104: 1172
5. Hiyama Y, Keiter EA, Brown TL (1986) J. Magn. Reson. 67: 202

EFNMR
Abbreviation of the Effects of an Electric Field on Nuclear Magnetic Resonance Spectra

When an external electric field is applied to a polar liquid, the molecular dipoles acquire a small average orientation parallel to the field. As a result the NMR spectrum of the liquid may be changed because anisotropic interactions, such as the direct dipolar spin-spin couplings and the quadrupole couplings, are not averaged out as in normal isotropic liquids. The anisotropic interactions will be seen in the NMR spectrum as splittings or shifts of the resonance lines, proportional to the molecular alignment.

EFNMR of dipolar molecules has been reviewed in two papers [1,2]. EFNMR has been used to measure quadrupole couplings of nitrogen-14 [3] and oxygen-17 [4], to determine pretransitional phenomena in isotropic liquid crystals [5], dipole moment orientations in asymmetric molecules [6], and dielectric saturation [7].

One of the serious limitations on the resolution of EFNMR spectra, especially of crowded proton magnetic resonance spectra, is the inhomogeneity of the magnetic field introduced by the non-spinning electric-field sample cell. A number of spin-echo procedures have been published which give high-resolution information from inhomogeneously broadened samples [8]. More recently [9] the use of zero-quantum transition (see ZQT) ^{1}H NMR spectroscopy in this field has been reported. This technique is of interest because the linewidths of the ZQT's are insensitive to the magnetic field inhomogeneity and only depend on the proton spin-spin relaxation time, T_2.

References

1. Hilbers CW, MacLean C (1972) NMR, basic principles and progress, Springer, Berlin Heidelberg New York, vol 7
2. van Zijl PCM, Ruessink BH, Bulthuis J, MacLean C (1984) Acc. Chem. Res. 17: 172
3. Plantenga TM, Bulsink H, MacLean C (1981) Chem. Phys. Lett. 82: 439
4. Ruessink BH, MacLean C (1984) Mol. Phys. 53: 421
5. Ruessink BH, Barnhoorn JBS, MacLean C (1984) Mol. Phys. 52: 939
6. Plantenga TM, Ruessink BH, MacLean C (1980) Chem. Phys. 48: 359; Plantenga TM, MacLean C (1980) Chem. Phys. Lett. 75: 294
7. Plantenga TM, de Kanter FJJ, Bulsink H, MacLean C (1982) Chem. Phys. 65: 77
8. Freeman R, Hill HDW (1971) J. Chem. Phys. 54: 301; Weitekamp DP, Garbow JR, Murdoch JB, Pines A (1981) J. Am. Chem. Soc. 103: 3578
9. Ruessink BH, de Kanter FJJ, MacLean C (1985) J. Magn. Reson. 62: 226

EIRFT
Acronym for **E**xtended **I**nversion-**R**ecovery **F**ourier Transform

Since the results of the methods used so far for the measurement of spin-lattice relaxation times T_1 are not always correct, a new sequence, the extended inversion-recovery method has been proposed [1]. The procedure involves addition of a variable perturbing pulse to the conventional two-pulse sequence (see IRFT) and can be implemented in modern NMR spectrometers without difficulties. The method may be a key to more accurate determination of spin-lattice relaxation times (see T_1) in the future.

Reference

1. Ejchart A, Oleski P, Wroblewski K (1986) J. Magn. Reson. 68: 207

ELDOR
Acronym for **E**lectron-**E**lectron **D**ouble **R**esonance

ELDOR is a continuous wave (see CW) method used in electron spin or paramagnetic resonance (see ESR or EPR). In an extended series of papers starting with a density matrix formalism, Freed [1] has developed the theory of CW-ELDOR and of saturation recovery electron spin resonance-abbreviated as SR-ESR-[2,3], with particular respect to nitroxide spin-labels. An alternative approach has been given by Stanley and Vaughn [4]. Within an analog theoretical framework not only saturation recovery and CW-ELDOR but also pulsed ELDOR [5] have been discussed [6].

References

1. Hyde JS, Chien JCW, Freed JH (1968) J. Chem. Phys. 48: 4211
2. Freed JH (1974) J. Phys. Chem. 78: 1155
3. Freed JH (1979) In: Kevan L, Schwartz RN (eds) Time domain electron spin resonance, Wiley, New York, p 32
4. Stanley KJ, Vaughn RA (1969) Electron spin relaxation phenomena in solids, Plenum, New York
5. Hyde JS, Froncisz W, Mottley C (1984) Chem. Phys. Lett. 110: 621
6. Yien J-J, Hyde JS (1987) J. Magn. Reson. 74: 82

ELSA
Acronym for Ellipse-Saddle

ELSA is the name of a new radio-frequency coil [1,2] proposed for the application in nuclear magnetic resonance imaging (see NMRI). Its geometry is a hybrid between the saddle-shaped and the crossed-ellipse coil [3–5]. This hybrid coil has a distributed capacitance and a proximity effect smaller than the saddle coil with the same number of turns and an equivalent radio-frequency field distribution and current-field efficiency. The signal-to-noise ratio (see SNR) showed that ESLA is better than the normally used two-turn saddle coil (see DS), also when B_1 distribution is included in the SNR evaluation.

References

1. DeLuca F, Luzzi M, Crescenzi A, DeSimone BC, Campanella R, Casieri C, Maraviglia B (1986) J. Magn. Reson. 67: 7
2. Patent No. 48375A85, Italy, 12 June 1985
3. Mansfield P, Morris PG (1982) NMR imaging in biomedicine, Academic, New York
4. Bottomley PA (1981) Proceedings of an International Symposium on NMR Imaging, Winston-Salem NC, p 25
5. Redpath TW, Selbic RD (1984) Phys. Med. Biol. 29: 739

EMR
Abbreviation of Electromagnetic Radiation

This abbreviation is sometimes used in the fields of nuclear magnetic resonance (see NMR) and electron spin resonance (see ESR or EPR) spectroscopy [1].

Reference

1. Hornak JP, Freed JH (1986) J. Magn. Reson. 67: 501

ENDOR
Acronym for Electron Nuclear Double Resonance

Electron nuclear double resonance, a new method in electron spin or paramagnetic resonance (see ESR or EPR) was originally introduced by G. Feher [1]. Over a long period ENDOR became a domain in solid state physics. In the meantime this kind of double resonance has attracted more and more interest in chemistry and biology [2–4]. Application of the method was perfectly demonstrated in a review article by A. Schweiger [5] and will be explained under special items of this dictionary (see CW-ENDOR and ESE-ENDOR), too. Sometimes FOURIER transform methods (see FT, FFT) can be directly used for analysis [6,7]. Recently modern approaches such as MEM (see p 173) or LPSVD (see p 163) have become fruitful analysis methods [8,9].

References
1. Feher G (1956) Phys. Rev. 103: 834
2. Kevan L, Lispert LD (1976) Electron spin double resonance spectroscopy, Wiley, New York
3. Dorio MM, Freed JH (eds) (1979) Multiple electron resonance spectroscopy, Plenum, New York
4. Schweiger A (1982) Struct. Bonding (Berlin) 51: 1
5. Schweiger A (1986) Chimia 40: 111
6. Narayana PA, Kevan L (1983) Magn. Reson. Rev. 7: 239
7. Merks RPJ, de Beer R (1980) J. Magn. Reson. 37: 305
8. van Ormondt D, Nederveen K (1981) Chem. Phys. Lett. 82: 443
9. Barkhuijsen H, de Beer R, Bovée WHMJ, van Ordmondt D (1985) J. Magn. Reson. 61: 465

ENDORIESR
Abbreviation of **E**lectron **N**uclear **D**ouble **R**esonance **I**nduced **E**lectron **S**pin **R**esonance

Typically in the field of electron spin resonance (see ESR) we find overlapping of the different spectra of individual paramagnetic species or compounds. Analysis of the ESR spectra of molecular crystals with different positions or orientations of the paramagnetic species were thought of as difficult or incorrect. ENDORIESR now permits separation of these overlapped ESR spectra in a simple way [1,2].

References
1. Hyde JS (1965) J. Chem. Phys. 43: 1806
2. Niklas JR, Spaeth JM (1980) Phys. Status Solidi B 101: 221

EPI
Acronym for **E**cho-**P**lanar **I**maging

High-speed nuclear magnetic resonance imaging (see NMRI) methods are now beginning to be recognized as very important factors in the economics of clinical imaging. In many cases where superconductive magnets are in operation, the existing imaging techniques, which could benefit from faster gradient switching, are limited by gradient rise times of about 1 ms.

The ultra-high-speed echo-planar imaging (EPI) technique [1,2] operating with unscreened gradient coils has, until recently, been limited to an intermediate-size probe for adult heads and small children in order to reduce the gradient interactions with the magnet. But, using single active screens, adult whole-body EPI has now been implemented at video rates of up to 20 frames per second [3].

While achieving the desired decoupling, it can incur spatial distortions and a reduction in the primary field per unit current [4-6]. The resulting field loss of the order of 50 per cent has to be compensated with more powerful gradient drivers.

Recently [7] it has been shown, that the selective reflective and absorptive properties of active screens [5] may be exploited in double-shield configurations to restore the value and the character of the primary field without sacrificing screening efficacy. A preliminary report of some aspects of this work was made in 1986 [8].

References

1. Mansfield P (1977) J. Phys. C 10: L 55
2. Mansfield P, Morris PG (1982) In: NMR imaging in bio-medicine, Academic, New York
3. Doyle M, Chapman B, Turner R, Ordidge RJ, Cawby M, Coxon R, Glover G, Coupland RE, Morris GK, Worthington BS, Mansfield P (1986) Lancet 2: 682
4. Mansfield P, Chapman B (1986) J. Magn. Reson. 66: 573
5. Mansfield P, Chapman B (1986) J. Phys. E 19: 540
6. Bangert V, Mansfield P (1982) J. Phys. E 15: 235
7. Mansfield P, Chapman B (1987) J. Magn. Reson. 72: 211
8. Chapman B, Mansfield P (1986) J. Phys. D 19: L 129

E

EPR
Abbreviation of Electron Paramagnetic Resonance

EPR or electron spin resonance (see ESR) is the most important spectroscopic method for the determination of structures of paramagnetic species and compounds in quite different fields of research such as solid state physics, organic and inorganic chemistry, materials and life sciences and medicine [1–4]. For example, new techniques such as ENDOR (see ENDOR) or ESE (see ESE) in electron paramagnetic resonance have shown new essential aspects and brought new insights into photosynthesis [5–7]. EPR or ESR spectra have been helpful in polymer chemistry, too.

References

1. Wertz JE, Bolton JR (1972) Electron spin resonance: Elementary theory and practical applications, McGraw-Hill, New York
2. Abragam A, Bleaney B (1970) Electron paramagnetic resonance of transition ions, Clarendon, Oxford
3. Gerson F (1970) Hochauflösende ESR-Spectroscopy, Verlag Chemie, Weinheim
4. Schweiger A (1986) Chimia 40: 111
5. Fajer J, Fujita I, Forman A, Hanson LK, Craig GW, Goff DA, Kehres LA, Smith KM (1983) J. Am. Chem. Soc. 105: 3837
6. Hoff AJ (1982) Biophys. Struct. Mech. 8: 107
7. Lubitz W, Lendzian F, Scheer H, Gottstein J, Plato M, Möbius K (1984) Proc. Natl. Acad. Sci. USA 81: 1401

Appendix

Much more recently [1] new application possibilities have been proposed in the *electrochemical generation of paramagnetic species* with simultaneous registration of the *EPR-signal* and electrochemical parameters. Figure 7 shows four different measurement modes, their applied parameters, and their recorded "answers". The first mode was described by Geske and Maki in 1960 [2]. Using a very simple three-electrode cell, Russian scientists obtained and studied series of unstable anion-radicals with lifetimes up to 1 s [3]. Modes 2 and 3 have been studied in

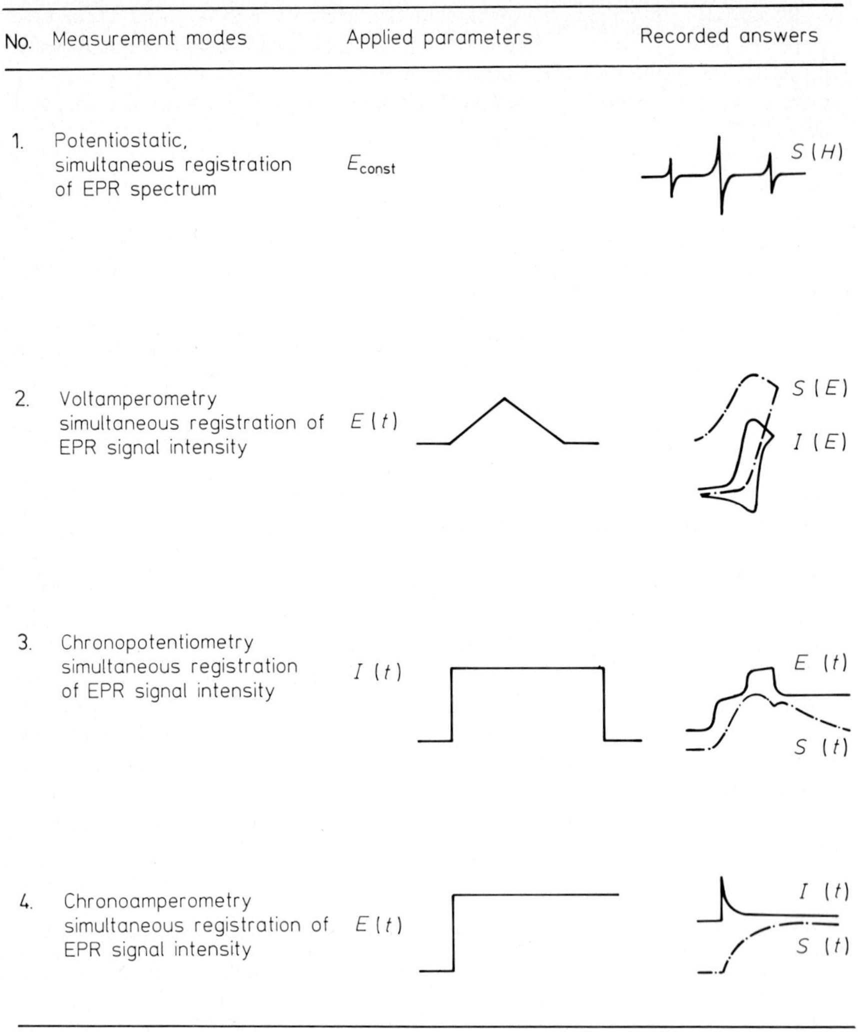

No.	Measurement modes	Applied parameters	Recorded answers
1.	Potentiostatic, simultaneous registration of EPR spectrum	E_{const}	$S(H)$
2.	Voltamperometry simultaneous registration of EPR signal intensity	$E(t)$	$S(E)$ $I(E)$
3.	Chronopotentiometry simultaneous registration of EPR signal intensity	$I(t)$	$E(t)$ $S(t)$
4.	Chronoamperometry simultaneous registration of EPR signal intensity	$E(t)$	$I(t)$ $S(t)$

Fig. 7

detail by Goldberg and Bard [4]. The principles of the generation of paramagnetic species on a special working electrode directly in the resonator system of an EPR spectrometer were described more than 10 years ago [5]. The fourth mode—chronoamperometry—allows one to study diffusion and kinetics of electrode processes [1].

References

1. Ilyasov AV, Kardirov MK, Kargin YM, Eichhoff U: Bruker Report 2/87, p 20
2. Geske DH, Maki AH (1960) J. Am. Chem. Soc. 82: 2671
3. Ilyasov AV, Kargin YM, Morozova ID (1980) Spectry EPR Organicheskikh Ion-Radikalov Izd., Nauka, Moscow

4. Goldberg IB, Bird AJ (1981) J. Phys. Chem. 75: 3281
5. Allendörfer DR, Martinchek GA, Bruckenstein S (1975) Anal. Chem. 47: 890

EPT
Acronym for **E**xclusive **P**olarization **T**ransfer

EPT is the name for a pulse sequence [1] used in nuclear magnetic resonance (see NMR) in analogy to INEPT or DEPT (see p 69) for assignment purposes. Figure 8 symbolizes the EPT pulse sequence where τ is equal to 1/2 J (J stands for the direct carbon-13/proton coupling constant, [see CC]) and **BB** is the abbreviation of proton broad-band decoupling which can be applied here in contrast to the simple INEPT pulse sequence (see INEPT).

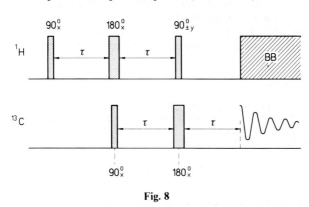

Fig. 8

The two 180° pulses in Fig. 8 are not essential for the distinct polarization transfer (see PT) and are therefore absent in the so-called "simplified version" of EPT given in Fig. 9 [1]. It must be pointed out that EPT can only be applied for CH fragments of a given molecule. The situations at the points (1) to (7) of Fig. 9 have been carefully analyzed [2] and understood on the basis of the product operator formalism (see POF).

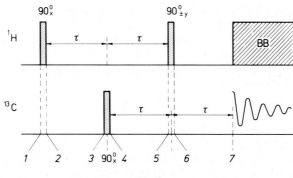

Fig. 9

References

1. Bendall MR, Pegg DT, Doddrell DM: J. Chem. Soc., Chem. Commun. 1982: 872
2. Wesener JR (1985) Dissertation, University of Siegen/Germany

ERS

Acronym for **E**xternal **R**eflection **S**pectroscopy

ERS is a technique used in infrared (see IR) spectroscopy. When surfaces are highly reflective, as in the case of metals, external reflection spectroscopy (ERS) can be applied with success [1]. In order to get optimum intensity of the reflection bands of thin films, angles of incidence near 88° are desirable. However, to avoid interference with the incoming beam, angles of incidence near 80° are used. However, for a highly reflective sample, in general, 50 to 60 percent of the energy is lost through the reflection optics [2]. If the optical constants of the film are known, the thickness of the film can be calculated on the basis of the optical theory [3].

References

1. Tompkins HG (1976) Appl. Spectrosc. 30: 377
2. Koenig JL (1984) Fourier transform infrared spectroscopy of polymers, Springer, Berlin Heidelberg New York (Advances in Polymer Science, vol 54)
3. Greenler RG (1966) J. Chem. Phys. 44: 310

ESCA

Abbreviation of **E**lectron **S**pectroscopy for **C**hemical **A**nalysis

Electron spectroscopy may be defined as the measurement of the kinetic energy distribution of electrons emitted by a sample on irradiation with ultraviolet (see UV), X-rays, or electrons [1]. In the special case of excitation with X-rays a new method was introduced by K. Siegbahn, who called it ESCA [2–7]. It has been shown, that ESCA can be a very helpful method in structure elucidation of nitrogen compounds [1]. On the other hand, ESCA has become a potent characterization discipline in polymer spectroscopy, especially in surface studies [8].

ESCA is based on the following general principle [2,6]: monochromatic short-wavelength radiation is used in order to detach electrons from the sample of interest, the kinetic energy of the electrons being measured by means of an electron analyzer. The difference in energy between exciting photons of the frequency v and the electrons detached (E_{kin}) is direct proportional to the binding energy of the electron (E_B):

$$E_B = hv - E_{kin} \quad . \tag{1}$$

ESCA spectra, therefore inform us about the binding energies or ionization energies of the electrons in the sample under investigation.

Core electron peaks are prominent in ESCA spectra since their E_B is normally nearer to the photon energy of the exciting X-rays than that of weakly bound valence electrons. Core electrons are influenced by the mode of binding, as evidenced by shifts in ionization energies relative to the free atom values. The free atom values are characteristic of the element and orbital considered. So-called shifts may be seen in analogy to nuclear magnetic resonance (see NMR) as "chemical shifts", and will depend on the electron densities near the nuclei on account of the shielding of the positive charge of the nuclei. Core electron peaks in ESCA spectra, therefore are useful in determination or identification of atoms, their measurement of binding mode, and their relative abundance, too.

References

1. Fitzky HG, Wendisch D, Holm R (1972) Angew. Chem. internat. Edit. 11: 979
2. Siegbahn K, et al. (1967) Nova Scota R. Soc. Sci. Uppsala, Ser. IV: 20
3. Siegbahn K, et al. (1967) ESCA: atomic, molecular, and solid state structure studied by means of electron spectroscopy, Almquist and Wiksell, Uppsala
4. Siegbahn K, et al. (1969) ESCA applied to free molecules, North Holland, Amsterdam
5. Nordling C (1972) Angew. Chem. internat. Edit. 11: 83
6. Siegbahn K (1974) In: Caudano R, Verbist J (eds) Electron spectroscopy. Progress in research and application, Elsevier, Amsterdam
7. Electron spectroscopy: theory, techniques and applications, Brundle CR, Baker AD (eds) (1977, 1978, 1979) vols 1–3, Academic, London
8. Klöpffer W (1984) Introduction to polymer spectroscopy, Springer, Berlin Heidelberg New York (Polymers/Properties and Applications, vol 7)

E

ESCORT

Acronym for an updated APT (see there) Experiment with Error Self-Compensation Reached by Tau-Scrambling

Procedure updates the normal attached proton test (see APT) under using pulse sequences in which error self-compensation is reached by tau-scrambling [1]. In the meantime the method has been incorporated into manufacturer's software for NMR spectrometers as the following parameter settings show [2]:

PW	:	$90°$ pulse on the observe channel.
P1	:	excitation pulse, normally $90°$; in order to allow faster repetition rate, P1 can be set greater than $90°$ (if P1 = 0, P1 is set equal to PW).
JCH	:	average of one-bond carbon-proton coupling.
D2	:	refocussing delay (should be greater than D3 and D4)
D3, D4	:	J-evolution delays (may be calculated automatically)

Flag		Basic Cycle
= 'YY'	:	conventional carbon-13 spectrum with all resonances,
= 'YN'	:	quart. C and CH_2 up; CH and CH_3 down,
= 'NY'	:	largest difference of CH and CH_3 signals,
= 'NN'	:	see 'YN' and 'NY' too,
QUAT = Y	:	for non-protonated carbons only

References

1. Madsen JC, Bildsøe H, Jakobson HJ, Sørenson OW (1986) J. Magn. Reson. 67: 243
2. See Bruker software manuals

ESE
Abbreviation of **E**lectron **S**pin-**E**cho

A new pulsed method in electron spin resonance (see ESR) like pulse Fourier transform spectroscopy (see PFT) in nuclear magnetic resonance. The procedure has become more and more popular in research fields of chemistry and biology [1–5]. Enhanced resolution is typical in the case of modulated electron spin-echos [6]. The determination of frequencies via modulation patterns is possible by Fourier transformation [7,8]. Recently modern procedures like the maximum entropy method [9] (see MEM) or the linear prediction singular value decomposition [10] (see LPSVD) have been used in analysis of modulated electron spin-echos, too.

References
1. Kevan L, Schwartz RN (eds) (1979) Time domain electron spin resonance, Wiley-Interscience, New York
2. Mims WB (1972) Electron spin echos. In: Geschwind S (ed) Electron paramagnetic resonance, Plenum, New York
3. Morris JR, Thurnauer MC, Bowman MK (1980) Adv. Biol. Med. Phys. 17: 316
4. Salikov KM, Semenov AG, Tsvetkov YD (1976) Electron spin echoes and their applications, Nauka, Novosibirsk
5. Lin TS (1984) Chem. Rev. 84: 1

ESEEM
Acronym for **E**lectron **S**pin-**E**cho **E**nvelope **M**odulation

A new method used in ESR or EPR spectroscopy (see p 89). The great power of this technique has been demonstrated in several studies of weakly coupled nuclei with a small magnetic moment in single crystals [1–4] and powders [5,6]. In these cases one has to deal with low-frequency nuclear spin transitions which are not easily accessible with the conventional continuous wave-ENDOR techniques (see CW-ENDOR) although spectacular improvements have been accomplished in this area [7–10].

ESEEM has very important advantages over classical CW-ENDOR. The modulation effect does not depend on a balance of relaxation times, very low frequency modulations can be observed, the signal-to-noise ratio (see SNR) does not depend on the frequency, the modulation intensity can be calculated directly via the spin HAMILTONian (see p 122) because it is not affected by relaxation phenomena [11], and with two-dimensional techniques one can get assignment information of spectral lines in single crystals [3,12].

In contrast, the modulation effect itself is strongly dependent on the existence of anisotropic interactions which indeed most not differ more than one order of magnitude from the usual nuclear ZEEMAN interaction (NZI). Therefore, modulation effects will be weaker than ENDOR effects for predominantly isotropic interactions as, for instance, for weakly coupled protons.

The application of the ESEEM spectroscopy to disordered systems still suffers from a lot of difficulties.

First, the anisotropic broadening results in destructive interference which makes the modulations disappear rapidly from the echo envelope. Thus, the so-called instrumental dead time is a serious limiting factor. Second, analysis of frequency-domain spectra is not straightforward due to the extreme angular dependence of the modulation intensity.

One way of analyzing ESEEM spectra is to fit the time domain itself to a model function following the spin HAMILTONian [13–15]. Impressive results have been obtained using this procedure in the case of systems with an isotropic g tensor [6]. But, for quadrupolar nuclei, this method is limited to such systems where a perturbation treatment of the nuclear quadrupole interaction (NQI) is allowed. Moreover, the hyperfine interaction (HFI) is always approximated as a (pseudo)-dipolar (hence axial) interaction. The approximations are limitated by the time neccessary for a computer fit. Recently [16] we gained knowledge about new model calculations of frequency-domain ESEEM spectra of disordered systems involving ESEEM, ENDOR and pulsed ENDOR (P-ENDOR) spectra of spin systems of one electron coupled with $I = 1/2$, $I = 1$, or $I = 3/2$ nucleus.

References

1. Isoya J, Bowman MK, Norris JR, Weil JA (1983) J. Chem. Phys. 78: 1735
2. Reijerse EJ, Paulissen MLH, Keijzers CP (1984) J. Magn. Reson. 60: 66
3. Barkhuijsen H, de Beer R, Deutz AF, van Ormondt D, Völkel G (1984) Solid State Commun. 49: 679
4. Single DJ, van de Poel WAJA, Schmidt J, van der Waals JH, de Beer R (1984) J. Chem. Phys. 81: 5453
5. Reijerse EJ, van Aerle NAJM, Keijzers CP, Böttcher R, Kirmse R, Stach J (1986) J. Magn. Reson. 67: 114
6. Kevan L (1979) In: Kevan L, Schwartz RN (eds) Time domain electron paramagnetic resonance, Wiley, New York, chap 8
7. Kirmse R, Abram U, Böttcher R (1982) Chem. Phys. Lett. 90: 9
8. Kirmse R, Stach J, Abram U, Dietzsch W, Böttcher R, Gribnau MCM, Keijzers CP (1985) Inorg. Chem. 23: 3333
9. Böttcher R, Kirmse R, Stach J, Keijzers CP (1985) Mol. Phys. 55: 1431
10. Böttcher R, Kirmse R, Stach J, Reijerse EJ, Keijzers CP (1986) Chem. Phys. 107: 145
11. Mims WB (1972) Phys. Rev. B 5: 2409, 3543
12. Merks RPJ, de Beer R (1980) J. Magn. Reson. 37: 305
13. Romanelli M, Narayana M, Kevan L (1984) J. Chem. Phys. 80: 4044
14. Hemig M, Narayana M, Kevan L (1985) J. Chem. Phys. 83: 1478
15. Iwasaki M, Toriyama K (1985) J. Chem. Phys. 82: 5415
16. Reijerse EJ, Keijzers CP (1987) J. Magn. Reson. 71: 83

ESR
Abbreviation of **E**lectron **S**pin **R**esonance

ESR or electron paramagnetic resonance (see EPR) is extensively used in various fields of research as a powerful tool for the determination of molecular structures of paramagnetic species and compounds. In particular, scientists working in solid state physics, organic and inorganic chemistry, materials science, biology, and environmental science apply this highly sensitive and non-destructive spectroscopic method with great success [1–3]. However, ESR often suffers from a poor spectral resolution which may severely limit its overall applicability. The demand for enhanced resolution stimulated the development of various novel experimental techniques [4]. Two modern methods, namely electron nuclear double resonance (see ENDOR) and the electron spin echo spectroscopy (see ESE) have become more and more favoured [4].

ESE spectroscopy has been used recently [5] to perform the first direct measurement of spin-spin relaxation times of a spin-labeled protein at physiological temperatures.

The experiments were recorded as two-dimensional (2D) contour plots which were found to be sensitive to different motions of proteins.

The considerably potential of 2D-ESE spectroscopy in the study of macromolecular motion has been illustrated by comparing 2D-ESE with the non-linear technique of saturation transfer electron paramagnetic resonance.

The accurate determination of the microwave field strength B_1 at the position of the sample is often an important prerequisite for any successful ESR experiments. Especially in pulse experiments, such as in extended time excitation [6,7], the rotation angles induced by the microwave pulses have to be set precisely to avoid artifacts. Recently [8] the determination of the microwave field strength by microwave-induced transitory oscillations in pulsed ESR has been proposed.

In 1986 [9], the setting up of a single-channel quadrature FT-ESR spectrometer was achieved. The spectrometer itself was built around the bimodal ESR probe head using the design proposed by Biehl and Schmalbein [10]. This setup was optimized with respect to spectrometer deadtime as shown by the Leiden group [11]. The authors used the quadrature phase detection (see QPD) scheme proposed by Pajer and Armitage [12] using sequential recording, which can also easily be implemented for the ESR regime [9].

It was demonstrated in 1987 [13] that electron-spin-echo-detected (see ESE) EPR, using an inversion-recovery three-pulse sequence, permits EPR imaging selectively based on electron spin longitudinal relaxation times. This is analogous to nuclear magnetic resonance imaging (see NMRI) where relaxation time differences have been widely exploited in the past [14].

EPR imaging with one spectral dimension and one spatial dimension can be achieved by obtaining spectra at a series of magnetic field gradients, followed by an approach of a convoluted back-projection image reconstruction algorithm [15]. The proposed concept has been demonstrated with samples of organic radicals, including those that exhibit hyperfine structure (see HFC) in the fluid solution EPR spectra.

In the field of *numerical analysis* of EPR spectra, recently [16] a method for determining the relative concentrations of paramagnetic species in mixtures by least-squares analysis has been described. The method is especially useful when the common double integration procedure cannot be used because of overlap between the component spectra. Most of the details of the analysis, including the use of the simplex algorithm to optimize the EPR parameters, have been published [17]. The fast FOURIER transform (see FFT) method [18] is used to calculate the spectra. For each of the free radicals which contribute to an EPR spectrum, the discrete FOURIER transform (see DFT) of the absorption spectrum is calculated. Suppose the real component of the first element of such a DFT is A, it may be shown that A is equal to the area under the corresponding absorption spectrum. For a relative concentration parameter of C, it is necessary to multiply all elements of the DFT by C/A. The DFTs calculated in this way for each of the free radicals involved are summed, and the resulting DFT modified to allow for first harmonic phase-sensitive detection [19]. Finally, the inverse FFT is used to recover the EPR spectrum. The relative concentration parameters represent integrated areas, corrected for modulation effects. These parameters will be reliable to the extent that the calculated spectral profile faithfully simulates the experimental profile, therefore a least-squares method of curve fitting is necessary [16].

References

1. Wertz JE, Bolton JR (1972) Electron spin resonance: Elementary theory and practical applications, McGraw-Hill, New York
2. Abragam A, Bleaney B (1970) Electron paramagnetic resonance of transition ions, Clarendon, Oxford
3. Gerson F (1970) Hochauflösende ESR-Spektroskopie, Verlag Chemie, Weinheim
4. Schweiger A (1986) Chimia 40: 111
5. Kar L, Johnson ME, Bowman MK (1987) J. Magn. Reson. 75: 397
6. Schweiger A, Braunschweiler L, Fauth JM, Ernst RR (1985) Phys. Rev. Lett. 54: 1241
7. Braunschweiler L (1985) Ph.D. thesis No. 7887, Swiss Federal Institute of Technology
8. Braunschweiler L, Schweiger A, Fauth JM, Ernst RR (1987) J. Magn. Reson. 72: 579
9. Dobbert O, Prisner T, Dinse KP (1986) J. Magn. Reson. 70: 173
10. Biehl R, Schmalbein D (1982) U.S. Patent 4, 312, 204
11. Barendswaard W, Disselhorst JAJM, Schmidt J (1984) J. Magn. Reson. 58: 477
12. Pajer RT, Armitage IM (1976) J. Magn. Reson. 21: 365
13. Eaton GR, Eaton SS (1987) J. Magn. Reson. 71: 271
14. Mansfield P, Morris PG (1982) NMR imaging in biomedicine, Academic, New York, chaps 2 and 3
15. Maltempo MM, Eaton SS, Eaton GR (1987) J. Magn. Reson. 72: 449
16. Beckwith ALJ, Brumby S (1987) J. Magn. Reson. 73: 260
17. Beckwith ALJ, Brumby S (1987) J. Magn. Reson. 73: 252
18. Brumby S (1984) Comput. Chem. 8: 75
19. Brumby S (1982) Chem. Phys. Lett. 87: 37

ETB-NOE
Acronym for **E**xchange-**T**ransferred **B**ound-Nuclear **OVERHAUSER** Effect (or **E**nhancement)

For more than 10 years it has been known that small molecules will display negative NOE's when bound to proteins. The use of ETB-NOE's [1][1] to test the conformations of drugs, enzyme substrates and cofactors, and other relatively small molecules at their receptors site is gaining increasing prominence as a proposed powerful tool in the investigation of several biorecognition phenomena. Some recent theoretical treatments of the ETB-NOE phenomenon [1–3] and practical demonstrations [3–5] have stressed the use of truncated radio-frequency driven NOE methods [6] for minimizing secondary cross-relaxation effects (=spin diffusion). But, these methods have to compete with a new procedure called SIR-ΔNOE [7] which is explained under this acronym in the dictionary.

References

1. Albrand JP, Birdsall B, Feeney J, Roberts GCK, Burgen ASV (1979) Int. J. Biol. Macromol. 1: 37

[1]Clore and Gronenborn [2,3] refer to these as TR-NOE's. In the opinion of some people [7] this abbreviation can bring confusions with those for both transient and truncated rf-driven NOE's. They [7] employ the term ETB-NOE throughout.

2. Clore GM, Gronenborn AM (1982) J. Magn. Reson. 48: 402
3. Clore GM, Gronenborn AM (1983) J. Magn. Reson. 53: 423
4. Gronenborn AM, Clore GM (1982) J. Mol. Biol. 157: 155
5. Levy HR, Ejchart A, Levy GC (1983) Biochemistry 22: 2792
6. Wagner G, Wüthrich K (1979) J. Magn. Reson. 33: 675
7. Anderson NH, Nguyen KT, Eaton HL (1985) J. Magn. Reson. 63: 365

Eu(fod)$_3$

Abbreviation of a commonly used lanthanide shift reagent in nuclear magnetic resonance (see LSR).

EXAFS

Acronym for **E**xtended **X**-Ray **A**bsorption **F**ine **S**tructure

EXAFS effects arise because of electron scattering atoms surrounding a particular atom of interest as that special atom absorbs x-rays and emits electrons. The atom of interest absorbs photons at a characteristic wavelength and the emitted electron, undergoing constructive or destructive interference as it is scattered by the surrounding atoms, modulates the absorption spectrum. The modulation frequency corresponds directly to the distance of the surrounding atoms while the amplitude itself is related to the type and number of atoms. An EXAFS spectrum can thus provide co-ordination numbers and distances for a selected atom, both critical structural parameters in analysis of molecular structures, too.

In the past EXAFS experiments have been associated with high intensity, large and most expensive x-ray generators such as synchrotron radiation sources. Recent advances in microcomputer control of monochromators, detectors and much more precise mechanical movements have brought routine EXAFS capability within the reach of a wider community, especially in the field of surface analysis.

Recently EXAFS studies of the B$_2$ subunit of the ribo-nucleotide reductase from E. coli have been published [1].

Reference

1. Scarrow RC, Maroney MJ, Palmer SM, Que L, Jr (1986) J. Am. Chem. Soc. 108: 6832

EXORCYCLE

is the name for a procedure used in two-dimensional nuclear magnetic resonance (see 2D-NMR) to "exorcise" two fundamental imperfections sometimes called "phantoms" or "ghost peaks":

a) the unwanted conversion of z-magnetization to detectable xy-magnetization, and
b) less than 100 per cent refocusing.

Fortunately, these two artifacts can be effectively suppressed by a simple phase-cycling technique that has been termed EXORCYCLE [1].

Experimental details of the application possibilities of EXORCYCLE in the field of 2D-NMR have been published recently [2].

References

1. Bodenhausen G, Freeman R, Turner DL (1977) J. Magn. Reson. 27: 511
2. Hull WE (1987) Experimental aspects of two-dimensional NMR. In: Croasmun WR, Carlson RMK (eds) Two-dimensional NMR spectroscopy: Applications for chemists and biochemists, Verlag Chemie, Weinheim

EXSY
Acronym for **Ex**change **S**pectroscopy

For EXSY essentially the same experimental procedure is applied as for nuclear OVERHAUSER effect spectroscopy (see NOESY). But, application of this procedure to the study of chemical exchange has been called EXSY or exchange spectroscopy [1].

In this case, however, it is desirable to extract accurate numerical values of rate constants in addition to exchange pathway information from the two-dimensional nuclear magnetic resonance (see 2D-NMR) spectra. This is relatively straightforward for cases of two-site exchange where explicit expressions have been devised [2] for peak intensities as a function of rate constant k, mixing time τ_m, and relaxation rate R. These expressions have been applied in a recent analysis of the carbon-13 2D EXSY spectra of carbonyl groups in the carbon-13 enriched complex $[Os_3H_2(CO)_{10}]$ by Hawkes and co-workers [3] who succeeded in measuring the rate of localized carbonyl group exchange.

The procedure becomes less straightforward when the number of exchanging sites, N, is greater than 2 leading to several exchange pathways, for example, in conformationally non-rigid or fluxional organometallic compounds.

One approach, used in a ^{119}Sn study of the kinetics of the halide redistribution reactions occurring in a mixture of $SnCl_4$ and $SnBr_4$ [4], is to repeat the experiment for several values of mixing time τ_m and plot the cross-peak intensities as functions of τ_m. Initial slopes of such curves then determine the exchange rates, but this procedure is extremely time consuming and impractical for precise kinetic studies. Randall et al. [5] developed a more satisfactory method to study three-site exchange of the CO groups in a ruthenium cluster by using a matrix formalism. Starting with initial trial values of the rate constants, it is possible to construct an exchange matrix diagonalization of which leads to the matrix of intensities, i.e. the two-dimensional spectrum. The latter is then compared with experimental intensities and an iterative procedure then varies rate constants and relaxation rates until the "best fit" of experimental and calculated 2D peak intensities is obtained. Note, that in this particular case, the problem is simplified, as the populations of the three sites are equal and the exchange matrix is symmetrical.

Recently [6] a general method for direct evaluating rate constants in complex exchange networks with N-sites from 2D EXSY NMR spectra has been proposed. A computer program capable of performing signal intensity to exchange rate calculations—and *vice versa*—based on a matrix formalism, has been described in detail [6]. This approach is similar to that briefly outlined by Perrin and Gipe [7] in their measurements of the acid-catalyzed proton exchange of acrylamide and thioacetamide. The new method has been applied to ^{195}Pt EXSY NMR of a $A \rightleftharpoons B \rightleftharpoons C$ three-site exchanging spin system with unequal populations as observed in complexes of trimethylplatinum (IV) halides with the ligand $MeSCH_2CH_2SMe$ [6]. Such a system has several advantages for demonstration purposes: First, the use of the ^{195}Pt nucleus eliminates interference from J cross-peaks [8] often present in proton spectra. There is only one ^{195}Pt nucleus per species and, therefore, no possibility of intramolecular NOE interactions which could contribute to the intensities of cross-peaks. Other advantages are associated with the spin-lattice relaxation times of platinum-195. Finally past experience of the laboratory group [6] with the above class of compounds [9] enabled the authors to reexamine them by conventional proton magnetic bandshape analysis and to make much more accurate quantitative comparisons of the 1D and 2D exchange NMR methods with some confidence.

A Japanese group of scientists [10] described, in 1987, a method for obtaining pure absorption phase spectra in four quadrants (complex method) in two-dimensional spin-exchange nuclear magnetic resonance spectroscopy. Modifying the phase-detection period proposed by States and co-workers [3], they have developed a new procedure which permits adjustment of the phases in both dimensions after double FOURIER transformation (see FT). Thus, phase-sensitive (pure absorption) spectra can be obtained with the homonuclear shift correlation, spin exchange, double-quantum-filtered, multiple-quantum coherence, hetero-nuclear shift coherence spectroscopy, etc. [10].

References

1. Wynants C, Van Binst G, Mügge C, Jurkschat K, Tzschach A, Pepermans H, Gielen M, Willem R (1985) Organometallics 4: 1906
2. Macura S, Ernst RR (1980) Mol. Phys. 41: 95
3. Hawkes GE, Lian LY, Randall EW, Sales KD, Aime S: J. Chem. Soc., Dalton Trans. 1985: 225
4. Ramachandran R, Knight CTG, Kirkpatrick RJ, Oldfield E (1985) J. Magn. Reson. 65: 136
5. Hawkes GE, Lian LY, Randall EW, Sales KD, Aime S (1985) J. Magn. Reson. 65: 173
6. Abel EW, Coston TPJ, Orrell KG, Šik V, Stephenson D (1986) J. Magn. Reson. 70: 34
7. Perrin CL, Gipe RK (1984) J. Am. Chem. Soc. 106: 4036
8. Macura S, Huang Y, Suter D, Ernst RR (1981) J. Magn. Reson. 43: 259
9. Abel EW, Khan AR, Kite K, Orrell KG, Šik V: J. Chem. Soc., Dalton Trans. 1980: 1175
10. Ohuchi M, Hosono M, Furihata K, Seto H (1987) J. Magn. Reson. 72: 279
11. States DJ, Haberkorn RA, Ruben DJ (1982) J. Magn. Reson. 48: 286

FC
Abbreviation of **FOURIER** Components

The fact that a given function f (t) may be expressed in a **FOURIER** series (see FS) of sines and cosines frequencies of 1/2T, 2/2T, 3/2T, etc., leads to the idea of describing the function itself only in terms of the coefficients of successive terms in these series [1]. Thus a function f(t) whose expansion gives rise to large values of A_1, A_2, and A_3 but small values of A_{30}, A_{31}, ... is said [1] to contain large contributions from the low frequency *FOURIER components* and small contributions from the large higher frequency FC.

Reference
1. See, for example, Farrar TC, Becker ED (1971) Pulse and Fourier transform NMR. Introduction to theory and methods, Academic, New York, Sect. 1.7

F

FELLGETT's Advantage
is one of the important advantages of a FT-IR (see p 116) spectrometer in comparison to a classical grating spectrometer [1]

FELLGETT's or the multiplex advantage of FT spectrometers may be expressed in the following way [2]: The signal-to-noise ratio (see SNR) of spectra measured on a FOURIER spectrometer will be much more greater than the SNR of a spectrum registered in the same time at the same resolution on a grating spectrometer with the same source, detector, optical throughput, optical efficiency, and modulation frequency by a factor equal to $M^{1/2}$, where M represents the number of resolution elements. In the case of mid-infrared spectra, with v_{max} = 4000 cm^{-1} and v_{min} = 400 cm^{-1}, measured in equal times at a resolution of 2 cm^{-1}, FELLGETT's advantage will be $1800^{1/2}$, or 42.

The time advantage of a FOURIER spectrometer should be even larger than the sensitivity, because of its direct proportionality to M. Thus, all other parameters being equal, a 2 cm^{-1}-resolution spectrum taking 30 min to measure on a grating spectrometer should be registered at equal sensitivity in 1 s on a FT spectrometer.

References
1. Fellgett PB (1958) J. Phy. Radium 19: 187, 237
2. Griffiths PR, de Haseth JA (1986) Fourier transform infrared spectrometry, Wiley, New York, p 275, 276 (Chemical analysis, vol 83)

FF
Abbreviation of **FOURIER** Frequencies

FF are modulation frequencies which have been take into account in infrared (see IR and FT-IR) rapid-scanning interferometry.

A typical mirror velocity commonly used for the measurement of IR spectra between 4000 and 400 cm^{-1} is 0.158 cm/sec. The modulations frequencies or the FF for the upper and lower limit of this range are, therefore, 1260 Hz (f_1 = 4000 cm^{-1}), respectively 126 Hz (f_2 = 400 cm^{-1}).

For further details the interested reader is referred to a monograph on FT-IR [1].

Reference

1. Griffiths PR, de Haseth JA (1986) Fourier transform infrared spectrometry, Wiley Interscience, New York

FFC-NMR

Acronym for **F**ast **F**ield-**C**ycling **N**uclear **M**agnetic **R**esonance

F. Noack and co-workers [1,2] have given the basic requirements and application possibilities of FFC-NMR.

Recently a field-cycling experiment has been reported which allows the detection of NOE's generated in a relatively weak magnetic field to be detected with the enhanced sensitivity and resolution of a stronger magnetic field [3]. This experiment is referred to as low-field NMR with the following basic scheme:

$$\begin{pmatrix} \text{high-field} \\ \text{equilibration} \\ \text{1. step} \end{pmatrix} \rightarrow \begin{pmatrix} \text{high-field} \\ \text{excitation} \\ \text{2. step} \end{pmatrix} \rightarrow \begin{pmatrix} \text{low-field} \\ \text{relaxation} \\ \text{3. step} \end{pmatrix} \rightarrow \begin{pmatrix} \text{high-field} \\ \text{detection} \\ \text{4. step} \end{pmatrix}$$

In the meantime a sample -shuttling device suitable for two-dimensional low-field NMR experiments has been constructed [4]. The design of this shuttling device meets all of the modern criteria and enhances the utility of low-field NMR. The described design of sample-shuttling should also be useful in other field-cycling experiments such as zero-field (see ZF) NMR [5] and the variety of new field-cycling experiments recently reviewed by Noack [1].

F. Noack et al. [6] have described much more recently a mathematical formalism which allows the design of the almost effective geometry for homo-generous, fast-switchable field-cycling magnets.

References

1. Noack F (1986) Prog. NMR Spectrosc. 18: 171
2. Rommel E, Mischker K, Osswald G, Schweikert KH, Noack F (1986) J. Magn. Reson. 70: 219
3. Kerwood DJ, Bolton PH (1986) J. Magn. Reson. 68: 588
4. Kerwood DJ, Bolton PH (1987) J. Magn. Reson. 75: 142
5. Zax DB, Bielcki A, Zilm KW, Pines A, Weitekamp DP (1985) J. Chem. Phys. 83: 4877
6. Schweikert KH, Krieg R, Noack F (1988) J. Magn. Reson. 78: 77

FFT

Abbreviation of **F**ast **FOURIER T**ransformation

Today discrete FOURIER transformations (see FT) are universally computed by an algorithm first elaborated by Cooley and Tukey [1]. Discrete FOURIER transformation can be defined by Eq. (1)

$$A_f = \sum_{t=0}^{N-1} X_t \exp - 2\pi i f t / N, \quad f = 0, 1, \ldots, N-1, \tag{1}$$

where A_f is the coefficient of the fth point in the transform (frequency domain), X_t is the value of the data (i.e. free induction signal or decay; see FID) at point t in the time domain, and N is the number of data points. From trigonometric identities, Eq. (2) is valid:

$$e^{-iy} = \cos y - i \sin y \ . \tag{2}$$

Hence the transform of Eq. (1) can be expressed as the complex sum of cosine and sine transforms [2]. The Cooley-Tukey algorithm [1] is an economical method of obtaining the A_f of Eq. (1) by combining in sequence progressively larger weighted sums of data points. This technique is often called FFT because of great saving in time over previously used methods of evaluating the A_f by direct multiplication and summation. The FFT method involves repetitive pairwise sorting of the data points; hence it works best for a number of data points, N, that is a power of 2. Saving in time arises from the smaller number of multiplications required by FFT compared with the direct method of transformation: $2N \log 2N$ multiplications for N points with FFT against N^2 multiplications with the direct method. A further important advantage of FFT comes from its efficient use of computer memory. Because the coefficients are evaluated by an iterative procedure, intermediate and final values can be written over original data values that are no longer needed. Thus $N(2^n)$ data points in computer memory can be transformed with only slightly more than N words of memory. A detailed and clear discussion of the rationale for the FFT method has been given by W.T. Cochran et al. [3].

References

1. Cooley JW, Tukey JW (1965) Math. Comput. 19: 297
2. Farrar TC, Becker ED (1971) Pulse and Fourier transform NMR: Introduction to theory and methods, Academic, New York
3. Cochran WT et al. (1967) Proc. IEEE 55: 1664

Addendum

Recently *alternatives to FFT* have attracted considerable interest [1]. The almost important among these procedures or proposals is the linear prediction singular value decomposition (see LPSVD) method of Kumaresan and Tufts [2]. In spectral analysis of nuclear magnetic resonance (see NMR) signals the work of Barkhuijsen and co-workers [3] is of particular interest and importance.

There are many *different alternatives to FFT* and one can classify them into two major categories:

a) the linear prediction methods (see LP) [2–7] and
b) the JAYNES maximum entropy method [8–17].

In a recent theoretical paper, Porat and Friedländer [18] suggested a modification of the Kumaresan-Tufts algorithm [2] used in LPSVD. Their modification was demonstrated to be more robust and efficient in the case of data with low signal-to-noise ratio (see SNR), without any important loss of computational economics. In a new paper Levy and co-workers [19] have shown that the Porat and Friedländer approach [18] is valid for NMR spectroscopy and can be used for an improvement of linear prediction processing of NMR spectra having very low SNR.

Recently Tang and Norris [20] have discussed the application of the linear prediction z-transform (see LPZ) method and the PADÉ rational approximation procedure [21,22]. These two approaches can yield phase-sensitive spectra with reduced noise and enhanced resolution while avoiding truncation artifacts. Similarities and differences among the LPZ method, the PADÉ approximation, and the BURG MEM (see p 34) have also been described [20].

The interested reader will bear in mind that all methods mentioned here are based on the LP theory developed by Prony [23] in the late 18th century.

References

1. Kay SM, Marple SL (1981) Proc. IEEE 69: 1380
2. Kumaresan R, Tufts DW (1982) IEEE Trans. Acoust. Speech Signal Process. 30: 833
3. Barkhuijsen J, de Beer J, Boveé WMMJ, van Ormondt D (1985) J. Magn. Reson. 61: 465
4. Tang J, Lin CP, Bowman MK, Norris JR (1985) J. Magn Reson. 62: 167
5. Tang J, Norris JR (1986) J. Chem. Phys. 84: 5210
6. Tang J, Norris JR (1986) J. Magn. Reson. 69: 180
7. Tang J, Norris JR (1986) Chem. Phys. Lett. 131: 252
8. Gull SF, Daniell GJ (1978) Nature (London) 272: 686
9. Collins DG (1982) Nature (London) 298: 49
10. Sibisi S (1983) Nature (London) 301: 134
11. Sibisi S, Skilling J, Brereton RG, Laue ED, Staunton J (1984) Nature (London) 311: 446
12. Laue ED, Skilling J, Staunton J, Sibisi S, Brereton RG (1985) J. Magn. Reson. 62: 437
13. Hore PJ (1985) J. Magn. Reson. 62: 561
14. Laue ED, Skilling J, Staunton J (1985) J. Magn. Reson. 63: 418
15. Martin JF (1985) J. Magn. Reson. 65: 291
16. Hore PJ, Daniell GJ (1986) J. Magn. Reson. 69: 386
17. Laue ED, Mayger MR, Skilling J, Staunton J (1986) J. Magn. Reson. 68: 14
18. Porat B, Friedländer B (1986) IEEE Trans. Acoust. Speech Signal Process. 34: 1336
19. Delsuc MA, Ni F, Levy GC (1987) J. Magn. Reson. 73: 548
20. Tang J, Norris JR (1988) J. Magn. Reson. 78: 23
21. Press WH, Flannery BP, Teukolsky SA, Vetterling WT (1986) Numerical recipes, Cambridge Univ. Press, Cambridge
22. Yeramian E, Claverie P (1987) Nature (London) 326: 169
23. Prony GRB (1795) J. L'Ecole Polytechnique 1: 22

New investigations in FT-NMR make it necessary to write the following *appendix*

Appendix

Phase anomalies in carbon-13 FT-NMR spectra have been discussed by Freeman and Hill [1] and some later on by Comisarow [2].

In repeated sequences of a transmitter pulse followed by data acquisition of the free induction decay (see FID), Freeman and Hill [1] have demonstrated that echoes are formed when the interval between successive pulses is short compared with the spin-spin relaxation time (T_2). As a result, the resonances in the FOURIER transformed spectrum exhibit a variable degree of dispersion mode character which cannot be removed by the commonly used frequency-independent and -dependent phase correction procedures. But, small random delays between pulses will prevent the accumulation of echoes and lead to properly phased carbon-13 spectra [1]. Comisarow [2] gave a quite different explanation for phase anomalies in carbon-13 FT-NMR spectra without any mention of the echo effect described earlier by Freeman and Hill [1,3].

Comisarow's explanation was based on the combination of a finite acquisition time and of the discrete character of the processing used to get the FT-NMR spectrum. He showed that for a *calculated continuous* pure absorption mode line shape, obtained by FT of a truncated signal, discrete sampling gives rise to an apparent misphasing when two conditions are satisfied: (a) the resonance frequency corresponding to the continuous absorption line shape of the actual signal does not fall onto one of the points in the discrete FT spectrum, and (b) the ratio of the acquisition time to the apparent relaxation time (here T_2^*) is less than about 2.

F.A.L. Anet and co-workers have discussed these controversial aspects on the origin(s) of phase anomalies observed in carbon-13 FT NMR spectra in detail [4]. They have shown experimentally that both these effects can occur separately or together and, in the latter case, the involved effects can interfere with one another. Procedures to ameliorate or remove phase anomalies in carbon-13 spectra have been briefly reviewed [4].

Current practice in FOURIER transform nuclear magnetic resonance (see FT, FFT, and NMR) involves multiplication of the sampled FID, f(t) by functions of time designed to accomplish three purposes:

a optimization of the signal-to-noise ratio (see SNR) [5],
b resolution enhancement [5], and
c phase corrections [6].

Recently [7] A. A. Bothner-By and J. Dadok have reported some additional useful manipulations of the FID. Both applications described by the authors [7] rest on the frequency-shift operation and seem to be interesting in the field of decoupling difference spectroscopy [8] or NOE difference spectroscopy (see NOE-Diff) [9].

References

1. Freeman R, Hill HDW (1971) J. Magn. Reson. 4: 366
2. Comisarow MB (1984) J. Magn. Reson. 58: 209
3. Freeman R, Hill HDW (1971) J. Chem. Phys. 54: 3367
4. Anet FAL, Zheng LK, Zhihong T (1987) Magn. Reson. Chem. 25: 439
5. Ernst RR (1966) Adv. Magn. Reson. 2: 1
6. Parks SI, Johannesen RB (1976) J. Magn. Reson. 22: 265
7. Bothner-By AA, Dadok J (1987) J. Magn. Reson. 75: 540
8. Gibbons WA, Beyer CF, Dadok J, Sprecher RF, Wyssbrod HR (1975) Biochemistry 14: 420
9. Pitner TP, Glickson JD, Dadok J, Marshall GR (1974) Nature (London) 250: 582

FGEPR
Acronym for **F**ield-**G**radient **E**lectron **P**aramagnetic **R**esonance

FGEPR is a new method used in electron spin resonance (see ESR or EPR), especially for *in vivo* studies. The method was first introduced by Karthe and Wehrdorfer [1], and has been reviewed recently [2].

In comparison with nuclear magnetic resonance imaging (see NMRI), the sample size in EPR or ESR imaging can presently be only up to 2 cm, which limits its application for *in vivo* problems. The spectra of spatially distributed molecules are broadened so that low signal-to-noise ratio (see SNR) restricts the resolution. Usually gradients up to 1 T/m seem to be close to the optimum [3].

Spatially separated samples of spin probes are detectable as separate peaks if their distance is 1 mm or more [4]. A concentration profile (or distribution function) of spin-labeled molecules within a sample [5] can be determined by a deconvolution procedure of the FGEPR spectra. Hornak and co-workers [6] studied the diffusion in anisotropic fluids by EPR imaging of concentration profiles. Diffusion of spin-labeled molecules in decalin has been measured by Galtseva et al. [7].

Recently [8] diffusion studies of spin probes in tissues have been published. In these studies a linear magnetic field gradient was superimposed on a main static field to resolve the spatial distribution of diffusing paramagnetic species. The diffusion coefficient and distribution function were evaluated from the FGEPR spectra by comparison with spectra calculated from a corresponding model. Swine subcutaneous adipose tissue penetration of spin-labeled methyl ester of palmitic acid was measured by this technique [8].

References

1. Karthe W, Wehrdorfer E (1979) J. Magn. Reson. 33: 107
2. Ewert U, Herrling T (1985) Proceedings of the 2nd International Conference on Modern methods in radiospectroscopy, Leipzig
3. Ohno H (1981) Japan J. Appl. Phys. 20: L179
4. Eaton SS, Eaton GR (1984) J. Magn. Reson. 59: 474
5. Hoch HJ (1981) J. Phys. C14: 5659
6. Hornak JP, Moscicki JK, Schneider DJ, Freed JH (1986) J. Chem. Phys. 84: 3387
7. Galtseva EV, Yakimochenko OY, Lebedov YS (1983) Chem. Phys. Lett. 99: 301
8. Demsar F, Cevc P, Schara M (1986) J. Magn. Reson. 69: 258

FGSE
Acronym for **F**ield-**G**radient **S**pin-**E**cho

FGSE is a technique used in nuclear magnetic resonance (see NMR) for diffusion measurements. Early in the history of NMR it was shown [1,2] that it is possible to produce a so-called "spin-echo", a refocusing of the nuclear magnetization which appears at a time τ after the second radio-frequency pulse in a two-sequence $90° - \tau - 180°$. This pulse sequence—or variations of it—is used to measure the spin-spin relaxation time (T_2). However, it was soon observed that in rapidly diffusing specimens the measured "T_2" became shortened in proportion to the inhomogeneities of the magnetic field in the region of the sample [1,2]. But, this effect, once analyzed [2], became the basis of all field-gradient spin-echo experiments. Measuring self-diffusion, a set of determinations of the magnitude of the spin-echo as a function of the magnitude and duration of the calibrated field-gradient yields the diffusion constant of the species at resonance.

References

1. Hahn EL (1950) Phys. Rev. 80: 580
2. Carr HY, Purcell EM (1954) Phys. Rev. 94: 630

FID
Acronym for **F**ree **I**nduction **D**ecay

In nuclear magnetic resonance (see NMR) the pulse FOURIER transformation (see PFT) technique has become the most important method [1], especially for nuclei with low natural abundance.

Suppose a 90° (or $\pi/2$) pulse is applied along the x' axis in the frame rotating at the radio-frequency (rf). Following this pulse, the magnetization, M, lies entirely along the y' axis, as shown in Fig. 10a. Since the normal NMR spectrometer is usually arranged to detect signals induced in a coil along the fixed x or y axis, the magnitude of M_{xy} determines the strength of the observed signal, called a free induction signal or FI, since the nuclei under investigation precess "freely" without applied rf. As transverse relaxation occurs, the FI signal decays. The instrument response is therefore the FID.

In a perfectly homogeneous field the time constant of that decay would be T_2, the so-called transverse relaxation time or spin-spin relaxation time. But, in fact, the FI signals decay in a time T_2^* that is often governed predominantly by field inhomogeneties, because nuclei in different positions of the field precess at slightly different frequencies, hence rapidly get out of phase with each other.

Thus the FID must be described by Eq. (1) where ΔH_0 stands for the magnetic field inhomogeneity and γ for the magnetogyric ratio of the nucleus involved.

$$1/T_2^* = 1/T_2 + (\gamma \Delta H_0/2) \tag{1}$$

Figure 10b indicates decreasing of M as magnetic moments dephase. Figure 10c shows the input signal, the 90° pulse, corresponding to Fig. 10a. Figure 10d visualizes the exponential FID, corresponding to Fig. 10b.

FOURIER transformation (see FT and FFT) of the FID leads to the "normal" NMR spectrum. A single LORENTZian line in the frequency sweep of an "ordinary" spectrum corresponds to a simple exponential decay of the FI after 90° pulse. Mathematically we can write

$$T_2[(1 + T_2^2(\omega - \omega_0)^2] = \int_0^\infty \exp(-t/T_2)\cos(\omega - \omega_0)t\,dt. \tag{2}$$

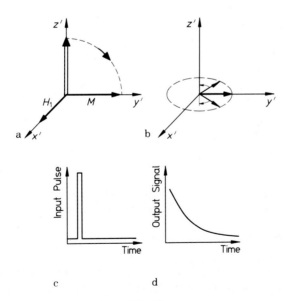

Fig. 10

On the left-hand side of Eq. (2) we find the LORENTZian function as expression for the commonly used absorption mode that comes from solution of BLOCH's equations (see p 29). The corresponding sine transformation of an exponential gives the dispersion mode, rather than the absorption mode. When several lines are present, the modulation or "ringing" frequencies interfere with each other to produce an *interferogram* as a complex FID.

Manipulation procedures applied to the FID before FT will be explained in this dictionary under their individual acronyms or abbreviations.

Unless $T_2 \ll (2/\gamma\Delta H_0)$ [see Eq. (1)], the contribution of inhomogeneity in H_0 to the FID precludes the use of this decay time, T_2^*, as a measure of T_2. An almost ingenious method to overcome the inhomogeneity problem was first proposed by Hahn [2], who called it the *spin-echo* method (see SE). The procedure consists of the application of a $90°$, τ, $180°$ pulse sequence and the observation at a time 2τ of a free induction *echo*, sometimes called the *Hahn spin-echo* experiment.

For alternative methods in the field of T_2 measurements see the acronyms CP-T, MG-M, and CPMG in this dictionary. Recently, non-iterative techniques have been proposed for a direct analysis of the FID in the time-domain [3]. In the presence of high signal-to-noise ratios (see SNR), these techniques can distinguish between closely spaced "signals" which would appear as overlapping peaks in the FOURIER-transformed spectrum. Although FT is a least-squares best fit, discriminating between small signal peaks and large noise peaks seems to be impossible. In the presence of increasing noise, the time-domain techniques become unstable and "true signals" will be ignored and treated as noise.

To overcome these problems, M. Shinnar and S.M. Eleff [4] have derived a new approach, which is based on the PISARENKO method [5] for harmonic retrieval [6] and uses information about the amount of noise in the "signal" to analyze the FID data directly and may be useful even in low SNR experiments.

References

1. Farrar TC, Becker ED (1971) Pulse and Fourier transform NMR: Introduction to theory and methods, Academic, New York, chap 2
2. Hahn EL (1950) Phys. Rev. 80: 580
3. Barkhuijsen H, de Beer R, Boveé MMMJ, van Ormondt DJJ (1985) J. Magn. Reson. 61: 465
4. Shinnar M, Eleff SM (1988) J. Magn. Reson. 76: 200
5. Pisarenko VF (1973) Geophys. J.R. Astron. Soc. 33: 347
6. Tang J, Lin CP, Bowman MK, Norris JR (1985) J. Magn. Reson. 62: 167

FILTERING

Such procedures have been used in almost all modern nuclear magnetic resonance (see NMR) methods

These techniques fall in two categories depending on whether the aim is to realize increased resolution or to improve a genuine low signal-to-noise ratio (see SNR) in NMR spectra.

Methods used for increased resolution include

a) GAUSSian transformation [1],
b) sine bell smoothing [2,3],
c) convolution difference (see CD) [4], or
d) apodization with the TRAF (see p 288) function [5].

For noise suppression, the commonly used procedures are matched filtering [6] and various maximum entropy methods (see ME or MEM) [7–12].

Recently both MEM alone and combinations of them have been claimed to do the two tasks [8,13,14].

All these problems have been discussed together, resulting in a method for automatic quantification of one-dimensional NMR spectra with low SNR [15].

References

1. Ferridge AG, Lindon JC (1978) J. Magn. Reson. 31: 337
2. De Marco A, Wüthrich K (1976) J. Magn. Reson. 24: 201
3. Brereton RG, Garson MJ, Staunton J (1981) J. Magn. Reson. 43: 224
4. Campbell ID, Dobson CM, Williams RJP, Xavier AV (1973) J. Magn. Reson. 11: 172
5. Traficante DD, Ziessow D (1986) J. Magn. Reson. 66: 182
6. Ernst RR (1966) Advances in magnetic resonance, Academic, New York, vol 2 pp 1–135
7. Hore PJ (1985) J. Magn. Reson. 62: 561
8. Laue ED, Skilling J, Staunton J, Sibisi S, Brereton RG (1985) J. Magn. Reson. 62: 437
9. Laue ED, Skilling J, Staunton J (1985) J. Magn. Reson. 63: 418
10. Martin JF (1985) J. Magn. Reson. 65: 291
11. Sibisi S (1983) Nature (London) 301: 134
12. Sibisi S, Skilling J, Brereton RG, Laue ED, Staunton J (1984) Nature (London) 311: 446
13. Ni F, Levy GC, Scheraga HA (1986) J. Magn. Reson. 66: 385
14. Lohman JAB (1980) J. Magn. Reson. 38: 163
15. Nelson SJ, Brown TR (1987) J. Magn. Reson. 75: 229

F

FIR
Abbreviation of **F**ar **I**nfrared (Spectroscopy)

The FIR spectral region is difficult to study because of energy limitations. The commonly used blackbody sources, such as NERNST's glowers or globars will only contribute less than one hundredth of one per cent of the total energy in the spectral region below 100 cm^{-1} with an peak output near 3000 cm^{-1}. In addition, in FIR, the elimination of unwanted energy from shorter wavelengths became a problem. Finally, detectors for the long wavelength region are poor, so the overall sensitivity in the FIR region is further reduced. But, with FOURIER transformed infrared (see FT-IR) this region is now accessible for additional studies.

FIRFT
Acronym for **F**ast **I**nversion-**R**ecovery **F**ourier **T**ransform

Measurements of spin-lattice relaxation times (T_1) in multiline NMR spectra are often achieved by the inversion-recovery method combined with FOURIER transform (see IRFT) [1]. The applicability of this method is limited by its duration which depends on the number of scans and on the waiting time, T. This difficulty is particularly stringent for low sensitivity nuclei such as ^{13}C or ^{15}N. It has been shown, however, that inversion-recovery with an arbitray short waiting time is comparable to SRFT (see p 268) and PSFT (see p 226) techniques in speed, and that such a sequence gives optimum dynamic range with respect to experimental time for each individual resonance line. For a set of T_1 spectra, dynamic range is essentially equivalent to sensitivity [2–4]. This new sequence got the acronym FIRFT [5] and does not share many of the disadvantages of the SRFT or PSFT sequences. The sequence utilizes a waiting time T much smaller than $5(T_1)_{max}$ but sufficient to allow the decay of the transverse magnetization before the next $180°$ pulse. This is actually achieved in a short time for carbon-13 measurements with broad-band proton decoupling [2]. The successive free induction decays (see FID) have amplitudes equal to:

$$S_\infty[1-2\exp(-\tau/T_1)]; S_\infty[1-(2-E_1)\exp(-\tau/T_1)]; \ldots;$$
$$S_\infty[1-(2-E_1)\exp(-\tau/T_1)] , \tag{1}$$

where $E_1=\exp(-T/T_1)$. The accumulated signal is consequently related to Eq. (2):

$$S_\tau=S_\infty[1-\alpha\exp(-\tau/T_1)] \tag{2}$$

with

$$\alpha=2-E_1((n-1)/n) . \tag{3}$$

If n is large or if the first FID is deleted, then S_τ can be written in terms of Eq. (4) and the usual logarithmic plot

$$S_\tau=S_\infty[1-(2-E_1)\exp(-\tau/T_1)] \tag{4}$$

allows the determination of the spin-lattice relaxation time T_1:

$$\ln((S_\infty-S_\tau)/S_\infty)=\ln\alpha-\tau/T_1 \tag{5}$$

In FIRFT τ values are chosen exactly as in a conventional IRFT experiment. Of course, the measurement of S_∞ requires that one spectrum be obtained where τ is $\gtrless 5(T_1)_{max}$. Inspection of Eq. (4) shows that instead of a dynamic range of 2 as in the common IRFT sequence, the new FIRFT sequence has a dynamic range of $(2-E_1)$, where E_1 is smaller than 1. The dynamic range or the overall sensitivity of this experiment is optimized for each resonance line since E_1 is very small when the waiting time, T, is longer than or comparable to T_1. In the case of $T \ll T_1$, E_1 approaches 1 and the dynamic range is consequently reduced to that of the SRFT and PSFT techniques. Thus the dynamic range will vary from a little over 1 (in the case of very long T_1 values) to exactly 2 (for very short T_1 values) for different nuclei during a single experiment, whose total experimental time can be up to an order of magnitude shorter than with the conventional IRFT sequence. The exact time shared depends on the discretionary value of the waiting time, T, in both cases (e.g. 3 to 5 times T_1 in IRFT; 1 to nearly 10 seconds in FIRFT) and on number of scans, n, since the IRFT method does not discard the first FID. The advantages of FIRFT can be summarized: (a) a short waiting time which could be equal to the acquisition time as in progressive saturation; (b) a better overall sensitivity than in other techniques ($E_1 < 1$); and (c) no limitation concerning the measurement of short relaxation times since τ can be set to any desired value [5].

References

1. Vold RL, Waugh JS, Klein MP, Phelps DE (1968) J. Chem. Phys. 48; 3831
2. Freeman R, Hill HDW (1971) J. Chem. Phys. 54: 3367
3. Freeman R, Hill HDW, Kaptein R (1972) J. Magn. Reson. 7: 82
4. Levy GC, Peat IR (1975) J. Magn. Reson. 18: 500
5. Canet D, Levy GC, Peat IR (1975) J. Magn. Reson. 18: 199

FLASH Imaging
Acronym for Fast Low Angle Shot NMR Imaging

Localized high-resolution nuclear magnetic resonance (see NMR) spectroscopy *in vivo* is gaining increasing interest for both research and medical applications, today. A variety of techniques and methods has been proposed to focus the NMR-sensitive region to a selected volume-of-interest (VOI). In particular, localization approaches (see the acronyms VSE, SPARS, ISIS, and STEAM in this dictionary on p 192) take full advantage of magnetic field gradients for the spatial selection, providing an almost high degree of flexibility with respect to location, size and shape of the VOI. In all these cases definition of the VOI is given by a set of frequencies or complex shift factors representing the spatial positions of a corresponding set of slice-selective excitation pulses or pulse packages to be applied in the presence of three orthogonal gradients. Of course, if high spectral resolution is wanted many imaging experiments have to performed, normally resulting in an unacceptably high measuring time from the medical point of view.

F

Here the introduction of a rapid NMR imaging sequence like FLASH was helpful [1]. This powerful sequence using low flip angles and gradient-recalled echoes [1,2] has been applied to the field of "spectroscopic images" (SFLASH) [3] and for the rapid image characterization for localized "*in vivo* spectroscopy" [4]. With measuring times of less than 10 seconds for images with a spatial resolution of 256×256 pixels, the procedure is almost ideally suited for a rapid characterization of the VOI and permits a direct correlation of the VOI to the surrounding tissues [4].

References
1. Haase A, Frahm J, Matthaei D, Merboldt KD, Hänicke W (1986) J. Magn. Reson. 67: 258
2. Frahm J, Hänicke W, Merboldt KD (in press) J. Magn. Reson.
3. Haase A, Matthaei D (1987) J. Magn. Reson. 71: 550
4. Frahm J (1987) J. Magn. Reson. 71: 568

FNMRI
Acronym for Flow Nuclear Magnetic Resonance Imaging

FNMRI is a new imaging method which allows a non-invasive determination of flow velocities via nuclear magnetic resonance (see NMR) signals [1].

The effect of motion on the NMR signal [2–5] has led to the development of various techniques for NMR flow measurements [6–11]. In principle, most of these experimental procedures can be adopted for NMR imaging (see NMRI) either to allow a much more qualitative discrimination between stationary and non-stationary magnetizations, e.g. to delineate vascular structures in the field of medical diagnosis, or to provide a spatially resolved determination of flow velocities or profiles in so-called rheotomography.

For example, Singer and Crooks [12] have demonstrated that a slice-selective saturation pulse (SSSP) preceding a conventional NMRI sequence may be used to measure flow velocities by varying the delay between the saturation and the common excitation pulse.

From F. Wehrli and co-workers [13] we have learned that to suppress the effects of stationary spins: saturation-recovery images (see SRFT) obtained with either a slice-selective or non-selective saturation pulse have to be digitally subtracted.

On the other hand, velocity-dependent phase-sensitive images of moving nuclear spins can be obtained by incorporation of self-refocusing magnetic field gradients into the imaging pulse sequence [14]. In time-of-flight type measurements the nuclear spins are excited within a certain spatial location, while their displacement is monitored by a typical "slice-selective read pulse" in a different position [15].

As mentioned above, FNMRI [1] is a new method in this field. The authors [1] describe, in their paper a related bolus-tracking technique which allows an extension of the common accessible range of flow velocities to much smaller values. Instead of using the conventional spin-echo (see SE) NMR signal, this technique employs stimulated-echo (STE) signals. The method itself is a time-of-flight measurement of spins between excitation and detection. For this purpose, the first and the third pulse of a stimulated-echo acquisition mode (STEAM) sequence $(90° - t_1 - 90° - t_2 - 90° - t_3)$ are taken to be slice-selective defining the tagging and target slice, respectively [1].

References

1. Merboldt KD, Hänicke W, Frahm J (1986) J. Magn. Reson. 67: 336
2. Suryan G (1958) Proc. Indian Acad. Sci. (A) 33: 580
3. Hahn EL (1950) Phys. Rev. 80: 580
4. Carr HY, Purcell EM (1954) Phys. Rev. 94: 630
5. Steijskal EO (1965) J. Chem. Phys. 43: 3597
6. Singer JR (1959) Science 130: 1652; Morse OC, Singer JR (1970) Science 170: 440; Grover T, Singer JR (1971) J. Appl. Phys. 42: 938
7. Packer KJ (1969) Mol. Phys. 17: 355
8. Hayward RJ, Packer KJ, Tomlinson DJ (1972) Mol. Phys. 23: 1083
9. Garroway AN (1974) J. Phys. D. 7: L159
10. Hemminga MA, De Jager PA (1977) J. Magn. Reson 27: 359
11. Battocletti JH, Halbach RE, Salles-Cunka SX, Sances A, Jr (1981) Med. Phys. 8: 435
12. Singer JR, Crooks LE (1983) Science 221: 654
13. Wehrli FW, MacFall JR, Axel L, Shutts D, Glover GH, Herfkens RJ (1984) Noninvasive Med. Imaging 1: 127
14. Moran PR (1982) Magn. Reson. Imaging 1: 197
15. Feinberg DA (1984) Magn. Reson. Med. 1: 151; Feinberg DA, Crooks LE, Hoenninger J III, Arakawa M, Watts J (1984) Radiology 153: 177

FOCSY

Acronym for **F**old**o**ver-**C**orrected **S**pectroscop**y**

A method introduced by R.R. Ernst et al. [1] to reduce the size of the data matrix without information loss in practical situations of two-dimensional nuclear magnetic resonance (see 2D-NMR) spectroscopy.

Reference

1. Nagayama K, Kumar A, Wüthrich K, Ernst RR (1980) J. Magn. Reson. 40: 321

FO-ESR
Abbreviation of **F**low **O**rientation-**E**lectron **S**pin **R**esonance

A flow-orientation technique for ESR experiments in the Q band region was described more than 20 years ago by Onishi and McConnell [1]. The major disadvantage of this technique was the restriction to Q-band ESR and the problems arising with experiments with flow parallel to the magnetic field, since Q-band cavities do not allow easy access from the sides.

Recently [2] a new flat cell, consisting of a long capillary folded within the outer dimensions of a regular TM_{110} flat cell, has been described. It can now be used for FO experiments in X-band ESR and can be oriented both parallel and perpendicular to the magnetic field of the spectrometer.

References
1. Onishi S, McConnell HM (1965) J. Am. Chem. Soc. 87: 2293
2. Stolze K, Mason RP (1987) J. Magn. Reson. 73: 287

F

FORMAN's Method
Name for a procedure involving FOURIER transformation and phase correction used in infrared (see IR) spectroscopy

The early steps of FORMAN'S method are identical with those of MERTZ's procedure (see p 175) because both these methods use a short double-sided interferogram to calculate the phase of the spectrum [1]. But later on, we find two different features in FORMAN's method. First, no apodization is applied to a double-sided interferogram and second, no interpolation of the phase curve is performed. The complete cosine and sine phase curves are calculated from the phase curve, and phase-undefined regions are left as zero. When inverse complex fast FOURIER transformation (see FFT) is applied to the cosine (real) and sine (imaginary) part of the phase curve, an interferometric signal results. The interferogram possesses maxima at both ends due to the fact that the phase is referenced to zero. The data are then shifted so that the centerbur is in the middle of the array. The interferogram of the phase is now apodized prior to convolution with a special apodization function $A(\delta)$, devised by Forman [1] and given with Eq. (1).

$$A(\delta) = [1 - (\delta/\Delta)^2]^2 \tag{1}$$

This kind of apodization function (see Apodization) has been used by Forman [1] in order to reduce the effects of the numerical filtering operation, which were introduced by default, by nulling the phase in the undefined regions. The interferogram that has been phase corrected exhibits no chirping and is by definition a real and even function. The remainder of the data is zero filled and the short side moved to the end prior to FT. The single-beam spectrum is calculated from the data by a cosine FFT. A reference spectrum is calculated in an identical manner and the two single-beam spectra are ratioed, as in MERTZ's method (see p 175), to produce a ratioed spectrum.

Reference
1. Forman ML, Steele WH, Vanasse GA (1966) J. Opt. Soc. Am. 56: 59

FORTRAN
Acronym for **For**mula **Tran**slation

FORTRAN in its different versions is the almost important programming language for scientific applications of digital computation. In the past, all versions of FORTRAN had to be compiled into more machine-oriented languages such as *assembler*, etc. for time sharing reasons.

FS
Abbreviation of **FOURIER** Series

It is well-known that a given function f (t) can usually be expressed as an FS, i.e. an infinite series of sines and cosines, as given in Eq. (1).

$$f(t) = \sum_{n=0}^{\infty} A_n \cos(n\pi/T)t + \sum_{n=1}^{\infty} B_n \sin(n\pi/T)t \qquad (1)$$

In standard mathematical textbooks [1] one can read that this expansion is valid over the region $-T \leq t \leq T$, and formulas are derived for determination of the coefficients A_n and B_n.

If f(t) is a symmetric or *even* function [i.e. $f(-t) = f(t)$], all B are equal to zero, and only the cosine series is needed; if f(t) is antisymmetric or *odd* [i.e. $f(-t) = -f(t)$], then only sine series is needed. A general, asymmetric function requires the complete series. For example, the representation of a simple square-wave function in a FOURIER sine series will show that inclusion of higher terms leads to a progressively better approximation [2].

The fact that a function may be expressed in a FS of sines and cosines with frequencies of 1/2T, 2/2T, 3/2T, etc., leads to an approach which describes the function merely in terms of the coefficients of successive terms. This approach is explained under the abbreviation FC in this dictionary.

References
1. See, for example, Margenau H, Murphy GM (1943) The mathematics of physics and chemistry, Van Nostrand, Princeton, chap 8
2. See, for example, Farrar TC, Becker ED (1971) Pulse and Fourier transform NMR. Introduction to theory and methods, Academic, New York, Sect. 1.7

FSD
Abbreviation of **FOURIER** Self-**D**econvolution

Most bands in the infrared (see IR) absorbance spectra have a LORENTZian profile. Since the LORENTZian broadening function is removed, this process has been called FSD. The technique was first described in 1962 by Stone [1] and more recently has been discussed in more detail in a series of three papers by Kauppinen and co-workers [2–4].

The type of weighting functions used normally in FSD are similar to those used in the generation of derivative spectra in that their magnitude is greatest for large retardations. It is again necessary for the noise level of the spectrum to be very low if self-deconvolution is to be successfully achieved. It is also important to realize that the amount by which the full width at the half-height (see FWHH) of bands can be reduced is determined by the resolution at which the spectrum was primarily measured. Since the selection of the parameters used in the FSD procedure is usually done empirically, it is always necessary to look for the appearance of side lobes, which indicates an "overdeconvolution". It has been pointed out [5], that the best way for checking that all features in deconvolution are real is to verify their existence by remeasuring the spectrum at a higher resolution and a better signal-to-noise ratio (SNR), if possible.

References

1. Stone H (1962) J. Opt. Soc. Am. 52: 998
2. Kauppinen JK, Moffat DJ, Mantsch HH, Cameron DG (1981) Appl. Spectrosc. 35: 271
3. Kauppinen JK, Moffat DJ, Mantsch HH, Cameron DG (1981) Anal. Chem. 53: 1454
4. Kauppinen JK, Moffat DJ, Cameron DG, Mantsch HM (1981) Appl. Opt. 20: 1866
5. Griffiths PR, de Haseth JA (1986) Fourier transform infrared spectrometry, J. Wiley, New York, pp 102–106 (Chemical analysis, vol 83)

F

FSW
Abbreviation of FOURIER Series Window, a method used for *in vivo* nuclear magnetic resonance (see NMR)

Among the different B_1 localization techniques, FSW provides a straightforward method to spatially restrict the delineated volume in *in vivo* NMR spectroscopy [1–4]. FSW localization is achieved by incrementing the spatially dependent nutation angle, θ, such that spatial information is encoded as the amplitude-modulation frequency of the transverse magnetization. FSW can maximize information for a single region of interest by ensuring that the incremented nutation angle at the centre of the window is always an odd multiple of $\pi/2$. FSW's on- and off-resonance using multiple coils and longitudinal modulation have been recently tested for *in vivo* ^{31}P spectroscopy [5].

References

1. Garwood M, Schleich T, Ross BD, Matson GB, Winters WD (1985) J. Magn. Reson. 65: 239
2. Garwood M, Schleich T, Bendall MR, Pegg DT (1985) J. Magn. Reson. 65: 510
3. Garwood M, Ugurbil K, Schleich T, Petein M, Sublett E, From AHL, Bache RJ (1986) J. Magn. Reson. 69: 576
4. Metz KR, Briggs RW (1985) J. Magn. Reson. 64: 172
5. Garwood M, Robitaille P-M, Ugurbil K (1987) J. Magn. Reson. 75: 244

FT-ESE
Acronym for FOURIER Transform-Electron Spin Echo Spectroscopy

This new technique [1] was born after the special features of FOURIER transformation were understood (see FT and FFT) in electron spin resonance (see ESR or EPR) and after the rapid developments of FT methods in other spectroscopic disciplines such as nuclear magnetic resonance (see NMR), infrared spectroscopy (see IR), etc.

But, it must be pointed out that ESE spectroscopy was a two-dimensional version [2,3] before FT was applied.

It is therefore realistic to expect that FT-ESE will increase the performance and speed of former 2D-ESE studies [4], since field-scanning is no longer necessary.

It has been estimated that a 2D spectrum recorded in 5 h by the older technique could take a few tens of minutes after implementation of the new procedures [1].

The requirements for rotating the magnetization associated with an ESR spectrum by a microwave pulse have also been considered [1].

References
1. Hornak JP, Freed JH (1986) J. Magn. Reson. 67: 501
2. Millhauser GL, Freed JH (1984) J. Chem. Phys. 81: 37
3. Merks PJ, de Beer R (1979) J. Phys. Chem. 83: 3319
4. Kar L, Millhauser GL, Freed JH (1984) J. Phys. Chem. 88: 3951

F

FTESR
Acronym for **F**ourier **T**ransform **E**lectron **S**pin **R**esonance

The technique of pulsed FOURIER transform nuclear magnetic resonance (see NMR) is well established and understood [1]. The equivalent technique for electron spin resonance (see ESR or EPR) has been slower to be developed, but recent dovelopments in ESR instrumentation have now made FTESR possible [2]. Modern FT (see FT or FFT) and two-dimensional techniques appear to be inadequately understood in ESR. However, such matters are becoming increasingly relevant to ESR with the development of low Q, high B_1 resonators [3–5] in electron spin-echo (see ESE) spectroscopy [2]. In fact, it is becoming practical to rotate the magnetization from an entire ESR spectrum with a microwave pulse [2,4].

References
1. Fukushima E, Roeder SBW (1981) Experimental pulse NMR. A nuts and bolts approach, Addison-Wesley, Reading/USA
2. Hornak JP, Freed JH (1986) J. Magn. Reson. 67: 501
3. Froncisz W, Hyde JS (1982) J. Magn. Reson. 47: 515
4. Hornak JP, Freed JH (1985) J. Magn. Reson. 62: 311
5. Papers presented at the 8th International EPR Symposium, Denver/USA 1985: (a) Froncisz W, Hyde JS: Topologies of the loop-gap resonator; (b) Grupp A, Seidel H, Hofer P, Moresh GG, Mehring M: Design and performance of slotted tube resonators for pulsed ESR and ENDOR; (c) Bo Brutto R, Leigh JS: Construction of a compact X-band microwave resonator for use with liquid helium flow systems; (d) Lin CT, Massoth RJ, Norris JR, Bowman MK: Microwave resonator performance in photosynthetic pulsed EPR experiments.

FT-IR
Abbreviation of **F**ourier **T**ransform-**I**nfrared (Spectroscopy)

The advent of FOURIER transform-infrared spectroscopy (FT-IR) has brought about a revival of interest in infrared spectroscopy (see IR) as a characterization technique. The increased speed and higher signal-to-noise ratio of FT-IR relative to dispersion infrared has lead to a substantially greater number of applications of infrared spectroscopy in research. Also the availability of a dedicated computer, which is required for the FT-IR instrumentation, has allowed the digitized spectra to be treated by sophisticated data processing techniques and has increased the utility of the infrared spectra for both qualitative and quantitative purposes. FT-IR has almost become a ubiquitous technique in physical and analytical chemistry and a number of sources can be consulted for further details [1-4].

References

1. Griffiths PR (1975) Chemical infrared transform spectroscopy, Wiley, New York
2. Ferraro JR et al. (1978) In: Ferraro JR, Basile LJ (eds) Fourier transform infrared spectroscopy applicaitons to chemical systems, Academic, New York
3. Koenig JL (1981) Acct. of Chem. Res. 14: 171
4. Durig J (ed) (1980) Biological and chemical applications of FT-IR, Reidel, Dordrecht

F

FT-RS
Abbreviation of **FOURIER** Transform-**RAMAN** Spectroscopy

The major limitation to the routine application of RAMAN spectroscopy (see p 233) is the interference caused be fluorescence, either of impurities or of the sample itself. In the past, several methods have been developed to circumvent this problem and the parallel process of thermal degradation. Among them are so-called time-resolved techniques [1], non-linear optical methods [2], ultraviolet RAMAN and resonance RAMAN measurements [3], and SERS methods [4] (see SERS). All these different techniques have shown some success for certain samples and problems, but none offer an universal approach to the problem. Initial studies by Hirschfeld and Chase [5] and Jennings and co-workers [6] demonstrated that reasonable RAMAN spectra could be obtained with a multiplexing spectrometer. Much more recently [7] it has been shown that FT-RS with a near-infrared laser (see NIR and LASER) can provide a universal solution to the problems of fluorescent interference in RAMAN spectroscopy as mentioned above. Data have been obtained on systems exhibiting extremely high fluorescent signals under visible excitation. RAYLEIGH's line filtering has been used for interferometric detection, and the possibilities for spectral subtraction (a common technique in FT-IR, see p 116) with RAMAN spectroscopic data have been demonstrated [7].

References

1. Watanabe J, Kinoshita S, Kushida T (1985) Rev. Sci. Instrum. 56: 1195 and additional references therein
2. Rahn L, Farrow RL, Maffern PL (1982) In: Lascombe J, Huong PV (eds) Linear and non-linear Raman spectroscopy, Wiley, New York, p 143
3. Asher S: private communication to D.B. Chase [7]
4. Metiu H (1984) Prog. Surf. Sci. 17: 162
5. Hirschfeld T, Chase DB (1986) Appl. Spectrosc. 40: 133
6. Jennings DE, Weber A, Braselt JW (1986) Appl. Opt 25: 284
7. Chase DB (1986) J. Am. Chem. Soc. 108: 7485

FWHH
Acronym for **F**ull **W**idth at the **H**alf-**H**eight

Several criteria have been used to define the resolution of infrared (see IR) spectrometers. The most popular are RAYLEIGH's criterion (see p 234) and the full width at the half-height (FWHH) of a line. The FWHH criterion is more useful in the case of spectrometers with a triangular slit function. Two triangularly shaped lines of equal intensity and half-width are not resolved until the spacing between the lines is greater than the FWHH of either line. The FWHH of a line whose shape is a sinc function given by Eq. (1) is $0.605/\Delta$, but two lines with sinc × line shapes are not resolved when they are separated by only this amount. In practice, a dip of nearly 20 per cent is found when the two lines with a sinc × instrument line shape function (see ILS) are separated by $0.73/\Delta$ [1].

$$f(\bar{v}) = \frac{2\Delta \sin(2\pi\bar{v}\Delta)}{2\pi v\Delta}$$

$$\equiv 2\Delta \text{ sinc } 2\pi\bar{v}\Delta \tag{1}$$

Reference
1. Griffiths PR, de Haseth JA (1986) Fourier transform infrared spectrometry, Wiley, New York, pp 9–15 (Chemical analysis, vol 83)

GARP
Acronym for **G**lobally **O**ptimized **A**lternating **P**hase **R**ectangular **P**ulse

The GARP sequence [1] has been applied in nuclear magnetic resonance (see NMR) for heteronuclear decoupling of nuclei with wide chemical shift (see CS) regions. Decoupling can be realized over a range of $\pm 2.5\gamma B_2$ which may be of interest in the field of *inverse* proton/carbon-13 NMR experiments [2].

References
1. Shaka AJ, Barker PB, Freemann R (1985) J. Magn. Reson. 64: 547
2. Kessler H, Gehrke M, Griesinger C (1988) Angew. Chem. 100: 507; (1988) Angew. Chem. Int. Ed. Engl. 27: 490

GASPE
Acronym for **Ga**ted **S**pin-**E**cho

G

GASPE is one of the various multi-pulse techniques for *spectral editing* of carbon-13 nuclear magnetic resonance (see NMR) spectra [1] and has been applied in the field of qualitative and quantitative analysis of fossil-fuel-derived oils [2,3].

References
1. Cookson DJ, Smith BE (1981) Org. Magn. Reson. 16: 111
2. Cookson DJ, Smith BE (1983) Fuel 62: 34
3. Cookson DJ, Smith BE (1983) Fuel 62: 987

GAUSSian Pulses
or GAUSSian-shaped pulses may be used insted of rectangular pulses with some advantages in high-resolution nuclear magnetic resonance (see NMR) spectroscopy [1].

Such pulses were already widely used in nuclear magnetic resonanace imaging (see NMRI) techniques [2,3].

Bauer et al. showed an application of one-dimensional COSY spectra to reveal proton-proton connectivities.

H. Kessler and co-workers [4] have extended the methodology of Bauer et al. [1] and have proposed a general procedure, which allows the transformation of almost all homonuclear two-dimensional NMR techniques into one-dimensional sequences by using semi-selective GAUSSian pulses.

These techniques are particularly advantageous in cases where only a limited amount of information is required for solving a chemical problem.

Kessler et al. [4] have derived nine homonuclear 1D pulse sequences for nine originally proposed two-dimensional experiments (COSY, MQF-COSY, COSY with z-filter, MQF-COSY with z-filter, NOESY, relayed COSY, relayed COSY with z-filter, relayed NOESY, and TOCSY) using semi-selective excitation with a GAUSSian pulse. Phase alternation of the first selective 90° pulse was applied to eliminate such signals which were not excited by the GAUSSian pulse.

Furthermore, a variant of the 1D-COSY, the refocused 1D-COSY with z-filtering has been suggested [4], which in combination with 1D-COSY allows the application of the DISCO (see p 74) technique for the determination of coupling constants from multiplets of crowded regions.

References

1. Bauer CJ, Freeman R, Frenkiel T, Keeler J, Shaka AJ (1984) J. Magn. Reson. 58: 442
2. Sutherland RJ, Hutchinson JM (1978) J. Phys. E 11: 79
3. Hutchinson JM, Sutherland RJ, Mallard JR (1978) J. Phys. E 11: 217
4. Kessler H, Oschkinat H, Griesinger C, Bermel W (1986) J. Magn. Reson. 70: 106

GD
Abbreviation of **G**ated **D**ecoupling

G

This is a technique used in carbon-13 nuclear magnetic resonance to obtain spectra with the full coupling information in reasonable measurement times [1–3]. Recording single-resonance proton-coupled carbon-13 spectra is a very tedious and time-consuming procedure. The drastically lower sensitivity found in such experiments has its origin in two factors: (a) the total intensity is distributed over a large number of multiplet components, and (b) the loss of the nuclear OVERHAUSER enhancement factor (see NOE). It would therefore be desirable to retain the NOE while eliminating the decoupling effect. This turns out to be possible by suitably "gating" the decoupler field, due to the fact that the NOE decays exponentially with a time constant in the order of the carbon-13 spin lattice relaxation time after the decoupling field has been turned off. By contrast, radio-frequency coherence vanishes almost instantaneously. In practice, this means that the decoupling effect disappears immediately while the NOE is retained when the decoupler is switched off. Hence by pulsing the decoupler so that it is on during the pulse interval except for the data acquisition period, the spin multiplets are retained. Maintaining a high duty cycle for the decoupler enables one to retain the NOE almost completely [2,3].

References

1. Feeney J, Shaw D: J. Chem. Soc. Chem. Commun. 1970: 554
2. Freeman R, Hill HDW (1971) J. Magn. Reson. 5: 278
3. Gansow OA, Schittenhelm W (1971) J. Am. Chem. Soc. 93: 4294

GHPD
Acronym for **G**ated **H**igh **P**ower **D**ecoupling

This technique has been introduced into solid-state nuclear magnetic resonance (see NMR) for the same reasons as GD (see p 120) in carbon-13 high-resolution resonance of liquids or solutions. High power decoupling is necessary in solid-state NMR to eliminate the dipole-dipole (see DD) interactions between carbons and protons in the solid in order to get less line-broadening. For further details see other acronyms such as 2D-MAS, MAS, MASS, MAR, CP, VASS, PASS, TOSS,

HAHA, 2D-HAHA, HOHAHA, WAHUHA, and PENIS dealing with solid state NMR in this dictionary.

GOE
Acronym for the General OVERHAUSER Effect

Intensity variations of individual lines in a given spin system in NMR spectra (see NMR) that are observed in connection with double resonance (DR) experiments are known as the general OVERHAUSER effect [1]. In this case, a very weak secondary field B_2 perturbs only the populations of those energy levels which are linked through the irradiated line. For the appearance of the GOE the condition that $\gamma B_2^2 T_1 T_2 \approx 1$ should be fulfilled. GOE has found versatile and valuable applications in INDOR spectroscopy (see INDOR).

Reference
1. Günther H (1980) NMR spectroscopy: An introduction, Wiley, Chichester, New York

G

GRE
Abbreviation of GAUSSian Resolution Enhancement

GRE was proposed over 20 years ago by Ernst [1], implemented much later by Ferrige and Lindon [2], and eventually found its way into commercially available FOURIER transform (see FT) nuclear magnetic resonance (see NMR) data-processing software.

Recently G.A. Pearson has published a critical paper dealing with the optimization of GRE [3] in practice. A comparison of the relative signal-to-noise ratios (see SNR) obtaining under using exponential and GAUSSian resolution enhancement apodization has been given.

The value of the relative linewidth p is uniquely determined by deciding how much of the obtained SNR one is willing to "pay" in order to "buy" resolution. Note, that this procedure directly and uniquely optimizes the GRE envelope for a given free induction decay (see FID) and is applicable to most high-resolution NMR spectra.

References
1. Ernst RR (1966) Adv. Magn. Reson. 2: 1
2. Ferrige AG, Lindon JC (1978) J. Magn. Reson. 31: 337
3. Pearson GA (1987) J. Magn. Reson. 74: 541

HAHA
Acronym for the **HARTMANN-HAHN** Condition or Match

The HAHA condition or match is used in solid-state nuclear magnetic resonance (see NMR) for a cross-polarization (see CP) process between high-abundant nuclei such as protons (with a lower spin temperature) and low-abundant nuclei like carbon-13 (with a higher spin temperature).

The CP process occurs when the smaller, hotter carbon-13 reservoir is brought into contact with larger, cooler proton reservoir.

One way this is accomplished is by irradiating the carbons with a radio-frequency field, H_{1C}, of a strength such that the so-called HAHA condition [1] is satisfied:

$$\gamma_H H_{1H} = \gamma_C H_{1C} \ . \tag{1}$$

When this condition is fulfilled the original proton and carbon-13 levels, which did not match, are brought to lower, matching levels. The energy-conserving spin flips can now occur between the carbon-13 and proton spins instead of just within their individual reservoirs. In other words, the rotating frames of both spin systems are made to rotate at the same rate.

The larger, cooler proton reservoir cools the carbon-13 reservoir; that is, a carbon-13 signal is created along H_{1C}. Under ideal conditions, the carbon signal produced by CP is four times the original carbon-13 magnetization [2].

After a reasonably long contact, H_{1C} is turned off and the carbon-13 signal is detected and stored in a computer as in conventional NMR spectrometers operating in the pulse FOURIER transform (see PFT) mode.

References
1. Hartmann SR, Hahn EL (1962) Phys. Rev. 128: 2042
2. Pines A, Gibby MG, Waugh JS (1973) J. Chem. Phys. 59: 569

HAMILTONian
Abbreviation of HAMILTON's operator used in Nuclear Magnetic Resonance (see NMR)

Quantum-mechanical description of the NMR phenomenon starts with HAMILTON's function

$$\mathscr{H}(x, y, z; \not p_x, \not p_y, \not p_z) \ .$$

SCHROEDINGER's equation can be written under using the HAMILTON operator $\mathscr{H}$ in the form of Eq. (1).

$$\mathscr{H}\psi = E\psi \tag{1}$$

In the case of NMR the HAMILTONian has a very complex form, given in Eq. (2).

$$\mathscr{H} = \frac{1}{2m}\sum_{i=1}^{n}\left(\vec{\not p}_i + \frac{e}{c}\vec{\mathscr{A}}_i\right)_2 - e\varphi + \mathscr{V}_M + \mathscr{V}_D + \mathscr{V}_Q \tag{2}$$

In Eq. (2) terms and symbols have the following meaning:

$-\vec{p}_1$ impulse operator; in connection with the NABLA operator $\vec{\nabla}$ ($\vec{p} = \hbar/i \cdot \vec{\nabla}$) we can write here

$p_x = \hbar/i \cdot \partial/\partial x, \quad p_y = \hbar/i \cdot \partial/\partial y$, and

$p_z = \hbar/i \cdot \partial/\partial_z$.

$-\vec{\mathscr{A}}_1$ an operator coming from HAMILTON's function; in connection with the impulse operator we get the so-called generalized impulse

$$\vec{p} = mv - e/c \, \vec{\mathscr{A}} \, (\vec{\mathscr{H}} = \mathrm{rot} \, \vec{\mathscr{A}})^1 \quad .$$

$-e\varphi$ the potential energy of all electric interactions with the exception of the nuclear quadrupole interaction [see $\mathscr{V}_q$ in Eq. (2)].

$-\mathscr{V}_M$ the potential energy of the nuclear dipole moments $\vec{\mu}_K$ in the magnetic field H_o, therefore

$$\mathscr{V}_M = \sum_{K=1}^{N} (\vec{\mu}_K \cdot \vec{H}_o) \; .$$

$-\mathscr{V}_D$ the potential energy of the direct magnetic coupling of all nuclear dipoles.

In the case of NMR of compounds in condensed phase or solved in suitable solvents the terms $\mathscr{V}_D$ and $\mathscr{V}_Q$ of Eq. (2) will vanish.

For practical reasons, one can define a so-called empirical HAMILTON operator, given here with Eq. (3). This empirical HAMILTONian is valid in the case of high-resolution NMR spectra of compounds with nuclei of the spin quantum number $I = 1/2$ in form of liquids or in solution.

$$\mathscr{H}^{\mathrm{emp}} = \mathscr{H}^{(0)} + \mathscr{H}^{(1)} = \sum_{i=1}^{n} \nu_i I_{zi} + \sum_{i=1}^{\mu-1} \sum_{j=i+1}^{n} J_{ij} I_{(i)} I_{(j)} \tag{3}$$

With Eq. (3) we define the empirical HAMILTONian as the sum of $\mathscr{H}^{(0)}$ and $\mathscr{H}^{(1)}$, standing for a "shift"- and a "coupling" operator[2], given with the Eqs. (4) and (5).

$$\mathscr{H}^{(0)} = 1/2 \, \pi \sum_i \gamma_i H_i I_z(i) = \sum_{i=1}^{n} \gamma_i I_{zi} \tag{4}$$

In Eq. (4) γ_i stands for the magnetogyric ratio of the nucleus of interest. H_i depends on the electronic shielding against the external magnetic field H_o per definitionem: $H_i = H_o(1 - \sigma_i)$, where σ_i is the magnetic screening constant.

$$\mathscr{H}^{(1)} = \sum_{i=1}^{n-1} \sum_{j=i+1}^{n} J_{ij} I_{(i)} I_{(j)} \tag{5}$$

Equation (5) symbolizes the indirect (via electrons leading) spin-spin couplings of a molecule in solution under consideration of the scalar vector products off all possible pair combinations.

---

[1] Rotation of a vector field.

[2] Nota bene! The matrix of $\mathscr{H}^{(0)}$ has only diagonal elements, but $\mathscr{H}^{(1)}$ will have off-diagonal elements, too.

The notion of an *average* HAMILTONian, which represents the "average" motion of the spin system, provides a clear description of the effects of a time-dependent perturbation applied to the system under investigation. It was initially introduced into nuclear magnetic resonance by Waugh [1,2] to explain the effects of multiple-pulse sequences.

The *average* HAMILTONian can be defined either by exact calculation involving the diagonalization of the time evolution operator or by means of an expansion known as BAKER-CAMPBELL-HAUSDORFF or MAGNUS expansion [3].

References
1. Haeberlen U, Waugh JS (1968) Phys. Rev. 175: 453
2. Waugh JS (1976) Proc. 4th Ampère Int. Summer School, Pula/Yugoslavia
3. Ernst RR, Bodenhausen G, Wokaun A (1987) Principles of nuclear magnetic resonance in one and two dimensions, Clarendon, Oxford, chap 3

H

Suter and Pines [1] have discussed the recursive evaluation of interaction pictures for NMR purposes, recently. They pointed out that the interaction picture is a powerful tool in quantum mechanics, allowing one to study slow and possibly complex processes in the presence of fast motions whose behavior is well understood. In their opinion, important examples in NMR are the rotating-frame transformation [2] and the so-called "toggling-frame" picture used in *multiple-pulse NMR* [3,4] and *composite pulses* [5] (see p 51). All their [1] concepts are based on a almost correct description by the *distinct* HAMILTONian. For practical reasons, the authors [1] have calculated the *interaction* HAMILTONian for the WAHUHA-4 sequence (see p 300) and a BLEW sequence (see p 28).

References
1. Suter D, Pines A (1987) J. Magn. Reson. 75: 509
2. Rabi II, Ramsey NF, Schwinger J (1954) Rev. Mod. Phys. 26: 167
3. Haeberlen U, Waugh JS (1968) Phys. Rev. 175: 453
4. Haeberlen U (1976) In: Waugh JS (ed) Advances in magnetic resonance, Suppl. 1, High resolution NMR in solids, Academic, New York
5. Levitt MH (1986) Progr. NMR Spectrosc. 18: 61

Appendix

An important advantage of NMR compared with other molecular spectroscopies is the possibility of *manipulating* and/or *modifying* the nuclear spin HAMILTONian at will, almost without any important restriction, and "adapt" it to the individual requirements of the problem to be solved.

Many infrared (see IR) and ultraviolet (see UV) spectra cannot be analyzed because of their complexity of incompletely resolved signal patterns. But, in NMR very often it is possible to simplify complex spectra by *modifying* the HAMILTONian to an extent which allows a fruitful analysis.

Easy *manipulations* of the nuclear spin HAMILTONian are unique for purely practical reasons [1]. Because, nuclear interactions are normally weak, it is possible to introduce competitive perturbations of enough strength to "override" certain interactions. But, in optical spectroscopies, the relevant interactions are normally of much higher energy, and therefore similar manipulations seem to be impossible.

It has been pointed out [1] that in various applications of one-dimensional nuclear magnetic resonance (see 1D-NMR) spectroscopy, the *manipulation* or *modification* of the HAMILTONian will play an almost essential role. Today, spin decoupling, coherent averaging by multipulse techniques, sample spinning, and partial orientation in liquid-crystalline phases are standard experiments, either for the purpose of spectra simplification of for an enhancement of the structural information content.

In the context of 2D-NMR spectroscopy, the *manipulations* of spin HAMILTONians turns out to be of even greater value, because one can use several different average HAMILTONians in the course of a single experiment [1].

Manipulation of the HAMILTONian requires an external perturbation of the system under investigation, which may either be time-independent or time-dependent.

Time-independent perturbations change the parameters which govern the HAMILTONian. Changes in temperature, pressure, solvents, concentrations, and of the applied static magnetic field can be used for an alternation of the HAMILTONian. But, many of these perturbations cannot be applied or removed sufficiently rapid to be incorporated in a 2D-NMR experiment. An important exception is magnetic field-cycling (see MFC) where the sample is transported between evolution and detection periods from one field strength to another [2]. An interesting application is time domain zero-field magnetic resonance, which can be used to determine dipolar or quadrupolar coupling constants in powders [3,4].

The most important time-dependent perturbations of HAMILTONians are mechanical sample spinning and those coming from stationary or pulsed radio-frequency fields. Rapid spinning leads to a spatial averaging of the inhomogeneous or anisotropic terms of the HAMILTONian. Magnetic field inhomogeneities which give rise to a distribution of LARMOR frequencies can be averaged out, and anisotropic interactions—like dipolar or quadrupolar couplings and the anisotropic term of chemical shift—may be eliminated by high-speed spinning about a suitable rotation axis. The resulting NMR spectra may be understood by a *modified* HAMILTONian without time-dependent terms. But, for slow spinning rates, families of spinning side-bands coming up which can no longer be described by a modified time-independent HAMILTONian. In those situations the theory of FLOQUET may be used for an appropriate description [5–7].

A great variety of methods that employ radio-frequency fields to *manipulate* the HAMILTONian have been proposed so far, and are explained under their individual acronyms or abbreviations.

References

1. Ernst RR, Bodenhausen G, Wokaun A (1987) In: Principles of nuclear magnetic resonance in one and two dimensions, Clarendon, Oxford (The International Series of Monographs on Chemistry, vol 14)
2. Ramsey NF, Pound RV (1951) Phys. Rev. 81: 278
3. Weitekamp DP, Bielecki A, Zax D, Zilm KW, Pines A (1983) Phys. Rev. Lett. 50: 1807
4. Bielecki A, Murdoch JB, Weitekamp DP, Zax D, Zilm KW, Zimmermann H, Pines A (1984) J. Chem. Phys. 80: 2232

5. Shirley JH (1965) Phys. Rev. B138: 979
6. Barone SR, Narcowich MA, Narcowich FJ (1977) Phys. Rev. A15: 1109
7. Suwelack D, Waugh JS (1980) Phys. Rev. B22: 5110

H-C COSY

Abbreviation of ^{1}H-^{13}C Shift **Cor**relation **S**pectroscop**y** in Nuclear Magnetic Resonance (see NMR and 2D-NMR)

H. Kessler and co-workers [1] have recently published a detailed comparison of four different pulse sequences for the purpose of a routine application of proton/carbon-13 shift correlation. The four different pulse sequences for the $^1H-^{13}C$ *shift correlation via* $^1J(CH)$ *coupling* constants have been compared and applied to a macrolide as an practical example. These are the COLOC sequence, a new improved variant of COLOC (see p 49), the two-dimensional DEPT-COSY sequence and its new constant-time variant. It emerged that the 2D-DEPT-COSY and the modified COLOC sequence are the almost appropriate ones for routine proton-carbon shift correlation via direct couplings. The interested reader will find more details dealing with these problems in the literature [2,3] and its applications [4–8], too.

H

References

1. Kessler H, Griesinger C, Zimmermann G (1987) Magn. Reson. Chem. 25: 579
2. Reynolds WF, Hughes DW, Perpick-Dumont M, Enriques P (1985) J. Magn. Reson. 64: 304
3. Kessler H, Griesinger C, Zarbock J, Loosli HR (1984) J. Magn. Reson. 57: 331
4. Kessler H, Griesinger C, Lautz J (1984) Angew. Chem., Int. Ed. Engl. 23: 444
5. Kessler H, Bermel W, Griesinger C (1985) J. Am. Chem. Soc. 107: 1083; Kessler H, Bernd M: Justus Liebigs Ann. Chem. 1985: 1145; Kessler H, Bermel W, Müller A, Pook KH (1985) In: Hruby V (ed) The peptides, Academic, New York, vol 7 p 437
6. Kessler H, Bermel W, Griesinger C, Kolar C (1986) Angew. Chem., Int. Ed. Engl. 25: 342
7. Müller N (1985) Magn. Reson. Chem. 23: 688
8. Cham JA, Shultis EA, Dingerdissen JJ, De Brosse CW, Roberts GD, Snader KM (1985) J. Antibiot. 38: 139

HETCOR

Acronym for **Het**eronuclear Chemical Shift **Cor**relation Spectroscopy

HETCOR can be defined as a heteronuclear version of the COSY experiment (see COSY) [1–3]. The basic requirement for a heteronuclear shift-correlation experiment is a mixing process that transfers information about the chemical shift evolution of one nucleus to a second for detection; therefore, a heteronuclear coupling constant-scalar or dipolar-between the two nuclei will be required [4].

References

1. Maudsley AA, Ernst RR (1977) Chem. Phys. Lett. 50: 368
2. Bodenhausen G, Freeman R (1977) J. Magn. Reson. 28: 471
3. Bolton PH, Bodenhausen G (1979) J. Am. Chem. Soc. 101: 1080
4. Hull WE (1987) Experimental aspects of two-dimensional NMR. In: Croasmun WR, Carlson RMK (eds) Two-dimensional NMR spectroscopy: Applications for chemists and biochemists, Verlag Chemie, Weinheim

HFC
Abbreviation of **H**yper**f**ine **C**ouplings

As in nuclear magnetic resonance (see NMR), coupling constants (see CC) symbolize the different relationships of the nuclei involved in electron spin or paramagnetic resonance (see ESR or EPR) HFC became here a similar meaning as an almost interesting parameter.

HFT
Abreviation of **H**ypercomplex **FOURIER T**ransformation

The HFT approach has been used in the field of two-dimensional nuclear magnetic resonance spectroscopy (see 2D-NMR).

Normally, absorption mode 2D spectra can only be recorded if the transmitter is placed at either the low or high field side of the spectrum. Since data acquisition during t_2 (detection period) has to be done in quadrature (for sensitivity reasons), this type of experiment is very inefficient as far as data storage is concerned.

The so-called hypercomplex FOURIER transformation is a more efficient way to obtain a 2D absorption mode spectrum in 2D quadrature and was first introduced by Müller and Ernst [1] and by Freeman and his group [2]. Later on States and co-workers [3] also described this approach.

The detailed description of the HFT method starts with the signal of the first experiment given by Eq. (1), if quadrature detection is employed during the detection period.

$$s(t_1, t_2) = M_0 \cos(\Omega_1 t_1) \times \exp(i\Omega_2 t_2)\exp(-t_1/T_2^{(1)})\exp(-t_2/T_2^{(2)}) \quad (1)$$

In Eq. (1) Ω_1 and Ω_2 denote the angular frequencies of the magnetization during the times t_1 and t_2, respectively. $T_2^{(1)}$ and $T_2^{(2)}$ are the decay constants of the magnetization during the times t_1 and t_2.

A second experiment, $90°_{-x} - t_1 - 90°_x$-acquire (t_2), is then performed. The signal in the second experiment is initially $(t_2 = 0)$ also along the x axis, but is modulated in amplitude by $\sin(\Omega_1 t_1)$. The free induction decays (see FID) obtained in the two experiments, $s(t_1, t_2)$ and $s'(t_1, t_2)$, are *not* co-added this time, but stored in separate locations in the memory. Hence, for every t_1, two "spectra" are recorded, one modulated by a cosine and the other one by sine function. FT of these two sets of "spectra" gives:

$$S(t_1, \omega_2) = M_0 \cos(\Omega_1 t_1)[A_2(\omega_2) + i\,D_2(\omega_2)] \quad (2)$$

$$S'(t_1, \omega_2) = M_0 \sin(\Omega_1 t_1)[A_2(\omega_2) + iD_2(\omega_2)] \quad (3)$$

In Eqs. (2) and (3), $A_2(\omega_2)$ and $D_2(\omega_2)$ are the absorptive and dispersive parts of the resonance.

The simple but crucial trick is to replace the imaginary part of $S(t_1, \omega_2)$ by the real part of $S'(t_1, \omega_2)$, yielding in:

$$S^r(t_1, \omega_2) = M_0 \exp(i\Omega_1 t_1)A_2(\omega_2) \quad (4)$$

Complex FOURIER transformation with respect to t_1 of Eq. (4) gives for the real part:

$$S^t(t_1, \omega_2) = M_0 A_1(\omega_1) A_2(\omega_2) \tag{5}$$

which represents a 2D absorption mode resonance.

In the opinion of Ad Bax [4], HFT undoubtedly will become the accepted way of data processing for most 2D-NMR experiments in the near future. For a slightly different approach to the same problem of two-dimensional quadrature and absorption mode see the acronym TPPI in this dictionary.

References

1. Müller L, Ernst RR (1979) Mol. Phys. 38: 963
2. Freeman R, Kempsell SP, Levitt MH (1979) J. Magn. Reson. 34: 663
3. States DJ, Haberkorn RA, Ruben DJ (1982) J. Magn. Reson. 48: 286
4. Bax A (1985) Bull. Magn. Reson. 7: 167

H, H-COSY

H

Acronym for Proton, Proton-Correlation Spectroscopy

H, H-COSY (see COSY, too) is one of various methods have been developed for evaluating coupling constants from larger proton spin systems in the field of nuclear magnetic resonance (see NMR). Normally, double resonance methods [1] (DR or NMDR) and two-dimensional J, δ-spectroscopy [2] are the methods of choice for obtaining accurate values of proton, proton coupling constants. However, both methods often fail when applied to complex spin systems with pattern resulting from a large number of couplings. Resonances which appear as broad multiplets are not fully decoupled when irradiated with the normal decoupling power. In J, δ-spectroscopy [2] the couplings are often difficult to assign or to extract. But, it is well known that cross-peaks in H, H-COSY spectra [3,4] contain important information about the size of a coupling constant, which can be evaluated if the spectral resolution is sufficient. Therefore, H, H-COSY is one of the most powerful techniques for the analysis of complex proton spin systems in NMR [5]. The modern availability of larger data storage facilities and faster computers allows one now to record routinely phase-sensitive NMR spectra with nearly perfect absorptional line shapes and sufficient resolution, at least in the ω_2 domain when using new procedures [6,7]. Problems involved in H, H-COSY in the case of evaluation of coupling constants have been discussed in detail [5]. The principle of a new method using differences and sums between the traces of cross-peaks and diagonal peaks has already been described [8]. But, the extension of the procedure has been shown by using co-addition of traces of a single cross-peak or diagonal peak, and by using differences and sums of cross-peaks alone [5]. This new procedure was given the acronym DISCO by the authors [5] and will be explained under the acronym DISCO in this dictionary.

References

1. Hoffman RA, Forsén S (1966) Prog. Nucl. Magn. Reson. Spectrosc. 1: 15
2. a) Aue WP, Karhan J, Ernst RR (1976) J. Chem. Phys. 64: 4226
   b) Bodenhausen G, Freeman R, Morris GA, Turner DL (1978) J. Magn. Reson. 31: 75
3. Aue WP, Bartoldi E, Ernst RR (1978) J. Chem. Phys. 64: 2229

4. a) Bax A, Freeman R, Morris GA (1981) J. Magn. Reson. 43: 333;
   b) Wagner G, Wüthrich K, Tschesche H (1978) Eur. J. Biochem. 86: 67;
   c) Bax A, Freeman R (1981) J. Magn. Reson. 44: 542
5. Kessler H, Müller A, Oschkinat H (1985) Magn. Reson. Chem. 23: 844
6. States DJ, Haberkorn RA, Ruben DJ (1982) J. Magn. Reson. 48: 286
7. Marion D, Wüthrich K (1983) Biochem. Biophys. Res. Commun. 113: 967
8. Oschkinat H, Freeman R (1984) J. Magn. Reson. 60: 164

HMBC
Acronym for ^{1}H-detected Multiple-bond Correlation

HMBC provides an extremely effective method for *long-range proton-carbon-13 shift correlation* [1] in nuclear magnetic resonance (see NMR) spectroscopy. The applied sequence is sketched in Fig. 11.

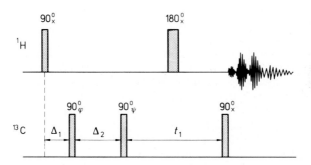

Fig. 11. Pulse scheme of the HMBC procedure

The delay Δ_1 in Fig. 11 is set to $1/2\, J_{CH}$ and Δ_2 to $1/2\, {}^{lr}J_{CH}$, where J_{CH} stands for the one-bond and ${}^{lr}J_{CH}$ for the long-range proton-carbon-13 coupling constants (see CC). In order to minimize relaxation effects, a shorter compromise value of about 60 ms is typically used for Δ_2. The phase cycling used is $\phi = x, x, x, x, -x, -x, -x, -x;\ \psi = x, y, -x, -y;\ acq. = x, x, -x, -x$. The data for odd- and evennumbered transients are stored in separate locations of the computer memory.

Recently Bax and Marion [2] have studied procedures to improve resolution and sensitivity in HMBC spectroscopy. They have demonstrated, that a more effective way of handling the acquired data is to treat them in the known *hypercomplex format* [3,4]. The data then can be processed to be absorptive in the carbon-13 dimension, followed by an absolute-value mode calculation in the dimension of the protons. This approach yields a gain in sensitivity of $2^{1/2}$ plus a significant resolution enhancement of the carbons. Mixed-mode representations as mentioned above, have been used before, mainly for reasons of convenience [5,7]. Bax and Marion [2] have pointed out, that pure two-dimensional absorption spectra cannot be obtained in the HMBC experiment and therefore the mixed-mode representation is the best alternative.

References

1. Bax A, Summers MF (1986) J. Am. Chem. Soc. 108: 2093
2. Bax A, Marion D (1988) J. Magn. Reson. 78: 186
3. Müller L, Ernst RR (1979) Mol. Phys. 38: 963
4. States DJ, Haberkorn RA, Ruben DJ (1982) J. Magn. Reson. 48: 286
5. Sklenář V, Miyashiro H, Zon G, Miles HT, Bax A (1986) FEBS Lett. 208: 94
6. Nagayama K (1986) J. Magn. Reson. 69: 508
7. Lerner L, Bax A (1986) J. Magn. Reson. 69: 375

HMDS

Abbreviation of **Hexamethyldi**siloxane, a typical high-temperature internal standard used in nuclear magnetic resonance (see NMR) for referencing chemical shifts (see CS).

HMQC

Acronym for **H**eteronuclear **M**ultiple-**Q**uantum **C**oherence used in Nuclear Magnetic Resonance Spectroscopy

H

It has been demonstrated convincingly [1–3] that ^{1}H-detected chemical shift-correlation via heteronuclear multiple-quantum coherence (HMQC) [3–5] offers a substantial sensitivity advantage over the conventional shift-correlation procedure [6–9], in which the nucleus with a low magnetogyric ratio (γ) is detected directly during the data acquisition period. The ^{1}H-detected HMQC procedure, however, has not yet gained the popularity one might expect on the basis of its intrinsic advantages. Major problems of this experimental procedure are the dynamic range problem that is introduced by the presence of large signals from protons that are not coupled to the low-γ nucleus and the required suppression of these signals in a difference experiment. Principally, both problems can be solved by presaturating the proton signals and transferring the NOE-enhanced (see NOE) low-γ signal to the protons [10] but unfortunately, the sensitivity advantage of the proton-detected experiments is partially lost in this procedure. Recently [11], a simple method that alleviates the dynamic range problem and facilitates suppression of proton signals that are not coupled to the nucleus with the low magnetogyric ratio has been described. The new method is a two-dimensional experiment using bilinear (see BIRD) pulses and have been proved for carbon-13 nuclear magnetic resonance.

References

1. Bax A, Griffey RH, Hawkins BL (1983) J. Am. Chem. Soc. 105: 7188
2. Live DH, Davis DG, Agosta WC, Cowburn D (1984) J. Am. Chem. Soc. 106: 6104
3. Bax A, Griffey RH, Hawkins BL (1983) J. Magn. Reson. 55: 301
4. Müller L (1979) J. Am. Chem. Soc. 101: 4481
5. Bendall MR, Pegg DT, Doddrell DM (1983) J. Magn. Reson. 52: 81
6. Maudsley AA, Müller L, Ernst RR (1977) J. Magn. Reson. 28: 463
7. Bodenhausen G, Freeman R (1977) J. Magn. Reson. 28: 471
8. Bax A, Morris G (1981) J. Magn. Reson. 42: 501
9. Bax A, Sarkar SK (1984) J. Magn. Reson. 60: 170
10. Neuhaus D, Keeler J, Freeman R (1985) J. Magn. Reson. 61: 553
11. Bax A, Subramanian S (1986) J. Magn. Reson. 67: 565

HNOE
Acronym for **H**etero**n**uclear **O**VERHAUSER **E**ffect

Among the different nuclear magnetic resonance (see NMR) methods in the field of NOE measurements (see GOE, NOE, and NOESY) the HNOE measured under selective proton irradiation became an interesting procedure in structure elucidation [1].

In contrast to the basic experiments with simultaneous irradiation of all proton multiplet lines with a relatively high decoupler power, the individual lines are irradiated successively several times using a very low decoupler power. An improved selectivity in the proton domain combined with higher sensitivity in the carbon-13 domain are the main attributes of this new technique [1].

Reference
1. Bigler P, Kamber M (1986) Magn. Reson. Chem. 24: 972

HOESY
Acronym for **H**etero**n**uclear **O**VERHAUSER **E**ffect or **E**nhancement **S**pectroscopy

H

Like NOESY (see p 200) [1–4] Rinaldi [5], and Yu and Levy [6] independently proposed the corresponding 2D pulse sequences to measure NOE's between protons and heteronuclei a few years ago and consequently introduced the acronym HOESY.

In the meantime, a growing number of applications [7–9] and improvements of the originally given sequence [10] have demonstrated the usefulness of this kind of experiment and information.

In connection with the information from experiments designed to detect scalar interactions between quaternary heteronuclei and protons, as COLOC (see p 49) [11] and related sequences [12], or selective measurements of long-range coupling constants [13], resonances of the involved heteronuclei can be assigned unambiguously by HOESY as has been demonstrated recently by P. Bigler [14].

References
1. Kumar A, Ernst RR, Wüthrich K (1970) Biophys. Chem. Soc. 92: 1102
2. Macura S, Huang Y, Suter D, Ernst RR (1981) J. Magn. Reson. 43: 256
3. Macura S, Wüthrich K, Ernst RR (1982) J. Magn. Reson. 46: 269
4. Macura S, Ernst RR (1980) Mol. Phys. 41: 95
5. Rinaldi PL (1983) J. Am. Chem. Soc. 105: 5167
6. Yu C, Levy GC (1984) J. Am. Chem. Soc. 106: 6533
7. Hammond SJ: J. Chem. Soc. Commun. 1984: 712
8. Yu C, Levy GC (1983) J. Am. Chem. Soc. 105: 6994
9. Bauer W, Clark T, Schleyer PVR (1987) J. Am. Chem. Soc. 109: 970
10. Kövér KE, Batta G (1986) J. Magn. Reson. 69: 519
11. Kessler H, Griesinger C, Zarbock J, Loosli HR (1984) J. Magn. Reson. 57: 331
12. Reynolds WF, Hughes DW, Perpick-Dumont M, Enriquez RG (1985) J. Magn. Reson. 63: 413
13. Bauer C, Freeman R, Wimperis S (1984) J. Magn. Reson. 58: 526
14. Bigler P (1988) Helv. Chim. Acta 71: 446

HOHAHA
Acronym for **Ho**monuclear **Hartmann-Ha**hn Spectroscopy

A new type of a two-dimensional nuclear magnetic resonance (see 2D-NMR) experiment for the determination of homonuclear scalar connectivities in complex molecules [1].

The method relies on the principle of cross polarization (see CP), first introduced by Hartmann and Hahn in 1962 [2] and commonly used for sensitivity enhancement in solid-state carbon-13 NMR [3].

In the new experiment homonuclear cross polarization is obtained by switching on a single coherent radio-frequency field. In cases where the effective radiofrequency field strengths experienced by two scalar coupled protons are identical, a perfect Hartmann-Hahn match is established and gives rise to an oscillatory exchange (with period $1/J$) of the spin-locked magnetization. For a simple two-spin system AX complete exchange of magnetization is obtained for spin-lock time equal to $1/2J_{AX}$. In the case of larger spin systems the time dependence of magnetization exchange follows a more complex pattern.

The HOHAHA experiment relies on quite the same basic principles as the heteronuclear CP experiment [2–5] and is also related to the TOCSY experiment (see p 284) [6] and to other experiments concerning the disappearance of J modulation in the presence of very rapid pulsing [7–9].

References
1. Davis DG, Bax A (1985) J. Am. Chem. Soc. 107: 2820
2. Hartmann SR, Hahn EL (1962) Phys. Rev. 128: 2042
3. Pines A, Gibby MG, Waugh JS (1973) J. Chem. Phys. 59: 569
4. Müller L, Ernst RR (1980) Mol. Phys. 38: 963
5. Chingas GC, Garroway AN, Bertrand RD, Moniz WB (1981) J. Chem. Phys. 74: 127
6. Braunschweiler L, Ernst RR (1983) J. Magn. Reson. 53: 521
7. Wells EJ, Gutowsky HS (1965) J. Chem. Phys. 43: 3414
8. Allerhand A, Gutowsky HS (1965) J. Chem. Phys. 42: 4203
9. Allerhand A (1966) J. Chem. Phys. 44: 1

HOMCOR
Acronym for **Hom**onuclear **Corr**elation

HOMCOR is the simplest experiment in the field of two-dimensional correlation methods in nuclear magnetic resonance (see 2D-NMR) based on coherence transfer, originally proposed by Jeener [1] and first put into NMR practice and analyzed in detail by Aue et al. [2]. Equation (1) symbolizes the used pulse sequence.

$$(\pi/2) - t_1 - (\beta) - t_2 \qquad\qquad (1)$$

R.R. Ernst and his group refer to this experiment of JEENER's as homonuclear 2D correlation spectroscopy, or COSY (see p 53) [3].

In the basic experiment, the coherence transfer between single-quantum transitions is induced by a single non-selective radio-frequency pulse with the rotation angle β. In the case of weakly coupled spin systems, the existence of a non-vanishing coupling between two nuclei is a necessary condition for coherence transfer to occur. Thus the appearance of "cross-peaks" in two-dimensional correlation spectra constitutes a proof of the real existence of resolved scalar or dipolar couplings, and allows the spectroscopist to identify the chemical shifts of the coupling partners. R.R. Ernst has interpreted such a 2D spectrum as a "map" which traces out coherence transfer in the mixing process [3].

Although the basic COSY experiment has proven to be quite a powerful tool for the interpretation of complex NMR spectra [4–6], some ambiguities in their assignments may remain. But, these problems can often be solved by using a different consecutive coherence transfer experiment such as relayed magnetization transfer (see MT) [7,8] in multiple-quantum NMR (MQ) [2,9].

References

1. Jeener J (1971) Ampère International Summer School, Basco Polje/Yugoslavia
2. Aue WP, Bartholdi E, Ernst RR (1976) J. Chem. Phys. 64: 2229
3. Ernst RR, Bodenhausen G, Wokaun A (1987) Principles of nuclear magnetic resonance in one and two dimensions, Clarendon, Oxford
4. Nagayama K, Wüthrich K, Ernst RR (1979) Biochem. Biophys. Res. Commun. 90: 305
5. Wagner G, Kumar A, Wüthrich K (1981) Eur. J. Biochem. 114: 375
6. Bax A, Freeman R (1981) J. Magn. Reson. 44: 542
7. Eich GW, Bodenhausen G, Ernst RR (1982) J. Am. Chem. Soc. 104: 3731
8. Bolton PH, Bodenhausen G (1982) Chem. Phys. Lett. 89: 139
9. Braunschweiler L, Bodenhausen G, Ernst RR (1983) Mol. Phys. 48: 535

H

HSS
Abbreviation of **H**omogeneity **S**poil **S**pectroscopy

HSS has been introduced as a tool of spectrum localization for *in vivo* nuclear magnetic resonance (see NMR) spectroscopy used in the field of medicine [1].

In vivo spectroscopy of phosphorus-31 is one of the most important NMR applications in biochemistry, biology, and clinical medicine, because of its insight into the energy metabolism of living tissue. ^{31}P—*in vivo*—NMR is, therefore a powerful tool in the assessment of the metabolic state and offers interesting applications ranging from the diagnosis of muscle diseases [2–4] to studies of brain metabolism of new-born infants [5] and possibilities of a therapy control in tumor treatment [6].

Since T_2 (spin-spin relaxation time) of ^{31}P *in vivo* is normally very short (several tens of milliseconds), application of localization techniques, which essentially require even a short time interval between excitation and acquisition, is gravely restricted. This fact and the low signal-to-noise ratio (see SNR) of *in vivo* phosphorus-31 spectra arising from the low tissue concentration of ^{31}P have made surface coil spectroscopy [7] to the commonly used procedure in the field of clinical spectroscopy.

Several kinds of approaches have been suggested dealing with the "ill-defined" sensitive volume caused by the complicated profile of the exciting radio-frequency field around the coil [8,9].

The basic concept of HSS [1] is related to the well-known principle of topical magnetic resonance (see TMR) [10] and to that proposed by Crowley and his group [11], using field inhomogeneities for destroying the coherence of unwanted signals. For hard- and software details the interested reader is referred to the original paper dealing with HSS [1].

References

1. Hennig J, Boesch C, Gruetter R, Martin E (1987) J. Magn. Reson. 75: 179
2. Ross BD, Radda GR, Gadian DG, Rocker G, Eseri M, Falconer-Smith J (1981) N. Engl. J. Med. 304: 1338
3. Edwards RT, Dawson MJ, Wilkie DR, Cordon RE, Shaw D (1982) Lancet 1: 725
4. Radda GK, Bove PJ, Rajagopalan B (1983) Brit. Med. Bull. 40: 155
5. Hope PL, Costello AM, Cady EB, Delphy DT, Tofts PS, Chu A, Hamilton PA, Reynolds EOR, Wilkie DR (1984) Lancet 2: 366
6. Williams SR, Gadian DG (1986) Q. J. Exp. Physiol. 71: 335
7. Ackerman JJH, Grove TH, Wong GG, Gadian DG, Radda GK (1980) Nature (London) 283: 167
8. Bendall MR (1984) J. Magn. Reson. 59: 406
9. Shaka AJ, Freeman R (1985) J. Magn. Reson. 62: 340
10. Gordon RE, Hanley PE, Shaw D, Gadian DG, Radda GK, Styles P, Bove PJ, Chan L (1980) Nature (London) 287: 763
11. Crowley MG, Ackerman JJH (1985) J. Magn. Reson. 65: 522

HSVD
Acronym for **HANKEL's Singular Value Decomposition**

In contrast to LPSVD (see p 163), this new method implicitly preserves the HANKEL structure of the data matrix. Therefore some authors have indicated it by the name HSVD [1,2] where H stands for HANKEL, although other names, referring to the state space formalism or to matrix factorization, are also conceivable.

Recently an improved algorithm for a non-iterative time-domain model fitting to exponentially damped magnetic resonance signals via HSVD has been proposed and tested for quantitative analysis [3].

References

1. Barkhuijsen H, de Beer R, Bovée WMMJ, van de Brink AM, Drogendijk AC, van Ormondt D, van der Veen JW (1986) Signal Processing III: Theory and Application Young IT et al. (eds) Elsevier, The Hague, pp 1359–1362
2. Barkhuijsen H (1986) Ph.D. thesis, Delft/NL
3. Barkhuijsen H, de Beer R, van Ormondt D (1987) J. Magn. Reson. 75: 553

HSVPS
Abbreviation of **High Speed Vector Processor System**

The first array processor developed for FOURIER transform infrared (see FT-IR) spectrometers was manufactured by Bowmen [1,2] and abbreviated as HSVPS. Bowmen's HSVPS performs high-speed calculations on floating point data. 16-bit floating point arrays are the input to the HSVPS, where they are converted to fixed points.

For further details the reader of this dictionary is referred to the literature [3]. This system is a general-purpose programmable array processor, and software is currently available to perform phase corrections using FORMAN's method (see p 113 and Ref [4]), numerical filtering, fast FOURIER transform (see FFT), and discrete FOURIER transform (see DFT).

References

1. Bowmen Inc., Vanier, Quebec/Canada
2. Berbe JN, Buijs HL (1977) ACS Symposium Series 57: 106
3. Griffiths PR, de Haseth JA (1986) Fourier transform infrared spectrometry, Wiley, New York
4. Foreman ML, Steele WH, Vanasse GA (1966) J. Opt. Soc. Am. 56: 59

HT

Abbreviation of **HADAMARD**'s or **HILBERT**'s Transformation used in addition to the FOURIER Transformation (see FT and FFT) both in infrared (see IR) and nuclear magnetic resonance (see NMR) spectroscopy. For further details the reader is referred to an excellent written monograph [1].

Reference

1. Marshall AG (ed) (1982) Fourier, Hadamard, and Hilbert transforms in chemistry, Plenum, New York

HT-IR

Abbreviation of **HADAMARD** Transform-Infrared Spectroscopy (see IR)

H

HT-IR spectroscopy was investigated in the later 1960s and early 1970s as a spatially multiplexed alternative to conventional IR spectroscopy [1]. A spectrum was spatially dispersed in a conventional monochromator and the bands were observed simultaneously through a series of masks. A multiplex advantage was so obtained. But, this technique was abandoned in favor of FT-IR (see p 116) spectroscopy [2,3], which offered a much more larger multiplex advantage. HADAMARD's encoding has recently been employed in a spectrometer based on an array of light-emitting diodes and designed for measurements of trace atmospheric species [4]. The performance of a HADAMARD transform photothermal deflection imager with continuous wave laser (see LASER) illumination has been described more recently [5].

References

1. Harwitt M, Sloane NJA (1980) Hadamard transform optics, Academic, New York
2. Vanasse GA (1982) Appl. Opt. 21: 189
3. Hirschfeld T, Wyntjes G (1973) Appl. Opt. 12: 2876
4. Sugimoto N (1986) Appl. Opt. 25: 863
5. Fotiou FK, Morris MD (1987) Anal. Chem. 59: 185

HYPERCOMPLEX NUMBERS

Hypercomplex numbers have been used recently [1] for a new spectral representation of phase-sensitive two-dimensional nuclear magnetic resonance (see 2D-NMR) experiments. With this representation, it has been shown that the FOURIER transformation (see FT and FFT) and the phasing of phase-sensitive 2D-NMR spectra (see p 10) can take relatively simple expressions.

It may be useful here to take in mind that *hypercomplex algebras* are linear algebras [2] which are presented as a generalization of *complex-number algebra*.

Hypercomplex algebras have already been successfully used in gravitation theory [3,4], in particle classification theory [5], and in NMR for the representation of composite pulses (see p 51) [6]. In addition, the product operator decomposition [7], which has proven to be very useful in multi-pulse NMR, can be expressed as the direct product of hypercomplex algebras.

References

1. Delsuc MA (1981) J. Magn. Reson. 77: 119
2. Abian A (1971) Linear associative algebras, Pergamon, New York
3. Shpilker GP (1984) Dokl. Akad. Nauk SSSR 282: 1090
4. Mann RB, Moffat JW (1985) Phys. Rev. D 31: 2488
5. Dixon G (1986) J. Phys. G 12: 561
6. Blümich B, Spiess HW (1985) J. Magn. Reson. 61: 356
7. Ernst RR, Bodenhausen G, Wokaun A (1987) Principles of nuclear magnetic resonance in one and two dimensions, Clarendon, Oxford

HZQC

Abbreviation of **H**omonuclear Proton **Z**ero-**Q**uantum **C**oherence Nuclear Magnetic Resonance (see NMR)

H

HZQC was first introduced by Müller [1] under using a well-defined pulse sequence:A $90°_x - \tau - 180°_y - \tau - 45°_y$ sequence creates the zero-quantum coherence (see ZQC) which is then allowed to evolve during the evolution period, t_1; chemical shift terms are suppressed by the $180°_y$ pulse midway through the excitation sequence.

At this point, one can use two alternative approaches: (a) utilization of a homogeneity spoiling pulse (HSP) with a minimal four-step phase cycle, or (b) more extensive phase cycling (16 up to 64 steps) without a HSP. When applied, the HSP at the beginning of the evolution period serves to destroy higher quantum coherence as suggested by Wokaun and Ernst [2]. Following the evolution, ZQC is converted back into an observable single-quantum coherence with the α-pulse. Shortening the α-pulse from 90° has several beneficial effects [1,3,4]: (a) transfer of ZQC to passive spins is suppressed, (b) quadrature phase detection (see QPD) is now possible, and (c) the number of responses contained in the data matrix is now halved. It should be pointed out finally that ZQC is insensitive to magnetic field inhomogenities [2].

It has been shown recently [5] on phenanthro [3, 4 : 3′, 4′] phenanthro [2, 1-b] thiophene as a model system that HZQC is a powerful method to establish vicinal proton-proton connectivities. The HZQC experiment has been compared with the commonly used autocorrelated proton-proton experiment (see COSY). In the case of a molecule as mentioned above, where the proton chemical shifts are so highly congested, the HZQC experiment has provided data not obtainable via the COSY experiment [5].

References

1. Müller L (1984) J. Magn. Reson. 59: 326
2. Wokaun A, Ernst RR (1977) Chem. Phys. Lett. 52: 407
3. Mareci TH, Freeman R (1983) J. Magn. Reson. 51: 531
4. Braunschweiler L, Bodenhausen G, Ernst RR (1983) Mol. Phys. 48: 535
5. Zektzer AS, Martin GE, Castle RN (1987) J. Heterocyclic Chem. 24: 879

ID

Abbreviation of **I**ndirect **D**etection used in Nuclear Magnetic Resonance (see NMR) Spectroscopy

Indirect detection pulse schemes have proven extremely valuable for the observation of low natural abundance and sensitivity nuclei (X nuclei) [1–6].

Different approaches have been provided by Müller [2] and by Bodenhausen and Ruben [1]. In the first case two-spin (^{1}H–X) multiple-quantum polarization is established and detected via the proton so that the sensitivity of the experiment is independent of the gyromagnetic ratio, γ, of the nucleus X. In the latter case proton polarization is transfered to X using the INEPT (see p 143) pulse sequence [7]. After a fixed evolution time coherence of X is indirectly detected through the more sensitive ^{1}H by transferring back to ^{1}H via a reverse INEPT sequence [8].

Most applications of the above mentioned sequences have been focused on chemical-shift measurements or correlation of ^{1}H and X shifts (see HETCOR) for assignment purposes.

Recently [9] indirect detection principles have been used in the field of *carbon-13 relaxation studies*. Carbon-13 NMR relaxation measurements can provide a detailed description of molecular dynamics [10]. This is possible since carbon-13 relaxation is dominated by dipolar interactions between ^{13}C and attached protons and since ^{1}H–^{13}C bond distance is known precisely [11,12]. But, the inherent insensitivity of carbon-13 requires large sample volumes, substantial concentrations, and large acquisition times. Several pulse sequences have been developed which improve the sensitivity of ^{13}C relaxation experiments [9]. These sequences are modification of the above mentioned Bodenhausen and Ruben sequence [1] in that carbon-13 magnetization is indirectly detected through bonded protons via polarization transfer. It has been anticipated that these pulse sequences will be useful for carbon-13 T_1 measurements of specifically labeled residues in macromolecules, where the nuclear OVERHAUSER enhancement (see NOE) effect is negligible [13] and where conventional methods require concentrations and aquisition times that are often impractical. But, it is necessary to come back to the problems arising with carbon-13 T_1-measurements.

Accurate measurements of carbon-13 longitudinal relaxation times (T_1) are often very time consuming because carbon-13 has a low magnetogyric ratio and correspondingly low NMR sensitivity. Moreover, carbon-13 relaxation times are often quite long, requiring long delay times between scans. It has been demonstrated now, that the sensitivity of the measurement can be increased significantly by using the much more sensitive proton resonance to monitor changes in carbon-13 magnetization [14]. This indirect—for the direct methods see acronyms like IRFT, PSFT, SRFT, and modifications of these T_1 measurement procedures—carbon-13 T_1 measurement utilizes the sensitive hydrogen nucleus in a way similar to the recently proposed proton-detected heteronuclear chemical-shift correlation experiments [15–17]. However, for the T_1 experiment *net* magnetization transfer (see MT) between proton and carbon-13 nuclei is required, which makes the experimental schemes more complex than those normally used for heteronuclear chemical-shift correlation spectroscopy.

Two different schemes for proton-detected measurement of carbon-13 T_1's have been reported [14]:

1. In the simplest scheme, the amount of carbon-13 z-magnetization present is measured by transferring this magnetization to the protons by using a reverse DEPT [18,19] transfer (see DEPT). This experiment can be performed with or without presaturation of the proton resonances, i.e. with or without an nuclear OVERHAUSER enhancement effect (see NOE and NOESY) on the carbon-13. In order to minimize systematic errors and to make it possible to fit the experimental data via a two parameter fit a difference experiment is used [20], with and without inversion of the carbon-13 magnetization at the beginning of the variable relaxation delay. For optimal sensitivity, a delay time between scans of about twice the carbon-13 T_1 value is recommended [21]. Note, however, that a non-optimal delay time will only reduce the sensitivity without introducing systematic errors.
2. In a more complex scheme, a double DEPT transfer is used to generate carbon-13 longitudinal magnetization. To minimize dynamic range problems and the problem arising from suppressing signals from protons not attached to carbon-13, proton saturation can be used during the variable carbon-13 recovery delay. Proton saturation is accomplished by a long (5 ms) non-selective pulse followed by a so-called "homogeneity spoiling pulse" (5 ms), and followed by a series of 90° pulses spaced at 10 ms intervals. Note, however, this scheme can yield erroneous T_1 values when applied to methyl and methylene sites.

Both methods have been compared (reverse and double DEPT) in their absolute values of T_1's, in their root-mean-square (see RMS) errors, and in their reproducibility (that is the largest difference between T_1 values measured in three consecutive experiments) with common IRFT experiments.

Recently [22] the application of the DEPT (see p 69) technique for T_1 measurements of insensitive nuclei like carbon-13 has been reported.

References

1. Bodenhausen G, Ruben DJ (1980) Chem. Phys. Lett. 69: 1
2. Müller L (1979) J. Am. Chem. Soc. 101: 16
3. Summers MF, Marzilli LG, Bax A (1986) J. Am. Chem. Soc. 108: 4285
4. Sklenář V, Bax A (1987) J. Chem. Reson. 71: 379
5. Wagner G, Brühwiler D (1986) Biochemistry 25: 20
6. Live D, Davis DG, Agosta WC, Cowburn D (1984) J. Am. Chem. Soc. 106: 6104
7. Morris GA, Freeman R (1979) J. Am. Chem. Soc. 101: 769
8. Freeman R, Mareci TH, Morris GA (1981) J. Magn. Reson. 42: 341
9. Kay LE, Jue TL, Bangerter B, Demou PC (1987) J. Magn. Reson. 73: 558
10. Richarz R, Nagayama K, Wüthrich K (1980) Biochemistry 19: 5189
11. Allerhand A, Doddrell D, Komoroski R (1971) J. Chem. Phys. 55: 189
12. Dill K, Allerhand A (1979) J. Am. Chem. Soc. 101: 15
13. Doddrell D, Glushko V, Allerhand A (1972) J. Chem. Phys. 56: 3683
14. Sklenář V, Torchia D, Bax A (1987) J. Magn. Reson. 73: 375
15. Müller L (1979) J. Am. Chem. Soc. 101: 4481
16. Bodenhausen G, Ruben DJ (1980) Chem. Phys. Lett. 69: 185
17. Bax A, Griffey RH, Hawkins BL (1983) J. Magn. Reson. 55: 301
18. Doddrell DM, Pegg DT, Bendall MR (1982) J. Magn. Reson. 48: 323
19. Bendall MR, Pegg DT, Doddrell DM, Field J (1983) J. Magn. Reson. 51: 520
20. Freeman R, Hill HDW (1971) J. Chem. Phys. 54: 3367
21. Becker ED, Ferretti JA, Gupta RK, Weiss GH (1980) J. Magn. Reson. 37: 381
22. Zhao H (1987) JEOL NEWS Vol. 23A: 13

IETS
Acronym for Inelastic Electron Tunnelling Spectroscopy

IETS was first developed by Lambe and Jaklevics in 1966 [1]. Over the last twenty years, applications of IETS range from the study of biochemicals to transition metal complexes. Several reviews and texts are recommended [2–4]. IETS is a non-optical method for measuring vibrational and electronic spectra. It is a very sensitive technique, providing the spectrum of a sample containing less than 10^{13} molecules. It has been extensively used for the study of chemisorbed species, as a molecular spectroscopy to infrared (see IR) and Raman spectroscopy, and for the observation of transitions which are "silent" (forbidden) in photon spectroscopy [1–9]. The most usual feature of this method is that the sample of interest must be incorporated into a metal-insulator-metal thin film device called a tunnel diode. Thus, sample preparation is more of an art than with IR and Raman spectroscopy. Recently [10] the use of IETS has been discussed in more detail for the observation of optically forbidden transitions. We have learned, that through IETS it is now possible to start probing electronic states which have been theoretically predicted but never observed [10]. This situation should lead to a much better understanding of distinct phenomena, such as magnetic susceptibility, thermal properties, photochemistry, and bonding.

IJ

References
1. Jaklevics RC, Lambe J (1966) Phys. Rev. Lett. 17: 1139
2. Hansma PK (1982) Tunnelling spectroscopy, Plenum, New York
3. Walmsley DG, Tomlin JL (1985) Prog. Surf. Sci. 18: 247
4. Hipps KW (1983) J. Electron. Spectrosc. 30: 275
5. Hipps KW (1978) Rev. Sci. Instrum. 58: 265
6. Hipps KW, Susla B (1987) Chem. Phys. Lett. 132: 505
7. Hipps KW, Mazur U, Knochenmuss R (1986) J. phys. Chem. 90: 1755
8. Hipps KW, Aplin AT (1985) J. phys. Chem. 83: 5459
9. Hipps KW, Williams SD, Mazur U (1984) Inorg. Chem. 23: 3500
10. Hipps KW (1987) Nature 326: 107

IGD
Abbreviation of Inverse Gated Decoupling

A technique used in carbon-13 nuclear magnetic resonance to obtain precise values of the nuclear OVERHAUSER enhancement factor (see NOE) [1]. The NOE can be obtained, in principle, from a comparison of the integrated intensities of the single resonance and the proton decoupled spectra. Particular attention is due to the choice of the pulse interval in order to avoid partial saturation. This method turns out to be inconvenient because of the multiplet overlaps encountered in the coupled spectra. In order to circumvent this problem, pulse-modulated decoupling techniques may be applied which allow one to suppress the NOE in proton decoupled carbon-13 spectra. By "gating" the decoupler, so that it is on only during the data acquisition period but off during a pulse delay (see PD), we have the inverse gated decoupling (see GD) situation. The spectra thus obtained are decoupled with the NOE eliminated [1]. This is possible since the spin coupling multiplets will collapse almost instantaneously when the decoupling field is turned on whilst the NOE builds up with a time constant in the order of the spin-lattice relaxation time T_1 (see T_1). In order to minimize errors resulting from sample temperature differences in the two experiments one can combine them to only one acquisition sequence. But, in this version of the method, the free induction decays (see FID) corresponding to the NOE-enhanced and NOE-suppressed spectra are collected alternately and stored on a peripheral storage device. It should be pointed out that the use of a sufficiently long pulse delay is mandatory (up to $10T_1$) [2–4].

References
1. Freeman R, Hill HDW, Kaptein R (1972) J. Magn. Reson. 7: 327
2. Canet D (1976) J. Magn. Reson. 23: 361
3. Harris RK, Newman RH (1976) J. Magn. Reson. 24: 449
4. Opella SJ, Nelson DJ, Jardetzky O (1976) J. Chem. Phys. 64: 2533

ILS
Abbreviation of Instrument Line Shape (Function)

IJ

Since the FOURIER Transform-Infrared (see FT-IR) spectrometer does not have any slits, the sinc function $f(\tilde{v})$ has been variously called the instrument line shape (ILS) function, the instrument function, or the apparatus function, of which some authors [1] prefer the term ILS function. For further details, especially in comparison to the classic version of IR see the literature [1–3].

References
1. Griffiths PR, de Haseth JA (1986) Fourier transform infrared spectrometry, Wiley, New York (Chemical analysis, vol 83)
2. Hirschfeld T (1979) In: Ferraro JR, Basile LJ (eds) Fourier transform infrared spectroscopy: Applications to chemical systems, Academic, New York, vol 2 chap 6
3. Griffiths PR (1975) Chemical infrared Fourier transform spectroscopy, Wiley, New York (Chemical analysis, vol 43)

IM
Abbreviation of Isotropic Mixing

IM is a process which is driven by the full scalar coupling between nuclear spins and leads to net coherence transfer (see CT) in nuclear magnetic resonance (see NMR) spectra [1–3]. Such processes have also been termed SHRIMP (see p 255) in the heteronuclear context [4], and homonuclear HARTMANN-HAHN mixing [5] otherwise. At first, there were papers presented dealing with detailed studies on cross-polarization (see CP) and IM which showed a number of essential differences between these two types of collective mode processes [2,3,6,7], employing spin 1/2 and spin-1 operator algebra formalism [8,9]. Recently [10] it has been shown that the evolution frequencies under IM in *any* A_MX_N system are identical to the frequencies operative in the corresponding AX_{M+N-1} spin system. Evolution under IM conditions is therefore periodic in *any* A_MX_N system [10].

References

1. a) Braunschweiler L, Ernst RR (1983) J. Magn. Reson. 53: 521;
   b) Caravatti P, Braunschweiler L, Ernst RR (1983) Chem. Phys. Lett. 100: 305
2. Chandrakumar N, Subramanian S (1985) J. Magn. Reson. 62: 346
3. Chandrakumar N, Visalakshi GV, Ramaswamy D, Subramanian S (1986) J. Magn. Reson. 67: 307
4. Weitekamp DP, Garbow JR, Pines A (1982) J. Chem. Phys. 77: 2870
5. Davis DG, Bax A (1985) J. Am. Soc. 107: 2820
6. Chandrakumar N (1985) J. Magn. Reson. 63: 202
7. Chandrakumar N (1986) J. Magn. Reson. 67: 457
8. Chandrakumar N (1984) J. Magn. Reson. 60: 28
9. Chandrakumar N (1985) J. Magn. Reson. 63: 174
10. Chandrakumar N (1987) J. Magn. Reson. 71: 322

INADEQUATE
Acronym for **I**ncredible **N**atural **A**bundance **D**ouble **Qua**ntum **T**ransfer **E**xperiment

Via this experiment, measurements of $^{13}C-^{13}C$ coupling constants in natural abundance may be possible in times, which can be tolerated economically. It is possible to register unperturbed satellite spectra by quenching the strong signals of molecules with a normal carbon-13 nucleus [1]. By assignment of satellite pairs with equal distance for each carbon-carbon bond in a given molecule we will get the so-called carbon-carbon connectivity (CCC). The procedure—normally given as a one-dimensional procedure—has been automatized using computer programs (see COSMIC). In the meantime two-dimensional applications (see 2D-INADEQUATE) have become much more advantageous [2].

References

1. Bax A, Freeman R, Kempsell SP (1980) J. Am. Chem. Soc. 102: 4849; (1980) J. Magn. Reson. 41: 349
2. Bax A, Freeman R, Frenkiel TA (1981) J. Am. Chem. Soc. 103: 2102

INCH
Acronym for **In**directly-Bonded **C**arbon-**H**ydrogen Shift Correlation

This acronym has been proposed [1] for the whole class of nuclear magnetic resonance (see NMR) experiments dealing with indirectly bonded carbon-13 and hydrogen atoms, whereas several other pulse sequences [2–5], and Kessler's COLOC sequence [6] (see COLOC)—probably being the most widely used—have been abbreviated or given an acronym individually. The acronym INCH [1] is equally applicable to the original type of experiment or alternative procedure based on proton detection [7].

A detailed analysis of the factors, relevant to INCH has been published [1]. It has been shown that the use of composite pulses (see p 51) for carbon-13, the application of a selective $180°$ (see BIRD) pulse at the midpoint of the final refocusing-delay, the optimization of fixed delays under using modified INEPT sequences (see INEPT), and finding out appropriate acquisition parameters can all lead together to a significant sensitivity enhancement. In addition, the application of composite pulses also minimizes artifacts in the case of wide spectral windows.

References

1. Perpick-Dumont M, Enriquez RG, McLean S, Puzzuoli FV, Reynolds WF (1987) J. Magn. Reson. 75: 414
2. Hallenga K, van Binst G (1980) Bull. Magn. Reson. 2: 343
3. Jacobs H, Ramdayal F, Reynolds WF, McLean S (1986) Tetrahedron Lett. 27: 1453
4. Reynolds WF, Hughes DW, Perpick-Dumont M, Enriquez RG (1985) J. Magn. Reson. 63: 413
5. Quast M, Zektzer AS, Martin GE, Castle RN (1987) J. Magn. Reson. 71: 554
6. Kessler H, Griesinger C, Zarbock J, Loosli HR (1984) J. Magn. Reson. 57: 331
7. Bax A, Summers MF (1986) J. Am. Soc. 108: 2093

INDOR

Acronym for **I**nter**n**uclear **Do**uble **R**esonance used in nuclear magnetic resonance spectroscopy

INDOR represents an interesting variation of nuclear magnetic double resonance (NMDR) that can be successfully employed for both homonuclear and hetero-nuclear spin systems [1]. The method belongs to the so-called non-linear effects (NLE) in NMR. Like "spin tickling" [2] (see p 264) and "selective population transfer" [3] (see SPT), INDOR is very helpful in the construction of energy niveau schemes and their relationship to structure elucidation. The physical basis of the INDOR technique is formed by the general OVERHAUSER effect (see GOE) in which the BOLTZMANN distribution for specific transitions of a given spin system is perturbed by a secondary field B_2, resulting in the variation of the intensity of progressively or regressively linked lines. These variations are recorded with the primary field B_1. Experimentally, one proceeds by recording the intensity of an A line in an AMX spin system of proton resonance, for example, with a weak B_1 field (in order to avoid saturation). The pen of the recorder is placed on top of this so-called monitor signal. The spectrum of the nuclei M and X is then scanned with the B_2 field. Each time a line is encountered that has an eigenvalue in common with the A line in question an increase or a decrease in the intensity of the A line is observed as a result of a GOE. For recording the INDOR spectrum, the pen of the spectrometer and the B_1 field must be decoupled. The horizontal movement of the pen is then synchronized with the frequency sweep of the B_2 field [4].

Figure 12.a shows an AMX system with $v_A > v_M > v_X$ and $J_{AM} > J_{MX} > J_{AX} > 0$ as a model case. This AMX system with line A_1 as the monitor line gives rise to the INDOR spectrum shown in Fig. 12.b. Figure 13 shows the corresponding energy level diagram. In Fig. 12.b, the intensity of A_1 is recorded as a function of the frequency v_2. If $v_2 = M_1$, then a perturbation of the BOLTZMANN distribution between the eigenstates $\alpha\beta\alpha$ and $\alpha\alpha\alpha$ occurs (see Fig. 13). Spin population is transferred from $\alpha\beta\alpha$ to $\alpha\alpha\alpha$, a process that can be described as "spin pumping". Consequently, the intensity of line A_1 decreases since now the state $\alpha\alpha\alpha$ has too high a population relative to the $\beta\alpha\alpha$ state. "Spin pumping" at M_3 gives rise to an overpopulation of the $\beta\alpha\alpha$ state and therefore an increased intensity of A_1. Quite analogous considerations apply for line X_1 and X_2 also linked with A_1. Pro-gressively linked transitions thus cause an INDOR "absorption line" while

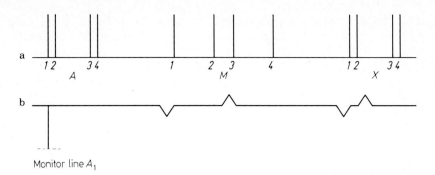

a

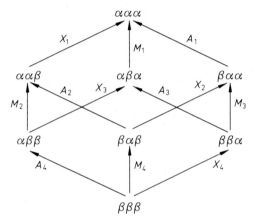

Monitor line A_1

Fig. 12. a) AMX spin system with $\nu_A > \nu_M > \nu_X$ and $J_{AM} > J_{MX} > J_{AX} > 0$. **b)** Schematic INDOR spectrum of the AMX spin system with A_1 as monitor line.

Fig. 13. Energy level diagram for an AMX spin system with $\nu_A > \nu_M > \nu_X$ and $J_{AM} > J_{MX} > J_{AX} > 0$.

regressively linked transitions give rise to an INDOR 'emission line". To avoid perturbation of the intensity of the monitor line by other unlinked transitions, the amplitude of the B_2 field should not be too high [4].

References

1. Becker E (1962) J. Chem. Phys. 37: 911
2. Freeman R, Anderson WA (1964) J. Chem. Phys. 37: 2053
3. Sørensen S, Hansen RS, Jakobson HJ (1974) J. Magn. Reson. 14: 243
4. Günther H (1980) NMR spectroscopy: An introduction, Wiley, Chichester

INEPT
Acronym for **I**nsensitive **N**uclei **e**nhanced by **P**olarization **T**ransfer

This method is useful in intensity enhancement of NMR signals from nuclei with small magnetogyric ratios [1]. The procedure is independent of different kinds of relaxation mechanisms—in contrast to the nuclear OVERHAUSER effect (NOE)—and its application for ^{15}N and ^{29}Si is indicated (negative gyromagnetic ratios). The following pulse sequence is used: $90_S^\circ(X) - t - 180_S^\circ(X)$, $180_I^\circ - t - 90_S^\circ(Y)$, 90_I° – acquisition (I stand for nucleus of interest, whereas S is a proton in most cases). Amplification of signals depends on the nuclear spin polarization and transfer from S-spins to the I-spins. Procedure can be explained as an alternative to cross-polarization experiments (see CP) in the rotating frame of nuclear magnetic resonance on solids, which have been adapted for condensed phases in the meantime.

Reference
1. Morris GA, Freeman R (1979) J. Am. Chem. Soc. 101: 760

INEPT CR
Acronym for **I**nsensitive **N**uclei **E**nhanced by **P**olarization **T**ransfer under **C**omposite **R**efocusing used in Nuclear Magnetic Resonance (see NMR)

INEPT CR is an improved refocused INEPT (see p 143) experiment [1] with simultaneous complete refocusing for all groups CH, CH_2, and CH_3. Figure 14 shows the pulse sequence for polarization transfer parity editing in carbon-13 NMR ($\tau = (2J)^{-1}$). Two experiments A ($\tau' = (4j)^{-1}$, $\phi = y$) and B ($\tau' = 3(4J)^{-1}$, $\phi = -y$) can be additively or subtractively combined to select CH_2 or $CH + CH_3$ groups, respectively.

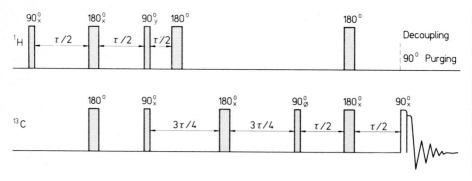

Fig. 14

Reference
1. Sørensen OW, Madsen JC, Nielsen NC, Bildsøe H, Jakobsen HJ (1988) J. Magn. Reson. 77: 170

INEPT-INADEQUATE
This acronym stands for **I**nsensitive **N**uclei **E**nhanced by **P**olarization **T**ransfer-**I**ncredible **N**atural **A**bundance **D**ouble **Qua**ntum **T**ransfer **E**xperiment

In this hybrid nuclear magnetic resonance (see NMR) experiment the INEPT polarization-transfer technique [1] has been combined with the INADEQUATE method [2] in order to partly circumvent severe sensitivity problems of the INADEQUATE experiment [3] (see INEPT and INADEQUATE in this dictionary).

References

1. Morris GA, Freeman R (1979) J. Am. Chem. Soc. 101: 761; Morris GA (1980) J. Am. Chem. Soc. 102: 428
2. Bax A, Freeman R, Frenkiel TA, Levitt MH (1981) J. Magn. Reson. 43: 478
3. Sørensen OW, Freeman R, Frenkiel TA, Mareci TH, Schuck R (1982) J. Magn. Reson. 46: 180

INMR
Acronym for **I**nverse **N**uclear **M**agnetic **R**esonance

Normally, in heteronuclear magnetic resonance $(X \neq H)$ one excites the magnetization of the proton and observes the magnetization of the heteronuclei. When estimating the sensitivity of a given pulse sequence one has to bear in mind a sequence independent factor of $\gamma_H(\gamma_X)^{3/2}$. INMR starts with the excitation of the proton magnetization but detects them too, yielding a sequence independent sensitivity factor of $(\gamma_H)^{5/2}$.

Because of this sensitivity improvement, INMR will become an attractive method in NMR of heteronuclei [1] as has been shown in a comparison of sensitivities achieved by different kinds of excitation and detection.

Kessler and co-workers[1] have discussed the possibilities of inverse H, C-correlations, BIRD (see p 27) pulses, inverse correlations via long-range couplings, and inverse relayed spectroscopy.

Reference

1. Kessler H, Gehrke M, Griesinger C (1988) Angew. Chem. 100: 507; (1988) Angew. Chem. Int. Ed. Engl. 27: 460

INSIPID
Acronym for **In**adequate **S**ensitivity **I**mprovement by **P**roton **I**ndirect **D**etection

INSIPID is a new approach for sensitivity enhancement of INADEQUATE (see p 141) by proton monitoring [1].

Reference

1. Keller PJ, Vogele KE (1986) J. Magn. Reson. 68: 389

in vivo-NMR
Abbreviation of **in vivo**-**N**uclear **M**agnetic **R**esonance (Spectroscopy)

Spatially resolved NMR, in particular localized high-resolution NMR spectroscopy *in vivo*, will become an important tool not only for non-invasive biochemical and biophysical research but also for medical diagnosis.

From the technical point of view one may generally distinguish between *global* and *local* methods of acquiring both *spatial* and *spectral* information in a combined imaging (see NMRI)/spectroscopy experiment.

Global methods can be defined as four- (or three) dimensional chemical-shift (see CS) imaging techniques yielding data sets with three (or two) *spatial* dimensions and one CS dimension.

Essential problems today, are the long measuring times, the processing and storage of extremely large data matrices, and the fact that there is not very often a real interest in truly *global* information. In addition, *spectroscopic imaging* approaches very often sacrifice *spectral* resolution and quantification in order to keep imaging times as short as possible.

In vivo NMR spectroscopy, so far, has made use of the following NMR-active nuclei: ^{1}H, ^{23}Na, ^{31}P.

Among these nuclei the *in vivo*-^{31}P-NMR spectroscopy has become the most important one because its clear differentiation between inorganic and organic phosphates in qualitative and quantitative measurements of the well shift-separated signals. One of the different applications of ^{31}P-NMR *in vivo* is the field of studies on ischaemic diseases [1]. In 1986, J.W. Prichard and R.G. Shulman published a very exciting paper about NMR spectroscopy *in vivo* and brain metabolism [2].

But, it must be pointed out that measuring and imaging (or visualization) of flow by NMR [3] will be surely applied in the field of blood flow as an additional tool to sonographic methods and in cardiological studies.

Before discussing *in vivo* phosphorus-31 spectroscopy much more in detail (see Progress), we want to present a paper dealing with the localized proton spectroscopy which uses stimulated echo signals [4]. The method itself has been given the acronym STEAM (see Table 1 on page 192 of this dictionary) and is a single-step procedure, minimizing radio-frequency power requirements and gradient switches. It further permits precise measurements of localized T_1 (spin-lattice relaxation) and T_2 (spin-spin relaxation) relaxation times relatively simply by varying the length of the corresponding intervals of the STEAM sequence. It will be of great interest to hear whether this approach can be applied to water/lipid-suppressed metabolic spectroscopy as proposed [4].

References
1. Roth K (1984) NMR-Tomographie und Spektroskopie in der Medizin, Springer, Berlin
2. Prichard JW, Shulman RG (1986) Ann. Rev. of Neurosciences 9: 61
3. Stepišnik J (1985) Progr. NMR Spectroscop. 17: 187
4. Frahm J, Merboldt K-D, Hänicke W (1987) J. Magn. Reson. 72: 502

Progress

In the field of *in vivo* NMR spectroscopy, one of the more promising localization techniques involves the use of phase-encoding pulsed field gradients applied prior to the registration of the free precession signal. The distance and frequency domain information can be reconstructed from the raw data matrix under using multi-dimensional FOURIER transformations [1–4]. Routinely the processed matrix is displayed in the absolute value mode [5] because conventional phase corrections methods cannot produce pure absorption lineshapes [6,7]. This may be satisfactory enough in many cases, but in some applications where resolution is at a premium, this technique may cause a significant line broadening to occur. Several proposals to overcome this problem have been published in the field of conventional 2D-NMR spectroscopy [7–10]. Two of these methods have been analyzed and applied [11] to the case of the chemical-shift imaging experiment first proposed by Brown and co-workers [12]. This experiment has been adapted by several groups [13–15], and is also currently being used in clinical applications [16].

An improved depth-selective single surface-coil ^{31}P-NMR spectroscopy using a combination of B_1 and B_0 selection techniques has been described and dubbed IDESSS-standing for *i*mproved *dep*th-selective *s*ingle *s*urface-coil *s*pectroscopy [17]. This method has been tested by measuring phosphorus-31 NMR spectra of the human liver [17].

References

1. Jeener J, Alewaters G (1971) Ampère International Summer Conference, Basko Polje/Yugoslavia
2. Aue WP, Bartholdi E, Ernst RR (1976) J. Chem. Phys. 64: 2229
3. Kumar A, Welti D, Ernst RR (1975) J. Magn. Reson. 18: 69
4. Edelstein WA, Hutchinson JMS, Johnson G, Redpath TW (1980) Phys. Med. Biol. 25: 751
5. Freeman R (1980) Proc. Royal. Soc. London. Ser. A 373: 149
6. Levitt MH, Freeman R (1979) J. Magn. Reson. 34: 675
7. Keeler J, Neuhaus D (1984) J. Magn. Reson. 63: 454
8. Bachmann P, Aue WP, Müler L, Ernst RR (1979) J. Magn. Reson. 28: 29
9. States DJ, Haberkorn RA, Ruben DJ (1982) J. Magn. Reson. 48: 286
10. Marion D, Wüthrich K (1983) Biochem. Biophys. Res. Commun. 113: 967
11. Barker PB, Ross BD (1987) J. Magn. Reson. 75: 467
12. Brown TR, Kincaid BM, Ugurbil K (1982) Proc. Natl. Acad. Sci. USA 79: 3523
13. Haselgrove JC, Subramanian VH, Leigh JS, Gyulal L, Chance B (1983) Science 220: 1170
14. Mareci TH, Brooker HR (1984) J. Magn. Reson. 57: 157
15. Pykett IL, Rosen BR (1983) Radiology 149: 197
16. Bailes DR et al. (1987) J. Magn. Reson. 74: 158
17. Segebarth C, Luyten PR, den Hollander JA (1987) J. Magn. Reson. 75: 345

IJ

IR
Abbrevation of Infrared (Spectroscopy)

IR spectroscopy—in former times called ultrared spectroscopy—in its classical form on grating spectrometers was until the mid 1950s the almost powerful tool in structure elucidation.

With the rapid development of nuclear magnetic resonance (see NMR) spectroscopy after 1955 IR spectroscopy lost its leading position in the field of the different molecular spectroscopies.

But, with the application of FOURIER techniques to IR after 1970 (see FT-IR) a renaissance started. New techniques on modern spectrometers led to almost new application possibilities in chemistry, polymer and material sciences, life sciences, environmental control, catalysis, and process engineering. A number of sources can be consulted for details [1–5].

Prior to FT-IR, infrared spectroscopy was carried out using a dispersive instrument utilizing gratings or prisms to disperse the IR radiation geometrically. By a scanning mechanism, the dispersed radiation was passed over a slit system which separated the frequency range falling on the detector. In this manner, the IR spectrum, that is, the energy transmitted through a sample as a function of frequency was obtained. This dispersive IR method is highly limited in sensitivity because most of the available energy is thrown away, i.e. does not fall on the open slits. Hence, to improve sensitivity of IR spectroscopy, a technique was sought which permits the examination of all of the transmitted energy all the time. In the MICHELSON interferometer such an optical device was found [6,7].

The interested reader will find in this dictionary a lot of acronyms or abbreviations for different IR methods and techniques. But, here we will give some basic comments on IR spectroscopy.

In polyatomic molecules, the atoms are normally not rigidly linked together and thus have the ability to vibrate from the residual position. For molecules with n atoms, there are, in space, 3n coordinates and $3n - 6$ degrees of freedom for the description of the fundamental internal vibrations since, of the 3n coordinates, 3 degrees of freedom must be deducted for translation and 3 for rotation.

In the case of linear molecules one has to take into account only 2 coordinates to describe the rotational mode and thus the number of degrees of freedom for the internal vibrations is equal to $3n - 5$.

The energy required to increase the amplitude of vibration of chemical bonds is of the order 5 kcal per mole and is provided by radiation in the IR region (400 to 5000 cm^{-1}) and the near infrared (see NIR) or so-called overtone region (5000 to 12500 cm^{-1}). Absorption of IR radiation leads to an excitation of the molecule to higher vibrational energy levels and is quantized. The spectrometer used measures this absorption at each frequency and one obtains an IR spectrum. From classical electrodynamics we have to learn that an excitation of a particular normal vibraion in a given molecule must obey the following two conditions: Firstly, the normal vibrations must have the same frequency as the electromagnetic radiation and secondly, a change in magnitude and direction of the dipole moment has also to occur.

Each different chemical bond in a given molecule has a different dipole moment and, therefore, absorption of radiation will occur over a relative wide range of frequencies. Thus, if IR radiation of successive frequencies is passed through the sample, series of absorption bands are recorded, the so-called *active fundamental modes of vibration*:

Symbols	Denominations
v	stretching
δ	bending, in-plane
γ	bending, out-of-plane
τ	twisting
ρ	rocking
ω	wagging

But, we have to bear in mind, that apart from the fundamental vibrations described above, harmonic and combination vibrations may also occur.

If the origin of the IR spectrum is kept in mind, it is apparent that the toral spectrum gives a wealth of information on the basic characteristics of the molecule under investigation, especially, functional groups and their spatial arrangements.

For this reason, the IR spectrum has been often referred to as the "finger print of a molecule."

Recently, J. Mink [5] has summarized the different possibilities of FT-IR in catalysis. Although hundred of papers are normally published per annum on this topic, nearly all of them are restricted to a small range of the IR spectrum. Support materials (such as silica, alumina, zeolites, etc.) do not permit IR radiation in the low-frequency region (below 1200 cm^{-1}). All vibrations involving the surface-adsorbate linkage or simply the metal-ligand motions are inaccessible. Alternative methods which do not suffer from this limitation are diffuse reflectance (DR), photoacoustic spectroscopy (see PAS), IR emission spectroscopy, and RAMAN spectroscopy.

References
1. Griffiths PR (1975) Chemical infrared transform spectrometry, Wiley, New York
2. Ferraro JR et al. (1978) In: Ferraro JR, Basile LJ (eds) Fourier transform infrared spectroscopy: Applications to chemical systems, Academic, New York
3. Koening JL (1981) Acc. of Chem. Res. 14: 171
4. Durig J (1980) Biological and chemical applications of FT-IR, D. Reidel, Dordrecht
5. Mink J (1987) Acta Phys. Hungarica 61: 71
6. Michelson AA (1981) Phil. Mag. 31: 256
7. Michelson AA (1982) Phil. Mag. 34: 280

IJ

IR-COSY
Abbreviation of Inversion Recovery-Correlation Spectroscopy

This procedure used in nuclear magnetic resonance (see NMR) may be described [1] as a combination of common inversion-recovery (see IR or IRFT) experiments [2] and homonuclear correlation spectroscopy (see COSY) pulse sequences [3]. The frequency-domain spectrum appears in this technique similar to the common COSY spectrum, but intensities of the so-called cross-peaks which correspond to two J-coupled nuclei are unequal due to differences in relaxation parameters of the nuclei of interest. In particular, IR-COSY might be useful for evaluation of molecular structure and surface accessibility to the environment by employing spin labels and paramagnetic probes in the most important field of biopolymers [1].

References
1. Arseniev AS, Sobol AG, Bystrov VF (1986) J. Magn. Reson. 70: 427
2. Vold RL, Waugh JS, Klein MP, Phelps DE (1968) J. Chem. Phys. 48: 3831
3. Aue WP, Bartholdi F, Ernst RR (1979) J. Chem. Phys. 64: 2229

IRFT
Acronym for Inversion-Recovery Fourier Transform

IRFT is one of the common methods for measuring spin-lattice relaxation times (T_1) in the field of nuclear magnetic resonance [1,2].

Normally, the macroscopic spin magnetization in thermal equilibrium is aligned along the $+z$ axis, i.e. the direction of the magnetic field H_0. Application of a 180° pulse nutates it onto $-z$. In terms of spin population, this situation corresponds to an exact inversion of the equilibrium populations of the two ZEEMAN levels. Since return to thermal equilibrium in general is exponential with time, and the equilibrium value of $M_z(t)$ is M_∞, the process obeys the differential equation Eq. (1)

$$\frac{dM_z}{dt} = -\frac{M_\infty - M_z}{T_1} \tag{1}$$

which after integration, yields

$$M_\infty - M_z = A\, e^{-t/T_1} \tag{2}$$

A in Eq. (2) is a constant and depends on the initial conditions. In the case of an inversion-recovery experiment, $M_z(0) = -M_\infty$, by definition, and thus A is equal to $2M_\infty$, which, upon insertion into Eq. (2), gives

$$M_\infty - M_z = 2M_\infty\, e^{-t/T_1} \tag{3}$$

At a time t $(t \lesssim T_1)$ after the initial perturbing pulse has elapsed, the magnetization has partially recovered. If a 90° observing pulse is applied and the resulting FID (see p 106) is collected and FOURIER transformed, the relative intensities of the various signals reflect their individual relaxation times. The inversion-recovery experiment is thus characterized by the following pulse sequence

$$(180° - t - 90° - T)_n \ .$$

In principle, only two different measurements suffice in order to determine T_1: one each to obtain $M_z(t)$ and M_∞. However, more precise data are retrieved from a set of spectra obtained by varying t. Usually, one plots a set of inversion-recovery spectra, three-dimensionally stacked in order of increasing t values. By plotting $\ln[M_\infty - M(t)]/M_\infty$ against t, the time interval between the perturbing and observing pulse, ideally a straight line with slope $-1/T_1$ should be obtained. Then data are refined by a least-squares regression analysis. Automatic determination of T_1 by inversion-recovery is possible on almost modern NMR spectrometers. In the meantime, however the traditional two-pulse inversion-recovery sequence [1,2] has been modified and extended: see FIRFT, MFIRFT and EIRFT. In comparison to other methods such as progressive-saturation (see PSFT) or saturation-recovery (see SRFT) a detailed error analysis [3] has shown that the inversion-recovery technique is least prone to systematic errors, which renders it particularly suitable as a routine method.

References

1. Vold RL, Waugh JS, Klein MP, Phelps DE (1968) J. Chem. Phys. 48: 3831
2. Freeman R, Hill HDW (1969) J. Chem. Phys. 51: 3140
3. Levy GC, Peat LR (1975) J. Magn. Reson. 18: 500

IRS
Acronym for **I**nternal **R**eflection **S**pectroscopy

IRS is a technique used in infrared (see IR) spectroscopy. In internal reflection spectroscopy the sample is in optical contact with another material (e.g. a prism) which is optically denser than the sample. Incoming light forms a standing wave pattern at the interface within the dense prism medium, whereas in the rare medium, the amplitude of the electric field falls off exponentially with the distance from the phase boundary. If the rare medium exhibits absorption, the penetrating wave becomes attenuated, so the reflectance can be written by Eq. (1)

$$R = 1 - kd_c \tag{1}$$

where d_c is the effective layer thickness. The resulting energy loss in the reflected wave is referred to as attenuated total reflection (see ATR). When multiple reflections are used to increase the sensitivity, the technique is often called multiple internal reflection (MIR). Thus qualitatively, a spectrum of IRS resembles a transmission spectrum. But, there are two adverse effects arising from the wavelength dependance of IRS. First, the long wavelength side of an absorption band tends to be distorted and second, bands of longer wavelengths will appear relatively stronger. J. Fahrenfort was the first to demonstrate the usefulness of the phenomenon of ATR [1,2] or total internal reflection, while N.J. Harrick developed the technique [3,4] and designed ATR cells for commercial use. With common FT-IR spectrometers, one does not achieve the same improvement in IRS as in transmission compared to the classical dispersion instruments because ATR attachments have not been redesigned yet for the larger round beam [5]. However, the signal averaging capability and the higher speed have increased the utility of IRS for polymers [6] especially for surface studies. ATR dichroism studies were helpful in characterizing polymer surfaces [7]. T. Hirschfeld [8,9] has generated the algorithms which are necessary for using the IRS method for determination of the optical constants of a sample from a pair of independent reflectivity measurements at each frequency. The optimum method is the one which determines the total reflectance at two polarizations at the same incidence angle.

References

1. Fahrenfort J (1961) Spectrochim. Acta 17: 698
2. Fahrenfort J, Visser WM (1962) Spectrochim. Acta 18: 1103
3. Harrick NJ (1962) Phys. Rev. 125: 1165
4. Harrick NJ (1967) Internal reflection spectroscopy, Wiley, New York
5. Koenig JL (1984) Fourier transform infrared spectroscopy of polymers, Springer, Berlin Heidelberg New York (Advances in polymer science, vol 54)
6. Jacobsen R: see Ref. 128 in Ref. [5] (ATR of biological systems)
7. Sung CSP (1981) Macromolecules 14: 591
8. Hirschfeld T (1970) Appl. Spectrosc. 24: 277
9. Hirschfeld T (1978) Appl. Spectrosc. 32: 160

ISC
Abbreviation of Inverse Shift Correlation

ISC is a new approach in nuclear magnetic resonance (see NMR) spectroscopy [1]. Inverse heteronuclear shift correlation via long-range couplings has been applied for carbonyl assignment and sequence analysis in polypeptides [1]. The pulse sequence involved is identical with that used by other authors [2]. It may be pointed out that this experiment has a higher sensitivity than the well-known COLOC sequence (see COLOC).

References

1. Bermel W, Griesinger C, Kessler H, Wagner K (1987) Magn. Reson. Chem. 25: 325
2. Bax A, Griffey R, Hawkins BL (1983) J. Magn. Reson. 55: 301

JACQUINOT's Advantage

This is one of the important advantages of FT-IR (see p 116) spectrometers in comparison to dispersive IR instruments [1]

The JACQUINOT's or throughput advantage [1] arises from the loss of energy in the dispersive system due to the gratings and slits. These losses can not occur in an FT-IR instrument which does not contain these elements. Basically, JACQUINOT's advantage means that the radiant power of the source is more effectively utilized in interferometers. The throughput for an FT-IR spectrometer is limited by the size of the mirrors. The JACQUINOT's advantage for an interferometer with a commercial dispersive spectrometer has been discussed in the past [2]. This higher throughput is particularly important in an IR region where the signals are weak since the IR sources are weak. The throughput advantage has been used successfully for studying strongly absorbing systems such as carbon-black filled rubbers [3] and emission from polymers [4].

IJ

References

1. Jacquinot P (1960) Rep. Prog. Phys. 13: 267
2. Griffiths PR, Sloane HJ, Hannah RW (1977) Appl. Spectrosc. 31: 485
3. Hart WW, Painter PC, Koenig JL (1977) Appl. Spectrosc. 31: 220
4. Jennings W (1976) Master Thesis, Case Western Reserve University, Cleveland, OH

JCP

Abbreviation **J** Cross-Polarization

In the case of liquids, the mixing HAMILTONian involves the indirect nuclear spin-spin coupling constant J of nuclear magnetic resonance (see NMR) spectra. One-dimensional as well as two-dimensional applications of this J cross-polarization process have been described [1–6], and detailed analyses of the coherence-transfer spectrum in AX_n spin systems $(n = 1, \ldots, 6)$ involving spin-1/2 nuclei [5,6] have been published.

In discussing coupled JCP spectra, the occurrence of phase and multiplet anomalies has been noted and a phase-corrected JCP sequence (see PCJCP) has been developed [3,5] to purge coupled spectra of phase anomalies.

Recently a modified JCP sequence has been proposed, which purges coupled spectra of both phase and multiplet anomalies, thereby restoring natural multiplet patterns [7]. This sequence has been termed CJCP (see p 48). For further details see Ref. [7].

References

1. Bertrand RD, Moniz WB, Garroway AN, Chingas GC (1978) J. Am. Chem. Soc. 100: 5227
2. Bertrand RD, Moniz WB, Garroway AN, Chingas GC (1978) J. Magn. Reson. 32: 465
3. Chingas GC, Bertrand RD, Garroway AN, Moniz WB (1979) J. Am. Chem. Soc. 101: 4058
4. Chingas GC, Garroway AN, Bertrand RD, Moniz WB (1979) J. Magn. Reson. 35: 283

5. Chingas GC, Garroway AN, Bertrand RD, Moniz WB (1981) J. Chem. Phys. 74: 127
6. Müller L, Ernst RR (1979) Mol. Phys. 38: 963
7. Chandrakumar N (1985) J. Magn. Reson. 63: 202

JMOD

Abbreviation of the **J-Mod**ulation pulse sequence used in carbon-13 nuclear magnetic resonance (see NMR)

The JMOD pulse sequence [1] may be explained with Eq. (1) and has been applied together with broad-band proton decoupling (see BB) in order to obtain ^{13}C-NMR spectra for structure elucidation.

$$D1 - 90° - D3 - 180° - D3 - FID \text{ (free induction decay)} \tag{1}$$

BB decoupling during D1, the relaxation delay (normally at lower power to avoid sample heating), builds up the nuclear OVERHAUSER effect (see NOE, 1D-NOE, 2D-NOE, and NOESY). With $D3 = 1/J_{CH}$, the spectra are recorded with CH and CH_3 signals in one direction and quarternary carbons and methylene carbons signals in the opposite direction.

The interested reader is referred to other or similar editing procedures in carbon-13 NMR, such as APT or DEPT.

IJ

Reference

1. Le Cocq C, Lallemand J-Y: J. Chem. Soc. Chem. Commun. 1981: 150

J-mod SE

Acronym for **J-mod**ulated **S**pin-**E**cho Technique used in Nuclear Magnetic Resonance (see NMR)

The so-called *J-modulated spin-echo* technique [1] has proved to be a valuable and simple alternative to off-resonance decoupling, INEPT (see p 143), DEPT (see p 69), and attached proton test (see APT) for the assignment of the different carbon species in carbon-13 NMR spectra. Recently [2] this method has been applied for spectral assignments in complicated spectral regions as demonstrated in a case of the β-lactam compound, known as a penicillin.

References

1. Le Cocq C, Lallemand J-Y: J. Chem. Soc. Chem. Commun. 1981: 150
2. Molin H (1987) Magn. Reson. Chem. 25: 606

J-RESIDE

Acronym for **J-Re**solved spectroscopy with **Si**multaneous **De**coupling

Nagayama [1] pointed out that problems due to overlapping in 2D-J spectra can be reduced by simultaneous spin decoupling and presented a theoretical analysis of experiments using continuous decoupling, decoupling during evolution, and decoupling during detection. A preliminary report on this technique as applied to

NMR of proteins was published [2], but the method has not been widely used. In the meantime it has been shown, however, that J-RESIDE is helpful in the analysis of complex proton spectra [3]. There are many cases in which spectral overlap does prevent the identification of the exact proton to which another proton is coupled.

References

1. Nagayama K (1979) J. Chem. Phys. 71: 4404
2. Nagayama K, Wüthrich K, Bachmann P, Ernst RR (1979) Biochem. Biophys. Res. Commun. 86: 218
3. Shoolery JN (1986) Magnetic Moments 2: 10

IJ

LAOCOON
Acronym for **L**east-squares **A**djustment **of C**alculated **on O**bserved NMR Spectra

LAOCOON I and II or the third revised version are FORTRAN programs (see FORTRAN) for the analysis of high-resolution NMR spectra (see NMR) which have been used for more than 20 years [1]. The program has two capabilities: 1. From an arbitrarily chosen set of chemical shifts and coupling constants for a given system of two to seven spin-1/2 nuclei (ABCDEFG), it can cause the computer to generate a table of frequencies and intensities of the absorption lines expected in the nuclear magnetic resonance spectrum. 2. If a spectrum thus calculated bears a recognizable resemblance to an observed NMR spectrum, the program can cause the computer to perform an iterative calculation process by means of which the calculated frequencies of assigned lines are brought as close as possible (by the least squares criterion) to the corresponding experimentally observed lines. The chemical shifts and coupling constants which yield this best fit are then printed out, together with information about the expected errors in them, and tables of observed and calculated line frequencies, calculated intensities, and errors in fitting the frequencies of observed lines.

References
1. Castellano S, Bothner-By AA (1964) J. Chem. Phys. 41: 3863; Bothner-By AA, Naar-Bolin C (1962) J. Amer. Chem. Soc. 84: 743; Bothner-By AA, Naar-Colin C, Günther H (1962) J. Amer. Chem. Soc. 84: 2748

Appendix
Comment of A.A. Bothner-By and S.M. Castellano about LAOCOON in 1966: ". . . The struggle of the Trojan priest Laokoon and his two sons with serpents seemed to the authors akin to their own struggles with keeping line assignments untangled. Fortunately, this difficulty has now disappeared". Detailed description of the program LAOCOON: Assignment of experimental frequencies to the calculated lines of the theoretically calculated spectrum from a trial parameter set is almost important. The fundamental idea here is that the best parameter set is the one that leads to the smallest sum of the squares of the errors, following Eq. (1):

$$\sum_{i=1}^{k} (f_{exp} - f_{calc})_i^2 \tag{1}$$

where k is the number of measured lines, $f_{exp} - f_{calc}$ is the frequency difference between observed and calculated transitions for the ith line.
    For each parameter p_j Eqs. (2) and (3) are valid.

$$\partial \sum_{i=1}^{k} (f_{exp} - f_{calc})_i^2 / \partial p_j = 0 \tag{2}$$

$$-2 \sum_{i=1}^{k} (f_{exp} - f_{calc})_i \left(\frac{\partial f_{calc}}{\partial p_j} \right)_i = 0 \tag{3}$$

If one assumes for only small parameter changes a linear dependence of the frequencies one can write Eq. (4):

$$\Delta f = \left(\frac{\partial f_i}{\partial p_j} \right) \Delta p_j \qquad (4)$$

Then for the best solution Eq. (5) is valid:

$$\left(\frac{\partial f_i}{\partial p_j} \right) \Delta p_j = (f_{exp} - f_{calc})_i \qquad (5)$$

The assignment of the experimental lines to the lines of the first calculated spectrum provides the computer with the information $f_{exp} - f_{calc}$. The partial derivatives are obtained from the eigenvalues of the trial spectrum. They are approximated just as the first parameter correction Δp_j. Various iteration cycles will lead finally to convergence toward the best solution. Because of its better convergence properties the LAOCOON procedure has been preferred in practice over a period more than 20 years. Like other methods or programs LAOCOON does not consider the intensities of the observed lines as input information.

LASER
Acronym for **L**ight **A**mplification by **S**timulated **E**mission of **R**adiation

The LASER is made possible by *stimulated emission*, an effect which is inherently related to the well-known effects of *spontaneous emission* and *absorption*, whereas all these effect are related to each other by EINSTEINS's probability coefficients.

A LASER, normally has three *basic components*:

− an optical amplifier,
− an excited-state pump, and
− an optical resonator or cavity.

The relevance of these basic LASER components has been described in detail in the literature [1].

LASER radiation is characterized by *extremely high values* of the following four *properties*:

− directionality,
− monochromaticity,
− coherence, and
− radiance.

These important properties of LASER's explain their *nearly universal application possibilities* in different fields of research and development.

The influence of the cavity upon LASER beam properties has been discussed [1] under the following *aspects*:

− spatial modes,
− concave mirror cavities,
− transverse electromagnetic modes (TEM),
− mode spacing and pulling,
− "real" cavities,
− pumping homogeneity,

−concentration effects,
−near- and far-field profiles,
−"focal spot",
−temporal modes such as continuous wave, normal pulsed Q-switched, cavity-dumped, and mode-locked,
−wavelength selection, and
−non-linear optical effects.

Normally, we can distinguish between *three types* of LASER's [1]:

−solid-state LASER's,
−dye LASER's, and
−gas LASER's.

Analytical applications of LASER's may be divided in *five classes* [2]:

−selected methods that use various detection schemes such as LASER-excited atomic ionic fluorescence in flames and plasmas, LASER-enhanced ionization in flames, detection of small numbers of atoms and molecules, optoacoustic spectroscopy, and infrared absorption spectroscopy (see IR);
−methods with enhanced spectral resolution such as cryogenic molecular fluorescence spectrometry and linear or non-linear site-selective LASER spectrometry;
−selected multiphoton and multiwavelength methods such as two-photon excited fluorescence and RAMAN spectroscopy (see p 233);
−methods using special characteristic of LASERs such as remote sensing with LASER'S, intracavity-enhanced spectroscopy, thermal lens effects, picosecond-spectroscopy, electrophoretic light-scattering, and LASER-flow cytometry;
−LASER's in combination with other methods such as LASER spectroscopy for detection purposes in chromatography, LASER ionization techniques in analytical mass spectrometry, and LASER ablation for atomic spectroscopy.

The interested reader will find in this dictionary only the applications of LASER's in molecular spectroscopy.

References
1. Piepmeier EH (1986) In: Elving PJ, Winefordner JD (eds) Analytical applications of lasers, chap 1, Wiley, New York (Chemical analysis, vol 87)
2. See chaps 2 to 19 (1986) In: Elving PJ, Winefordner JD (eds) Analytical applications of lasers, Wiley, New York (Chemical analysis, vol 87)

LDS
Abbreviation of **L**inear **D**ichroism **S**pectrometry

Like vibrational circular dichroism (see VCD) spectrometry used in the infrared region (see IR) of molecular spectroscopy LDS became interesting, too [1]. The photoelastic modulator (PEM) commonly used in the measurements of VCD spectra can also be applied in the field of LDS [2,3]. Although LDS has been applied to DNA using PEM in conjunction with a monochromator [4], a lot of scientist believe that by far the most important application of LDS will be the spectrometry of species absorbed onto flat metallic surfaces [1]. The combined use of a PEM and an FT-IR spectrometer for surface analysis was first demonstrated to be feasible by Dowrey and Marcott [5].

References

1. Griffiths PR, de Haseth JA (1986) Fourier transform infrared spectrometry, Wiley, New York (Chemical analysis, vol 83)
2. Nafie LA, Diem M (1979) Appl. Spectrosc. 33: 130
3. Nafie LA, Vidrine DW (1982) In: Ferraro JR, Basile LJ (eds) Fourier transform spectroscopy: Applications to chemical systems, vol 3, Academic, New York
4. Kusan T, Holzwarth G (1976) Biochemistry 15: 3352
5. Dowrey AE, Marcott C (1982) Appl. Spectrosc. 36: 414

LG function
Abbreviation of LORENTZian to GAUSSian Transformation Function

Undoubtedly one of the most important advantages in using a pulse FOURIER transform (see PFT) mode in nuclear magnetic resonance (see NMR) is that one can easily perform a *deconvolution* or *convolution* process by simply multiplying the accumulated free induction decays (see FID) by a relevant mathematical function to enhance the resolution or to improve the signal-to-noise ratio (see SNR) of the spectrum [1].

L

Among the various mathematical functions, *LG function* [see Eq. (1)] [2] and *sine-bell* functions [see Eq. (2)] [3] which include *shifted sine-bell* ($\phi \neq 0$) [4] and *shifted squared sine-bell* (n = 2) [5] functions have been widely used in 1D- and 2D-NMR [5,6].

$$g(t) = \exp(t/C1)\exp(-t^2/C2^2) \tag{1}$$

$$g(t) = \sin^n(\pi t/ACQ + \pi\phi/180) \tag{2}$$

In Eq (1), C1 and C2 and ϕ in Eq. (2) denote parameters which should be arbitrarily determined and ACQ stands for the acquisition time.

Recently [7] advantages and disadvantages of the LG function have been investigated by applying this function to the two-dimensional nuclear OVERHAUSER effect spectroscopy (see 2D-NOE and NOESY) of a drug-phosphatidylcholine vesicles solution. It was found [7] that, although this function is known to be useful in resolution enhancement of an NMR spectrum of a molecule of low molecular weight, the function often fails to yield a good line shape or contour plot in the case where there exist broad resonance peaks having different line widths as in the above mentioned sample solution. A careful choice of the relevant parameters is required. In conclusion, the LG function is inferior to a sine-bell function which has also been tested as an alternative approach [7].

The recently developed *phase-sensitive mode* of operation in 2D-NMR (see p 7) yields pure absorption phase spectra [8] and thus does not always require such vigorous "*reshaping*" of NMR signals as in the *absolute-value* two-dimensional experiments. But the phase-sensitive approach cannot be applied to such variations in 2D-NMR pulse sequences as in homonuclear 2D J-spectroscopy and in correlation spectroscopy (see COSY) with radio-frequency pulses not equal to 90° for the second (detection) pulse [6c]. In the cases, again, careful consideration of an appropriate *weighting function* is still required [7].

References

1. Shaw D (1976) Fourier transform NMR spectroscopy, Elsevier, Amsterdam, chap 3
2. a) Ferrige AG, Lindon JC (1978) J. Magn. Reson. 31: 337;
   b) Lindon JC, Ferrige AG (1979) J. Magn. Reson. 36: 277
3. de Marco A, Wüthrich K (1976) J. Magn. Reson. 24: 201
4. Guéron M (1978) J. Magn. Reson. 30: 515
5. Wider G, Macura S, Kumar A, Ernst RR, Wüthrich K (1984) J. Magn. Reson. 56: 207
6. a) Bax A (1982) Two-dimensional nuclear magnetic resonance in liquids, Reidel, Dordrecht,
   chap 1
   b) Morris GA (1986) Magn. Reson. Chem. 24: 371
   c) Ernst RR, Bodenhausen G, Wokaun A (1987) Principles of nuclear magnetic resonance in
   one and two dimensions, Clarendon Press, Oxford, chap 6
7. Kuroda Y, Fujiwara Y, Saito M, Shingu T (1988) Chem. Pharm. Bull. [Japan] 36: 849
8. a) States DJ, Haberkorn RA, Ruben DJ (1982) J. Magn. Reson. 48: 286
   b) Marion D, Wüthrich K (1983) Biochem. Biophys. Res. Commun. 113: 967

LGR
Abbreviation of **L**oop-**G**ap **R**esonator

Application of the LGR in pulsed electron spin resonance (see ESR and FT-ESR) has become very important since its high efficiency (in converting incident microwave power to B_1 in the resonator) allows the use of a lower Q, thus diminishing the ringdown time and the pulse distortion (note! the resonator bandwidth is equal to v_0/Q) [1].

L

Reference
1. Horwak JP, Freed JH (1986) J. Magn. Reson. 67: 501

LIS
Acronym for **L**anthanide **I**nduced **S**hifts in Nuclear Magnetic Resonance (see NMR)

Paramagnetic rare-earth complexes have been widely used in proton NMR as chemical shift dispersion agents [1–5]. The lanthanide induced shifts, brought about by secondary internal fields originating from unpaired electron spins, usually result in greatly simplified spectra (so-called *1st order spectra*). Necessary for the applicability of the lanthanide shift reagents (see LSR) is the presence of polar groups in the molecule under investigation allowing binding to the lanthanide ion (e.g. $NH_2 > OH > C=O > COOR > CN$, etc.). The data generated so far clearly indicate a significant predominance of a dipolar contribution. For a dipolar, or so-called *pseudocontact* interaction of the lanthanide ion with a nucleus i, the induced shift $\Delta H_i/H$ is expected to be proportional to the McCONNELL/ROBERTSON term [6]:

$$\Delta H_i/H = Kr_i^{-3} \ (3 \ cis^2 \ \chi_i - 1) \tag{1}$$

In Eq. (1), which is only valid in the case of axial symmetry of the complex involved, r_i stands for the distance of the i-th nucleus relative to the site of the lanthanide atom itself, and χ_i represents the angle between the principal magnetic axis of the complex and the distance vector r_i. Following Eq. (1), the incremental shifts are thus predicted to be proportional to the inverse cube of the internuclear distance. The relative sign, on the other hand, is given by χ_i. A critical angle ($\chi = 54°44'$) has been defined, at which sign reversal occurs. In the meantime this has been confirmed experimentally [7].

Quantitative information on the basis of Eq. (1) is difficult to obtain, because of uncertainties regarding the geometry of the complex formed. Moreover, the magnetic axis of the complex formed does not necessarily coincide with the metal-ligand bond axis [8]. Several attempts have been made to get a computer-fit between experimental and geometrical parameters [9,10] under minimizing deviations between experimental and calculated shift values. For the most commonly applied shift reagents see the acronym LSR in this dictionary.

Paramagnetic shifts induced at carbon-13 atoms, obtained for some typical ketones, alcohols, and amines have been published in the literature [10]. Although, in principle, the angular term in Eq. (1) cannot be neglected, in particular in conformationally rigid ring systems a quite near-proportionality between the observed shifts and $1/r^3$ has been found often [11]. Shift reagents have recently been used in conjunction with the total assignment of carbon-13 spectra in a series of natural products [12–14].

Based on the wellknown McCONNELL/ROBERTSON relationship [6] [see Eq. (1)], almost excellent agreement was only obtained in the case of proton magnetic resonance. However, corresponding carbon-13 data could not always be satisfactorily explained by the dipolar model [15,16]. A separation of the dipolar contribution was achieved as follows: a least-squares regressional analysis of proton data yielded the mutual space coordinates of the lanthanide ion involved; these, in turn, were used for computation of the induced carbon-13 shifts [16]. Finally, the difference between experimental and computed values was considered to be the contribution from a FERMI contact interaction [16].

In carbon-13 NMR spectra LIS can be helpful in the case of selective irradiation of better separated proton resonances in order to get more precise carbon assignments [17].

The LIS method has a lot of limitations [4]; these require quantitative analysis and a reliable significance test in order that they may be detected. The use of the widely applied *Hamilton R ratio test* [18] in LIS analysis has been questioned, and another significance test, the so-called *jacknife test*, has been proposed [19]. On the other hand, some authors have used weighting schemes, in spite of the lack of information in the literature about this approach. In order to evaluate the use of weighting schemes, and to compare the results obtained under using the *Hamilton significance test* [18] with those of a *jacknife test* [19], a group of Spanish scientists [20] has published a much more detailed significance testing program on LSR (see p 166) data. Their results [20]: the use of weighted agreement factors in the significance testing of LIS analysis is physically meaningless, and can manipulate the resulting significance values; the *R ratio test* is more easily applied, whereas the *jacknife test* does not show any clear advantage and will be more sensitive to changes resulting from the use of weighted agreement factors.

References

1. Hinkley CC (1969) J. Am. Chem. Soc. 91: 5760
2. Ammon RV, Fischer RD (1972) Angew. Chem. Int. Ed. 11: 675
3. Petersen MR, Wahl GH (1972) J. Chem. Educ. 49: 790

4. Hofer O (1976) The lanthanide induced shift technique. Application in conformational analysis, Wiley-Interscience, New York (Topics in stereochemistry, vol 9)
5. Sievers RE (ed) (1973) Nuclear magnetic resonance shift reagents, Academic, New York
6. McConnell HM, Robertson RE (1958) J. Chem. Phys. 29: 1361
7. Shapiro BL, Hlubucek JR, Sullivan GR, Johnson LF (1971) J. Am. Chem. Soc. 93: 3281
8. Hawkes GE, Leibfritz D, Roberts DW, Roberts JD (1973) J. Am. Chem. Soc. 95: 1659
9. See, e.g.: Briggs J, Hart FA, Moss GP, Randall EW: Chem. Commun. 1971: 364
10. Chadwick DJ, Williams DW: J. Chem. Soc., Perkin II, 1974: 1202
11. Gansow OA, Wilcott MR, Lenkinski RE (1971) J. Am. Chem. Soc. 93: 4295
12. Wehrli FW, Wirthlin T (1976) Interpretation of carbon-13 NMR spectra, Heyden, London, p 96
13. Almqvist SO, Enzell CR, Wehrli FW (1975) Acta Chem. Scand. B29: 695
14. See, e.g.: Wenkert R, Cochran DW, Hagamon EN, Burton Lewis R, Shell FM (1971) J. Am. Chem. Soc. 93: 6271
15. Hawkes GE, Marzin C, Johns SR, Roberts, JD (1973) J. Am. Chem. Soc. 95: 1661
16. Gansow OA, Loeffler PA, Davis RE, Wilcott MR, Lenkiski RE (1973) J. Am. Chem. Soc. 95: 3390
17. Birdsall B, Feeney J, Glasel JA, Williams RJP, Xavier AV: Chem. Commun. 1971: 1473
18. Hamilton WC (1965) Acta Crystallogr. 18: 502
19. a) Richardson MF, Rothstein SM, Li W-K (1979) J. Magn. Reson. 36: 69;
    b) Rothstein SM, Beel WD, Richardson MF (1980) J. Magn. Reson. 41: 310
20. Ribó JM, Valera G (1987) Magn. Reson. Chem. 25: 293

LOCK

An artificial name for field/frequency control and stabilization in nuclear magnetic resonance (see NMR) spectroscopy

L

There are different techniques for the purpose of field/frequency stabilization (1).

In the first method, a side-band produced by modulation of the frequency derived from the crystal oscillator is used to measure the NMR absorption of a control sample, usually water in proton magnetic resonance, which is located in a separate cell next to the position of the sample tube. This control or *lock signal* is then used for automatic regulation of the modulation frequency which is maintained so that the resonance condition for the water sample is always met. This arrangement is called an *"external lock system"*. The stability obtained with it limits the drift to about 1 part in 10^8 per hour.

In an *"internal lock system"*, the control signal used for regulating the magnetic field is generated directly in the analytical sample tube by means of a side-band of constant frequency. Usually the signal of the reference substance (see TMS) serves as the lock signal. This arrangement has been called *"homolock"*. Further, with a second and independent modulation oscillator, a side-band of variable frequency is generated which serves for observation of the spectrum in the frequency-sweep mode of a continuous wave experiment (see SWEEP and CW). Even a field-sweep experiment can be performed with this system if the lock frequency is varied while the field is swept. The internal lock system has the advantage that the analytical and the control substances are exposed to exactly the same magnetic field.

Other nuclei, such as deuterium or fluorine-19, have also been used to generate the lock signal for field/frequency stabilization purposes. In these cases the so-called *"heterolock"* is used. The stability produced limits the drift to about 1 part in 10^9 per hour.

On modern FOURIER Transform spectrometers (see PFT) data accumulation over long time intervals requires high field/frequency stability and an "*internal heterolock system*" that usually employs the deuterium resonance of a deuterated solvent ($CDCl_3$, C_6D_6, DMSO-D_6, etc.) is therefore essential. This "*internal deuterium lock*" may be a "*pulsed lock*" or a "*CW lock*".

The use of deuterated solvents for NMR field-shimming and -locking purposes at magnetic fields corresponding to a proton frequency between 300 and 400 MHz or higher has been evaluated recently [1]. It has been found that the appearance of quadrupolar splittings and the interference of scalar deuterium-deuterium couplings will influence "*lock*"-*intensity* and/or -*sensitivity* and -stability for many deuterated solvents commonly used in NMR spectroscopy. The effect of self-decoupling on this process and on the detectability of quadrupolar couplings in the deuterium-NMR spectrum have been carefully discussed presenting a list of problematic and suitable "*lock*" *solvents* [1].

The lock-signal itself can be used for a shim procedure (see SHIM), too.

References
1. van Zijl PCM (1987) J. Magn. Reson. 75: 335

LODESR
Acronym for **L**ongitudinal **D**etection of **E**lectron **S**pin **R**esonance

The phenomenon of the longitudinal detection in the field of electron spin or paramagnetic resonance spectroscopy (see ESR or EPR) was described 10 years ago [1–4]. Recently [5] variable frequency experiments in LODESR have been reported. The signal observed in a LODESR experiment has been studied, for the case where at a fixed static magnetic field the corresponding frequencies of the two irradiating waves are independently varied. The authors [5] have described the applied improved LODESR experimental procedure in detail. The linewidth dependence of the longitudinal relaxation time (T_1) and a new T_1 measurement method have been discussed [5].

References
1. Chiarini F, Martinelli M, Pardi L, Santucci S (1975) Phys. Rev. B 12: 847
2. Davis JJ (1976) Solid State Commun. 20: 433
3. Martinelli M, Pardi L, Pinzo C, Santucci S (1977) Phys. Rev. B 16: 164
4. Holmes BJ, Hagston WE (1979) Mol. Phys. 38: 25
5. Giordano M, Martinelli M, Pardi L, Santucci S, Umeton C (1985) J. Magn. Reson. 64: 47

LP
Abbreviation of **L**inearity **P**rinciple in Spectroscopy

Molecular spectroscopy in its classical form describes molecular systems in their linear approximation. Linearity is the essential problem or question. One can ask, whether input/output relations of a given physical system may be characterized correctly by LP. In the case of linear systems we can use the so-called *electronic system theory* [1,2]. Equivalence of expressions such as "*spectrum*", "*transfer function*" or "*frequency response function*" on one hand, and names such as "*interferogram*", "*free induction decay*" (see FID) or "*impulse response*" on the other, became very important in the past. These correspondencies became very fruitful in the development of FOURIER spectroscopy in NMR [3] and IR [4]. But, we have to have in mind, that linearity is often a simplification and not a real correspondence. In the radiofrequency field, linearity is only valid in the case of very weak perturbation fields. Strong perturbations will cause line broadening, line shifts and saturation as typical non-linear effects. These special cases will be explained under other items, abbreviations, acronyms etc. in this dictionary.

References

1. Goldman S (1968) Information theory, Dover, New York
2. Brown BM (1965) The mathematical theory of linear systems, Science, New York
3. a) Ernst RR, Anderson WA (1966) Rev. Sci. Instrum. 37: 93;
   b) Ernst RR (1966) Adv. in Magn. Reson. 2: 1;
   c) Farrar TC, Becker ED (1971) Pulse and Fourier transform NMR: Introduction to theory and methods, Academic, New York
4. a) Fellgett PB (1951) Thesis, University of Cambridge
   b) Vanasse GA, Sakai H (1967) Prog. Opt. 6: 259;
   c) Bell RJ (1972) Introductory Fourier transform spectroscopy, Academic, New York

LPQRD

Abbreviation of **L**inear **P**rediction Principle and the Householder triangularization decomposition (**QRD**) technique [1]

L

This method may be seen as an alternative to fast Fourier Transform (see FFT) spectral analysis with improved resolution [2]. In comparison to the new LPSVD technique (see p 163) LPQRD is an improvement in computational speed. LPQRD is based on the linear prediction principle and the Householder triangularization decomposition procedure. In common with LPSVD, this method is a linear least-square procedure and can yield accurate estimates of frequencies, decay rates, amplitudes, and phases of damped sinusoidal signals. It requires no guessing of starting trial values and eliminates the slow iterative execution of nonlinear curve fitting routines. The LPQRD method has a better resolution than the FFT and is much less subject to phase distortion arising from signal truncation.

References

1. Lawson CL, Hanson RJ (1974) Solving least square problems, Prentice-Hall, Englewood Cliffs, NJ
2. Tang J, Lin CP, Bowman MK, Norris JR (1985) J. Magn. Reson. 62: 167

LPSVD

Abbreviation of **L**inear **P**rediction Principle and **S**ingular **V**alue **D**ecomposition (see SVD)

This method was proposed as an alternative to the fast Fourier transformation (see FFT). It is a numerical technique using the linear prediction principle (see LP) and singular value decomposition (see SVD) to analyze time-domain signals of exponentially damped sinusoids [1]. This procedure has been applied recently to magnetic resonance [2]. The new LPSVD technique has many advantages over FFT, while the major drawback is considerably longer computational time. To improve computational speed and minimize this drawback a more efficient method (see LPQRD) has been developed [3].

References

1. Kumaresan R, Tufts DW (1982) IEEE Trans. ASSP-30: 833
2. Barkhuijsen H, De Beer J, Boveé WMMJ, Van Ormondt D (1985) J. Magn. Reson. 61: 465
3. Tang J, Lin CP, Bowman MK, Norris JR (1985) J. Magn. Reson. 62: 167

LPZ
Acronym for the combination of **L**inear **P**rediction (see LP) and **Z**-Transformation

To overcome difficulties associated with fast FOURIER transformation (see FFT) in nuclear magnetic resonance (see NMR) a method of spectral analysis has been developed using a combination of LP and the Z-transformation (LPZ) for which resolution enhancement and noise reduction can be achieved simultaneously [1].

Recently it has been demonstrated how LPZ can be used in two dimensions where one dimension permits the display of greatly enhanced resoultion [2].

Z-transformation is a combination of the discrete FOURIER transformation (see FT) and the discrete LAPLACE transformation [3].

References

1. Tang J, Norris JR (1986) J. Chem. Phys. 84: 5210
2. Tang J, Norris JR (1986) J. Magn. Reson. 69: 180
3. Oppenheim AV, Schafer RW (1975) Digital signal processing, Prentice Hall, Englewood Cliffs NJ

LR-HETCOR
Acronym for **L**ong-**R**ange **Het**eronuclear Shift-**Cor**relation

Recently a modified long-range heteronuclear chemical shift correlation pulse sequence which eliminates modulations of response intensity due to one-bond proton-carbon-13 spin couplings has been described [1].

The "decoupling" of the one-bond modulation has been realized by using a bilinear rotational decoupling (see BIRD) pulse of the following type

$$90_x^\circ(^1H) - \tau - 180_x^\circ(^1H)/180_x^\circ(^{13}C) - \tau - 90_{-x}^\circ(^1H)$$

midway through the second magnetization transfer (see MT) delay, Δ_2.

One-bond modulation of response intensities over a range of optimization values has been demonstrated for the alkaloid norharmane when long-range optimization of the conventional heteronuclear chemical shift correlation (see HETCOR) pulse sequence was employed. The modified pulse sequence eliminates one-bond modulations over the same range of optimization values. Slower modulations which arise due to long-range couplings other than that optimized for were present in both experiments.

Repression of the effects of one-bond modulation may be a significant improvement, making long-range HETCOR experiments a powerful tool in spectroscopic assignments and structure elucidation [1–3].

References

1. Zektzer AS, John BK, Martin GE (1987) Magn. Reson. Chem. 25: 752
2. Salazar M, Zektzer AS, Martin GE (1988) Magn. Reson. Chem. 26: 24
3. Salazar M, Zektzer AS, Martin GE (1988) Magn. Reson. Chem. 26: 28

LRJR
Acronym for a **L**ong-**R**ange **J**-**R**esolved (Pulse Sequence) used in Nuclear Magnetic Resonance (see NMR)

This new pulse sequence was originally given by Bax and Freeman [1] in order to observe selectively the carbon-13/proton long-range couplings of carbons with specific protons. The behaviour of carbon-13 magnetization following the spin echo displays modulation pattern for different carbon species as shown in the case of carvon [1]. Figure 15 shows the different situations of the new pulse sequence in the proton and carbon-13 channel.

This technique has been used for mixture analysis prior to the combination of the information with GC/MS data [2]. But, more recently an alternative mixture approach solely based upon LRJR and a carbon-13 chemical shift data bank has been reported [3].

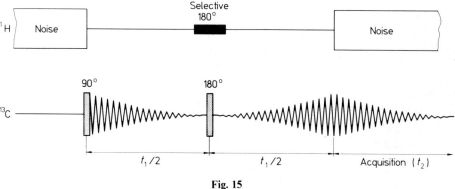

Fig. 15

References

1. Bax A, Freeman R (1982) J. Am. Chem. Soc. 104: 1099
2. Laude DA, Cooper JR, Wilkins CL (1986) Anal. Chem. 58: 1213
3. Steven TK, Lee RWK, Wilkins CL, paper presented at the Pittsburgh Conference, Atlantic City, NJ, March 9–13, 1987.

LROCSCM
Abbreviation of **L**ong-**R**ange **O**ptimized Heteronuclear Chemical Shift Correlation Method

LROCSCM is a new pulse sequence used in two-dimensional nuclear magnetic resonance (see 2D-NMR). A comparison of the results obtained using

LROCSCM with those from conventional long-range optimized heteronuclear chemical shift correlation has been presented [1].

Reference

1. Zektzer AS, Quast MJ, Linz GS, Martin GE, McKenney JD, Johnston MD Jr, Castle RN (1986) Magn. Reson. Chem. 24: 1803

LSE
Pulse sequence standing for the authors **Levitt**, **Suter**, and **Ernst**

LSE is a typical composite-pulse (see p 51) quadrupolar echo sequence [1], especially used in the field of nuclear magnetic resonance (see NMR) in the solid-state. This abbreviation or acronym was introduced by Barbara [2].

Calculations of the effects arising from finite pulse widths in solid-state deuterium NMR was given by Bloom and co-workers [3] where distortion factors coming in from the use of the quadrupolar echo pulse sequence were presented. More recent papers have addressed extensions of this effect to spin alignment echoes [4] and inversion recovery using a quadrupolar echo read pulse [5]. Barbara [2] has presented analytical expressions and lineshape simulations for LSE composite-pulse quadrupolar echo sequences. In this paper, it is shown that analytical expressions employing only relative phase shifts of 180^0 can be easily obtained from the usual relations pertaining to three-dimensional rotations. This is possible because the spin dynamics on the generally eight-dimensional space [6] decomposes into two spaces with four dimensions, allowing one to take advantage of the close relationship between four- and three-dimensional rotations [7]. The LSE composite sequences [1] can be analyzed in this fashion, which also provides a complementary picture to the similarities between composite-pulses that compensate for resonance offset and the problem of broadband excitation of quadrupolar-broadened resonances in solids [2].

References

1. Levitt MH, Suter D, Ernst RR (1983) J. Chem. Phys. 80: 3064
2. Barbara TM (1986) J. Magn. Reson. 67: 491
3. Bloom M, Davis JH, Valic MI (1980) Can. J. Phys. 58: 1510
4. Henrichs PM, Hewitt JM, Linder M (1984) J. Magn. Reson. 60: 280
5. Siminovitch DJ, Griffin RG (1985) J. Magn. Reson. 62: 99
6. Mehring M, Wolff KE, Stoll ME (1980) J. Magn. Reson. 37: 475
7. Hamermesh M (1962) Group theory, Addison-Wesley, Reading MA, p 308

LSR
Abbreviation of **L**anthanide **S**hift **R**eagents

Lanthanide shift reagents are used in practical nuclear magnetic resonance (see NMR) spectroscopy in different ways. The induced chemical shifts by these reagents (see LIS), originally investigated by Hinckley [1], have found a lot of application forms in NMR:

a) simplification of complex proton spectra;
b) simplification of complex proton and carbon-13 spectra of the same molecule, too;

c) analytic help in qualitative and quantitative assignments of product mixtures;
d) contributions to configurational and conformational studies of molecules in solution;
e) tests of optical or enantiomeric purity under using chiral LSR.

The most used lanthanide shift reagents, their anions, cations, chiral ligands, and their common abbreviations [2] are:

Anions (achiral and chiral)	Cations (lanthanides)

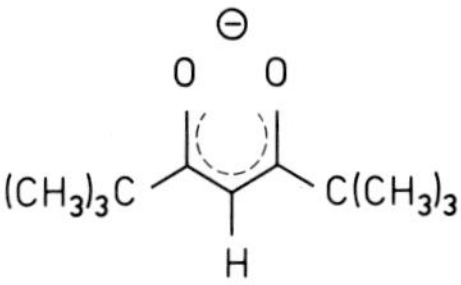

DPM$^{\ominus}$ (= dipivalvomethanato)

Eu^{3+}
Pr^{3+}
Yb^{3+}

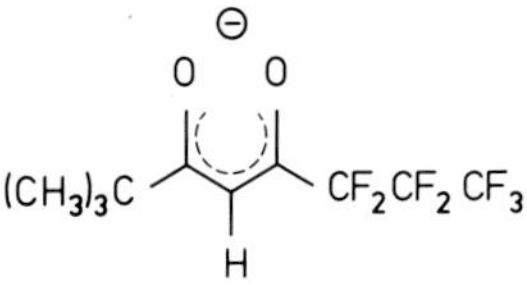

FOD$^{\ominus}$ (= 1, 1, 1, 2, 2, 3, 3, - heptafluoro-7,7-dimethyl-4,6-octandionato)

Eu^{3+}
Pr^{3+}
Yb^{3+}

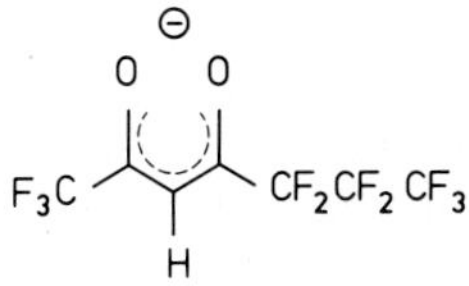

DFHD$^{\ominus}$ (= 1, 1, 1, 5, 5, 6, 6, 7, 7, 7, - decafluoro-2,4-heptandionato)

Eu^{3+}
Pr^{3+}
Yb^{3+}

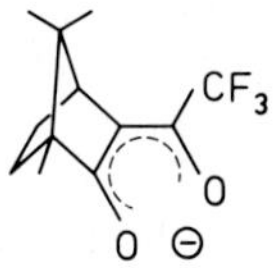

FACAM$^{\ominus}$ (= 3-trifluoracetyl-D-camphorato)

Eu^{3+}
Pr^{3+}
Yb^{3+}

Fig. 16

Since that time, binuclear (lanthanide/metal) shift reagents such as $AgYb(fod)_4$ or $AgPr(fod)_4$ have also been used [3]. On the other hand, mixed-ligand lanthanide complexes like $Eu(fod)_3pz$ or $Yb(fod)_3pz$ (with pz standing for pyrazole) have been considered as effective reagents in NMR [4].

Recently [5] we have learned, that it is possible to determine the enantiomeric purity of 1,2-diols by NMR spectroscopy without chiral auxiliaries.

Lanthanide complexes of the general formula $Ln(fod)_4^-$ are effective organic-soluble NMR shift reagents for ammonium [6], sulfonium [7], and isothio-uronium salts [7]. The reagent itself is formed in solution from $Ln(fod)_3$ and

Ag(fod) or K(fod). The selection of Ag(fod) or K(fod) in forming the shift reagent depends on the anion of the organic salt. Ag(fod) is more effective with halide salts, whereas K(fod) is preferred with tetrafluorobarate salts. These reagents can be employed in solvents such as chloroform and benzene [7]. Resolution of diastereotopic hydrogen atoms has been observed in the shifted spectra of certain substrates [7]. Enantiomeric resolution was obtained in the spectrum of *sec*-butylisothiouronium chloride with a chiral shift reagent [7].

All effects involved are explained under the acronym LIS (see LIS).

References

1. Hinckley CC (1969) J. Am. Chem. Soc. 91: 5160
2. Sievers RE (ed) (1973) Nuclear magnetic resonance shift reagents, Academic, New York
3. Hájek M (1986) Chemické listy 80: 795
4. Iftikhar (1986) Inorganica Chimica Acta 118: 177; (1986) Inorganica Chimica Acta 118: L53
5. Luchinat C, Roelens S (1986) J. Am. Chem. Soc. 108: 4873
6. Wenzel TJ, Zaia J (1985) J. Org. Chem. 50: 1322
7. Wenzel TJ, Zaia J (1987) Anal. Chem. 59: 562

LSRA

Abbreviation of **Least-Squares Regression Analysis**

One can find LSRA in almost all disciplines of molecular spectroscopy, applied for a wide range of purposes. For the sake of completeness we want to discuss here the use of LSRA for quantitative analysis in the field of infra-red spectroscopy (see IR and FT-IR).

Classically, quantitative IR analysis was carried out using a single analytical frequency characteristic of the molecule under investigation. LSRAs have been developed for utilization in the analysis of multi-component mixtures with *overlapping spectral features* [1]. It has been shown that inclusion of all the data in the spectral region of interest will substantially improve precision and accuracy of the results [2]. Today, newer programs allow the simultaneous determination of baselines as well as spectra. In addition, it is possible to define a *threshold value* below which the data will not be used, thus using only those spectral regions with spectral information content [3]. *Weighting factors* can be applied to make maximum use of the data available and minimize those ranges with high noise response [3]. One of the various advantages of LSRA is that no kind of assumption is made about spectral line shapes and this means the technique is particularly suited for polymer analysis where band shapes are asymmetric. Today it is possible to utilize LSRA to perform band shape analysis [4] and to establish much more non-subjective criteria for absorbance substraction [5].

References

1. Antoon MK, Koenig JH, Koenig JL (1977) Appl. Spectrosc. 31: 518
2. Haaland DM, Easterling RG (1980) Appl. Spectrosc. 34: 539
3. Haaland DM, Easterling RG (1982) Appl. Spectrosc. 36: 665
4. Gilette PC, Lando JB, Koenig JL (1982) Appl. Spectrosc. 36: 401
5. Koenig JL (1984) Fourier transform spectroscopy of polymers, Springer, Berlin Heidelberg New York (Advances in polymer science, vol 54)

MA
Abbreviation of **M**agnetic **A**lignment Effects in Nuclear Magnetic Resonance (see NMR)

High magnetic fields (e.g. 10 Tesla) can induce an observable partial magnetic alignment (MA) of diamagnetic molecules solved in ordinary isotropic solutions [1]. Because the MA effect is normally very small, detection by proton magnetic resonance has been realized so far in polycyclic molecules that have large anisotropic magnetic susceptibilies [2]. The NMR spectra of such compounds show a field dependence which permits a direct determination of direct coupling constants [3]. In deuterated species, deuterium resonance at a single field can provide valuable information on MA of even simple aromatics [2], but this method has some important limitations [4].

Strongly resolution-enhanced proton NMR spectra at a single high magnetic field (500 MHz) of *symmetrical* systems have now been shown, by A.L. Anet [5], to give information about extremely small direct coupling constants (down to a few millihertz), provided that NMR frequencies can be measured to a fraction of 1 mHz (see UHRNMR). He presented the interesting NMR parameters of *o*-dichlorobenzene measured in acetone-d$_6$ at 500 MHz [5].

References
1. For a review of MA in normal liquids see: Zijl PCM, Ruessink BH, Bulthuis J, MacLean C (1984) Acc. Chem. Res. 17: 172
2. a) Gayathri C, Bothner-By AA, Zijl PCM, MacLean C (1982) Chem. Phys. Lett. 87: 192
   b) Gayathri C (1982) Ph.D. Thesis, Carnegie-Mellon University, Pittsburgh
3. See Refs [1] and [2] for definition purposes
4. See Refs [1] and [2] for discussion
5. Anet FAL (1986) J. Am. Chem. Soc. 108: 1354

M

MAF
Acronym for **M**agic **A**ngle **F**lipping

The MAF technique is used as a special procedure in solid-state nuclear magnetic resonance (see NMR) and was first introduced by Bax et al. [1].

This kind of spinner axis flipping experiment creates 2D spectra (see 2D-NMR): along one axis the isotropic chemical shift frequencies can be measured, along the other static chemical shift anisotropy (see CSA).

The interested reader is referred to similar acronyms in solid-state NMR; see MAH, MAS, and VASS.

Reference
1. Bax A, Szeverenyi NM, Maciel G (1983) J. Magn. Reson. 55: 494

MAH
Abbreviation of **M**agic **A**ngle **H**opping

The MAH technique is used as a special method in solid-state nuclear magnetic resonance (see NMR) and was first introduced by Bax and his group [1].

MAH experiments create two-dimensional spectra (see 2D-NMR) with the isotropic chemical shift frequencies along one axis and the static chemical shift anisotropy along the other axis (see CSA).

For comparison, the interested reader is referred to other acronyms such as MAS, MAF, and VASS in this dictionary.

Reference
1. Bax A, Szeverenyi NM, Maciel GE (1983) J. Magn. Reson. 52: 147

MAR
Abbreviation of **Magic Angle Rotation**

MAR is sometimes used in nuclear magnetic resonance (see NMR) of samples in the solid-state instead of MAS or MASS (see below).

MAS
Abbreviation for **Magic Angle Spinning** (**Magic Angle Rotation** is used sometimes, too)

MAS is one of the important methods for the registration of high resolution NMR spectra of solids. In liquids rapid molecular motion causes line narrowing by averaging out the dipolar broadening and the chemical shift anisotropy (see CSA). MAS is a technique which, in effect, provides a substitute for this molecular motion in solid state and thereby allows direct observation of the isotropic chemical shift. The technique has been recognized for some 25 years [1–3] and more recent treatises deal with the quantitative aspects of the theory [4–6]. The expression magic-angle spinning arises from the fact that anisotropy is modulated during spinning by an equation containing the term $(3\cos^2 \theta - 1)$, were θ is the angle between the axis of rotation and the external magnetic field. This term vanishes, if θ is equal to $54°44'$ (the so-called magic angle). In this case the observed chemical shift is the isotropic value.

References
1. Andrew ER, Bradbury A, Eades RG (1958) Nature (London) 18: 1659
2. Andrew ER (1959) Arch. Sci. Geneva 12: 103
3. Lowe IJ (1959) Phys. Rev. Lett. 2: 285
4. Mehring M (1983) Principles of high resolution NMR in solids, Springer, Berlin Heidelberg New York
5. Haeberlen U (1976) High resolution NMR in solids; selective averaging, Adv. Magn. Reson., Suppl. 1, Academic, New York
6. Andrew ER (1972) Progr. Nucl. Magn. Reson. Spectrosc. 8: 1

MASS
Acronym for **Magic Angle Sample Spinning**

MASS is sometimes used in nuclear magnetic resonance (see NMR) instead of magic angle spinning (see MAS) or magic angle rotation (see MAR). Very often these acronyms are combined with the abbreviation of cross-polarization (see CP) and the nucleus of interest.

For example, recently, studies of the so-called "^{13}C CP-MASS NMR spectra" at 1.32 and 4.7 Tesla of solid α-D-glucose-1-d$_1$ have been presented [1]. The resonance of the anomeric carbon, directly bonded to deuterium, is split into an asymmetric doublet by the scalar and residual dipolar coupling between deuterium and carbon. Comparison of experimental and theoretical calculated (or simulated) spectra demonstrates the positive sign of the deuterium quadrupole coupling constant [1].

Reference

1. Swanson SD, Ganapathy S, Bryant RG (1987) J. Magn. Reson. 73: 239

MASS-DIFF
Acronym for **M**agic-**A**ngle-**S**ample-**S**pinning-**Diff**erence Spectroscopy

MASS-DIFF spectra can be used in nuclear magnetic resonance (see NMR) to selectively observe the contribution of an isotopic label to the total NMR spectrum [1]. Because the natural abundance background is removed, the measurement of center- and sideband intensities, and therefore chemical shift anisotropies (see CSA), is facilitated. The device which permits precise setting and regulation of the spinning speed has been described in detail [1]. The equipment may be useful in other MASS experiments where measurement and regulation of the spinning speed frequency (v_R) is required.

M

Reference

1. de Groot HJM, Copié V, Smith SO, Allen PJ, Winkel C, Lugtenburg J, Herzfeld J, Griffin RG (1988) J. Magn. Reson. 77: 251

MCD and MORD
stand for **M**agnetic **C**ircular **D**ichroism and **M**agneto-**O**ptical **R**otatory **D**ispersion

For both the effects or methods sometimes the general name *FARADAY-effect spectroscopy* is used [1].

These phenomena can be measured by means of a dichrometer in the case of MCD or a spectropolarimeter in the case of MORD, provided a magnetic field is induced in the sample compartment, parallel to the direction of the light beam. The origin of the *FARADAY-effect* lies basically in the so-called *helical symmetry* of the applied magnetic field. The sample under investigation thus responds differently to optical stimulation with right or left polarized light. Finally, the methods of *FARADAY-effect spectroscopy* are applicable—at least in principle— to any optically inactive or active compound.

The interested reader is here referred to a lot of papers [2–10], but it must be pointed out, that the originally discussed development possibilities of MCD and MORD and their advantages in application found their limitations.

Later on, however, MCD studies aroused new interest. Under chapter "progress in MCD" we will discuss applications in biochemistry separately.

References

1. Briat B (1973) In Ciardelli F, Salvadori P (eds) Fundamental aspects and recent developments in optical rotatory dispersion and circular dichroism, Heyden, London, Sect. 5.1
2. Rosenfield JS, Moscowitz A, Linder RE (1974) J. Chem. Phys. 61: 2427
3. Rosenfield JS (1977) J. Chem. Phys. 66: 921
4. Rosenfield JS (1976) Chem. Phys. Lett. 39: 391
5. Seamans L, Moscowitz A, Linder RE, Barth G, Bunnenberg E, Djerassi C (1976) Chem. Phys. 13: 135
6. Linder RE, Barth G, Bunnenberg E, Djerassi C, Seamans L, Moscowitz A (1976) Chem. Phys. Lett. 38: 28
7. Linder RE, Morrill K, Dixon JS, Barth G, Bunnenberg E, Djerassi C, Seamans L, Moscowitz A (1977) J. Am. Chem. Soc. 99: 727
8. Michl J, Josef Michl (1974) Tetrahedron 30: 4215
9. Kolc J, Josef Michl, Vogel E (1976) J. Am. Chem. Soc. 98: 3935, and references therein
10. Stephens PJ (1975) In: Dey P (ed) Electronic states of inorganic molecules, Reidel, Dordrecht

Progress in MCD

There are a lot of interesting MCD studies in the field of biochemistry, but we will focus our interest on those dealing with porphyrins.

M

Sign inversion in the MCD spectra of porphyrin derivatives is most commonly observed in those with an electron-withdrawing substituent at a peripheral position [1–7], although it can occur in some reduced porphyrins (8), some D_{4h} porphyrins [1, 9–12] and in a few symmetrically substituted porphyrins of lower symmetry [13]. While the occurrence of substituent-induced sign variation became well-established for specific cases, until recently there has been no experimental or theoretical study of the MCD and absorption spectra of series of porphyrins with peripheral substituents presenting a systematic variation of electron-donating or -accepting demand upon the porphyrin ring.

Much more recently, C. Djerassi and co-workers have published absorption and MCD spectra for a series of monosubstituted free-base alkylporphyrins possessing vinyl, oxime, cyano, ethoxycarbonyl, acetyl, and formyl substituents [14,15]. These porphyrins were alkylated at all the peripheral pyrrole positions except that adjacent to the π-substituent. This pattern of alkylation allows for the first time a direct comparison between the MCD of substituted benzenes and free-base porphyrins. Comparison with substituted benzenes, and with the homologous β-methylporphin series, now permits the coupled effects of central-proton tautomerism and substituent conformation on MCD to be disentangled and their separate contributions estimated. The occurrence of sign inversion has been found to be sensitive to small shifts in the toutomer equilibrium, as well as to changes in π-substituent conformation [15]. The interpretation of substituent and central-proton effects on MCD in the context of MICHL's perimeter model has been supplemented by explicit INDO-CI calculations of B terms [15].

References

1. Barth G, Linder RE, Bunnenberg E, Djerassi C (1973) Ann. N.Y. Acad., Sci. 206: 223
2. Haussier C, Sauer K (1970) J. Am. Chem. Soc. 92: 779

3. Gabriel M, Grange J, Niedercorn F, Selve C, Castro C (1981) Tetrahedron 37: 1913
4. Callahan PM, Babcock GT (1983) Biochemistry 22: 452
5. Kaito A, Nozawa T, Yamamoto T, Hatano M, Orii Y (1977) Chem. Phys. Lett. 52: 154
6. Woodruff WH, Kessler RJ, Ferris NS, Dallinger RF, Carter KR, Anitalis TM, Palmer G (1982) Adv. Chem. Ser. No. 201: 625
7. Thomson AJ, Englington DG, Hill BG, Greenwood C (1982) Biochem. J. 207: 176
8. Briat B, Schooley DA, Records R, Bunnenberg E, Djerassi C (1967) J. Am. Chem. Soc. 89: 6170
9. a) Linder RE, Barth G, Bunnenberg E, Djerassi C, Seamans L, Moscowitz A: J. Chem. Soc., Perkin Trans. II, 1974: 1712;
   b) Barth G, Linder RE, Waespe-Sarcevic N, Bunnenberg E, Djerassi C, Aronowitz YJ, Gouterman M: J. Chem. Soc., Perkin Trans. II, 1977: 337
10. Perrin MH, Gouterman M, Perrin CL (1969) J. Chem. Phys. 50: 4317
11. Kielman-van Luijte ECM, Dekkers HPSM, Canters GW (1976) Mol. Phys. 32: 899
12. a) Barth G, Linder RE, Bunnenberg E, Djerassi C: J. Chem. Soc., Perkin Trans. II, 1974: 696;
    b) Keegan JD, Bunnenberg E, Djerassi C, (1984) Spectrochim. Acta, Part A, 40: 287
13. Keegan JD, Bunnenberg E, Djerassi C (1983) Spectrosc. Lett. 16: 275
14. Djerassi C et al. (1984) J. Am. Chem. Soc. 106: 4241
15. Djerassi C et al. (1986) J. Am Chem. Soc. 108: 6449

MCT
Abbreviation of Mercury Cadmium Telluride

MCT, a mixture of two semiconductors, is the commonly used *quantum detector* for detecting infra-red radiation (see IR and FT-IR). MCT detectors must be maintained at liquid nitrogen temperatures (77 K), whereas others may require cooling to 4.2 K (temperature of liquid helium).

M

The specific detectivity, abbreviated as D^+, of MCT detectors is strongly dependent on their composition. MCT elements with a certain composition have a very high D^+, and their peak response is at about 800 cm^{-1}. At 750 cm^{-1}, however, the response approaches zero. If response is wished at longer wavelengths, some sacrifice must be made in the maximum value of D^+. MCT detectors are often classified as high-sensitivity (or narrow-band) or low-sensitivity (broadband, or wide-range) detectors. But, it should be realized that the composition of the HgTe/CdTe mixture can be varied almost continuously to manufacture detectors with the desired characteristics [1].

Reference
1. Griffiths PR, de Haseth JA, Azarraga LV (1983) Anal. Chem. 55: 1361A

ME or MEM
Acronym for Maximum Entropy or Maximum Entropy Method

A method used in nuclear magnetic resonance (see NMR) to explain non-linear effects in this field. In the past, much effort has been devoted to developing efficient non-linear methods in order to evaluate the so-called *power-spectra* of time-domain signals. Algorithms are generally based on a different class of models than that on which the usual fast FOURIER transform (see FFT) is based, and, therefore, they can be profitably used to process specific classes of signals [1].

Among the different non-linear methods, MEM has received more and more attention as a practical result of its good performances in the analysis of signals from very different sources. We have learned the different ways of applying MEM in data analysis [2]: One of these takes advantage of a constrained optimization of entropy over many spectra obtained via the common FFT: following the approach by Gull and Daniel, originally developed for image construction [3], this procedure has also been applied in nuclear magnetic resonance spectroscopy [4–7]. A second, developed by Burg [8] and others [9,10], is a procedure for extrapolating the estimated autocorrelation function after the period of data acquisition using the entropy concept. Via FOURIER transformation we will get an estimate of the "*power spectrum*". This kind of method has found important applications in the analysis of complex stationary as well as non-stationary signals in NMR [9–13]. In particular, it has been shown [9,10], that this latter MEM approach is a very powerful tool for processing NMR signals (free induction decays). Great advantages can be obtained with respect to FFT in terms of increased sensitivity and resolution when dealing with very noisy FID's and/or with short data records, thus allowing a more accurate determination of the interesting spectral parameters [1].

A property of MEM which has not been exploited in previous applications to NMR is its ability to produce spectral estimates from short data records, free of typical truncation artifacts which characterize discrete FOURIER transform (see DFT) spectra of zero-extended data records [14]. This property holds significance for 2D-NMR spectroscopy, allowing reduction of the time necessary to acquire a 2D data set. Following J.C. Hoch [14] MEM in 2D-NMR stidies can lead to sensitivity enhancements, reduction of instrumental artifacts, and shortened data acquisition times. But, it has been pointed out, that the non-linear nature of MEM may render it unsuitable for certain applications in nuclear magnetic resonance [14].

E.D. Laue and co-workers demonstrated that it is possible to reconstruct 2D-NMR spectra containing antiphase peaks by using MEM [15]. Later on, they found out that this very general approach can be used to reconstruct all phase-sensitive spectra [16]. The use of MEM in the analysis of electron spin resonance spectra (see ESR) has been described recently [1]. The different methodology complements the use of correlation approaches which have also been published in the analysis of weak and complex electron spin resonance spectra [2–5].

To apply MEM to ESR spectra, the experimental data are placed in a vector. A solution vector—initially filled with identical small positive numbers—is convolved with a GAUSSian or LORENTZian derivative function to calculate at each stage a "theoretical" spectrum for comparison with the experimental one. Ideally, at the end of the analysis for a noise-free spectrum, the solution vector should approximate a stick spectrum of the experimental data.

Recently P.J. Hore and G.J. Daniell [1] published a paper dealing with a ME reconstruction of rotating-frame zeugmatography data.

For special purpose the interested reader, will find additional two acronyms dealing with MEM: BURG's MEM and CAMBRIDGE MEM.

References

1. Viti V, Massaro E, Guidoni L, Barone P (1986) J. Magn. Reson. 70: 379
2. Jaynes ET (1982) Proc. IEEE 70: 939

 3. Gull SF, Daniell GJ (1978) Nature (London) 272: 686
 4. Sibisi S (1983) Nature (London) 301: 134
 5. Sibisi S, Skilling J, Brereton RG, Laue ED, Staunton J (1984) Nature (London) 311: 446
 6. Laue ED, Skilling J, Staunton J, Sibisi S (1985) J. Magn. Reson. 62: 437
 7. Martin JF (1985) J. Magn. Reson. 65: 291
 8. Burg JP (1967) Proc. 37th Meeting Soc. Expl. Geophys.
 9. Viti V, Barone P, Guidoni L, Massaro E (1986) J. Magn. Reson. 67: 91
10. Barone P, Massaro E, Guidone L, Viti V (1986) Techniques different from FFT for NMR data processing. In: Onori S, Tabet E (eds) Proc. Int. Conf., Physics in environmental and biomedical research, Rome, 5–7th November 1985, World Sci., Singapore, p 247
11. Radoski HR, Zawalick EJ, Fougere PF (1976) Phys. Earth. Plan. Int. 12: 208
12. Landers TE, Lacoss RT (1977) IEEE Trans. GE-15: 26
13. Jurkevics AJ, Ulrich TJ (1978) BSSA 68: 781
14. Hoch JC (1985) J. Magn. Reson. 64: 436
15. Laue EA, Skilling J, Staunton J (1985) J. Magn. Reson. 63: 418
16. Laue ED, Mayger MR, Skilling J, Staunton J (1986) J. Magn. Reson. 68: 14
17. Jackson RA (1987) J. Magn. Reson. 75: 174
18. Jackson RA: J. Chem. Soc., Perkin Trans. II, 1983: 523
19. Jackson RA, Rhodes CJ: J. Chem. Soc. Chem. Commun. 1984: 1278
20. Jackson RA, Rhodes CJ: J. Chem. Soc., Perkin Trans. II, 1985: 121.
21. Motton A, Schreiber J (1986) J. Magn. Reson. 67: 42
22. Hore PJ, Daniell GJ (1986) J. Magn. Reson. 69: 386

MERTZ's Method

This is the name of a procedure involving FOURIER transformation and phase correction used in infrared (see IR) spectroscopy.

M

This method was devised by Mertz twenty years ago [1,2]. The first step in this procedure is to extract a small double-sided interferogram from around the centerburst. Then a triangular apodization function (see Apodization) is used to apodize the double-sided interferogram. After the arrays are multiplied together, the data are shifted about the maximum value to reference the function as close to zero phase as possible. The resulting short double-sided interferogram is used to calculate the phase curve. Upon complex FOURIER transformation (see FT) of the function, both real and imaginary portions of the phase curve are produced, as the original interferogram has both cosine and sine components. The real and imaginary parts of the phase curve are even functions by virtue of the fact that the interferogram is even, or approximately so. In these curves, the central point is at the highest wavenumber of the bandpass whereas the two end points correspond to 0 cm^{-1}. The actual phase curve is then calculated. The final step in the preparation of the phase correction curve is an interpolation of the curve to the same resolution as the final spectrum. Once interpolated, the cosine and sine functions are calculated from the phase curve. Only one-half of each curve has been retained since each half is a mirror image of the other and no information is lost when one half is discarded. It should be noted that the phase curve has a value of zero where the phase is not defined. Regions of undefined phase are those outside the spectral bandwith, that is, at wavenumbers greater than 4400 cm^{-1}. The cosine and sine phase curves are not calculated in these undefined regions. At the end of all data manipulations one gets the single-beam phase corrected infrared spectrum.

References

1. Mertz L (1965) Transformations in optics, Wiley, New York
2. Mertz L (1967) Infrared Phys. 7: 17

MFIRFT

Acronym for **M**odified **F**ast **I**nversion-**R**ecovery **F**ourier **T**ransform

The selection of a technique is of considerable importance for minimizing experimental uncertainities within a fixed spectrometer time in measurements of spin-lattice relaxation (T_1) times [1,2]. There has been a continuing interest during recent years in arriving at an optimum technique for T_1 measurements, especially in situations where T_1 is relatively long. The known methods have been analyzed in the meantime and a modified version of the fast inversion-recovery (see FIRFT) was developed: MFIRFT [3]. Whereas the FIRFT method employs a fixed time T between successive 180°, τ, 90°, sequences, the MFIRFT version use a fixed value of $\Delta = T + \tau$, so that T decreases as τ increases. Both methods are shown to be capable of nearly equal precision under conditions of ideal 180 and 90° pulses. With imperfect pulses, however, the modified fast inversion-recovery method circumvents certain systematic errors inherent in the FIRFT procedure [3].

M

References

1. Weiss GH, Gupta RK, Ferretti JA, Becker ED (1980) J. Magn. Reson. 37: 369
2. Becker ED, Ferretti JA, Gupta RK, Weiss GH (1980) J. Magn. Reson. 37: 381
3. Gupta RK, Ferretti JA, Becker ED, Weiss GH (1980) J. Magn. Reson. 38: 447

MG-M

Abbreviation of the **M**eiboom and **G**ill-**M**ethod

In 1958, Meiboom and Gill [1] proposed for the field of nuclear magnetic resonance (see NMR) measurements of T_2, the spin-spin or transverse relaxation time, an often much easier experimentally realized approach than the famous *Hahn spin-echo* experiment from 1950 [2] and CP-T (see p 57) from 1954 [3]. This method uses the same pulse sequence as the CP-T [3], but the 180° pulses are applied along the positive y' axis, i.e. at 90° *phase* difference relative to the initial 90° pulse.

For further details the interested reader is referred to a critically written monograph [4] and to the combination of both the methods with the acronym CPMG (see p 57) in this dictionary.

References

1. Meiboom S, Gill D (1958) Rev. Sci. Instrum. 29: 688
2. Hahn EL (1950) Phys. Rev. 80: 580
3. Carr HY, Purcell EM (1954) Phys. Rev. 94: 630
4. Farrar TC, Becker ED (1971) Pulse and Fourier transform NMR: Introduction to theory and methods, Academic, New York, Sect 2.5

MINT
Acronym for **M**agnetic **Int**erference

Line shifts in nuclear magnetic resonance (see NMR) arising from larger instabilities of modern magnets with superconducting coil systems have been ascribed to MINT resulting from electric currents inside or outside the laboratory building [1]. MINT has been detected by recording the voltage generated in an electrically shielded air coil made from many windings of wire-wrap wire with standard laboratory equipment. Vertical magnets in basements seem to be prone to MINT whereas horizontal magnets are susceptible to vertical conductors. In view of MINT's easy detection and the annoyance stemming from improper site selection, it is probably useful to check for MINT prior to magnet installation in a laboratory [1].

Reference
1. Gewiese B, Jentschura U, Ziessow D (1986) paper presented at the 8th European Experimental NMR Conference, Spa/Belgium, June 3–6

MIS
Acronym for **M**anipulated **I**dentification of **S**pins

MIS is a new technique in the field of nuclear magnetic resonance (see NMR) for the identification of particular spins in congested homonuclear coupling networks [1].

M

Pulse sequences for MIS in special systems where distinctively large and uniform values of coupling constants define chains or pairs of spins are: (a) identification of terminal spins—abbreviated as MIS(T)—is accomplished, if 90° pulses in the "cluster" have the same phase (an appropriate phase cycle will suppress unwanted signals); (b) the complementary pulse sequence MIS(I) selectively reveals resonances of inner spins in the chain.

The pulse sequence MIS(T) incorporates the well-known procedure EXORCYCLE (see p 98) which eliminates resonances of single protons.

Bilinear rotation pulses—normally used in the BIRD technique (see BIRD)—have been implemented in the sequences and give rise to a lot of useful spin manipulations in special homonuclear spin systems.

The results of the pulse sequences MIS(T) and MIS(I) are not symmetric with respect to the main diagonal and symmetrization procedures cannot be used to improve the appearance. But, these pulse sequences impose less demanding requirements for large data storage capabilities. Since homonuclear coupling constants are mostly eliminated from the first frequency dimension (F_1), even a smaller number of data points allows separation of diagonal peaks and almost correct measurements of chemical shifts. Digital resolution along the F_2 dimension seems to be more important and may be increased at the expense of the F_1 dimension.

Reference
1. Rutar V, Wong TC (1987) J. Magn. Reson. 74: 275

MLEV
Acronym standing for **M**alcolm H. **Lev**itt who has established new techniques for broadband (see BB) decoupling in nuclear magnetic resonance (see NMR)

The efficiency of BB heteronuclear decoupling has been improved by innovations which use periodic sequences of composite pulses (see p 51), for example, the MLEV sequences [1,2] or the WALTZ sequences (see WALTZ).

Efficient decoupling over the entire band of proton resonance frequencies is a very important factor in determining resolution and sensitivity of carbon-13 NMR spectroscopy.

Continuous monochromatic irradiation [3] does not achieve this goal, but several modulation methods have been suggested such as coherent single-frequency modulation [4], noise modulation [5], square-wave phase modulation [6], and chirp frequency modulation [7]. Today, most of the commercially available carbon-13 spectrometers employ the proton noise decoupling (see PND) method introduced by Ernst [5], where the second radio-frequency field B_2 is applied continuously but with phase inversions at pseudo-random intervals. Normally, the B_2 intensity is set at its highest level compatible with the limitations on sample heating, leaving the frequency setting and the mean rate of phase inversion as adjustable parameters, the latter acting as a bandwidth control. Noise modulation can be transmitted to the observed resonances, interfering with carbon-13 spin-echo measurements [8] and giving rise to noise sidebands on the flanks of the signals, hindering the detection of carbon-13 satellites.

M

The concept of MLEV [1,2] can be formulated on the basis of three principles:

a) The strongly irradiated protons (S spins) should undergo a coherent cyclic perturbation at a rate fast compared with the largest coupling constant J_{IS} (I represents the observed carbon-13 nucleus).
b) This cycle should consist of four elements, RRR'R', where R stands for any perturbation known to invert the longitudinal magnetization, and R' is the same one with the radio-frequency pulse phase inverted (MLEV-4; other cycles: MLEV-16, MLEV-17, and MLEV-64).
c) The operation of R and R' should be insensitive to the offset parameter $\Delta B/B_2$ over the whole frequency range of interest.

The important innovation lies in the cycling principle. The theoretical justification for the choice of cycling comes from the average HAMILTONian theory [9] (see HAMILTONian).

Recently [10] the cycling sidebands in BB decoupling have been studied. A schematic diagram of a circuit used to implement the cycling permutation scheme for repetitive phase-alternated decoupling sequences has been presented. It has been used to generate cyclically permuted versions of WALTZ-16 (see WALTZ) and GARP (see p 119).

References
1. Levitt MH, Freeman R (1981) J. Magn. Reson. 43: 502
2. Levitt MH, Freeman R, Frenkiel T (1982) J. Magn. Reson. 47: 328
3. Bloch F (1954) Phys. Rev. 93: 944
4. Anderson WA, Nelson FA (1963) J. Chem. Phys. 39: 183
5. Ernst RR (1966) J. Chem. Phys. 45: 3845
6. Grutzner JB, Santini RE (1975) J. Magn. Reson. 19: 173

7. Basus VJ, Ellis PD, Hill HDW, Waugh JS (1979) J. Magn. Reson. 35: 19
8. Freeman R, Hill HDW (1971) J. Chem. Phys. 54: 3367
9. Haeberlein U, Waugh JS (1968) Phys. Rev. 175: 453
10. Shaka AJ, Barker PB, Bauer CJ, Freeman R (1986) J. Magn. Reson. 67: 396

MMT
Abbreviation of **M**ulti-site **M**agnetization **T**ransfer (Experiments)

In comparison to the common magnetization transfer (see MT) experiments in nuclear magnetic resonance (see NMR) spectroscopy, discussion seems, as yet, to be open for detailed information about the so-called multi-site approach. Two recent papers [1,2] have discussed the determination of the rates of multisite chemical exchange processes as studied by selective inversion-recovery experiments (see IRFT and other acronyms, too). These articles prompted other authors [3] to publish their own work, which had been started independently.

References
1. Gesmar H, Led JJ (1986) J. Magn. Reson. 68: 95
2. Grassi M, Mann BE, Pickup BT and Spencer CM (1986) J. Magn. Reson. 69: 92
3. Muhandiram DR, McClung RED (1987) J. Magn. Reson. 71: 187

MQC
Acronym for **M**ultiple-**Q**uantum **C**oherence

M

The notion *coherence* itself is of central importance in two-dimensional nuclear magnetic resonance (see 2D-NMR) spectroscopy [1].

One practically important application is 2D-INADEQUATE (see p 6) spectroscopy [2] to find out carbon-carbon connectivities (CCC) of a given molecule.

However, we will here discuss *coherence* and MQC much more principally. The preparation period, the first one of four successive time periods in 2D-NMR, usually consists of a *delay time*, during which the thermal equilibrium is attained, followed by one or several radio-frequency pulses to create *coherence*. *Coherence* can mean *phase coherence* among like or unlike spins. It includes the concept of transverse magnetization, corresponding to observable single-quantum transitions, where the majority of spins in an ensemble of like spins have the same phase in the $x'y'$ plane. In context of, not directly observable, zero-quantum, double-quantum, and/or *higher-order quantum coherence*, it means that in the majority of molecules two or several spins of different types will have the same phase relative to each other. In the following evolution period of the 2D-NMR experiment *coherence evolved*, and at the end of this time interval the system assumes a distinct state depending on the elapsed time t_1 and the HAMILTONian (see there) operator $\mathscr{H}^1$ effective during t_1. The following mixing period may include one or several radio-frequency pulses and delay intervals. During the mixing period, *coherence is transferred* between spins (e.g. in autocorrelation NMR experiments, the mixing process determines the frequency pairs with cross-peaks of non-vanishing intensity) [3].

References
1. Ernst RR, Bodenhausen G, Wokaun A (1987) Principles of nuclear magnetic resonance in one and two dimensions, Clarendon, Oxford
2. Mareci TH, Freeman R (1982) J. Magn. Reson. 44: 158
3. Wüthrich K (1986) NMR of proteins and nucleic acids, Wiley, New York

MQF-COSY
Acronym for **M**ultiple-**q**uantum-**f**iltered **C**orrelated **S**pectroscopy

A new pulse sequence which is used in novel editing and spin filtration techniques in two-dimensional NMR spectroscopy [1–3]. The identical pulse sequence is used in E. COSY (see p 84). Recently, the use of MQF-COSY has been investigated with regard to the identification of amino acid spin systems in proteins [4]. In addition to a simplification of the spectra when using multiple-quantum filters, multiplet structures and symmetry properties of the "cross-peaks" in MQF-COSY can give new information, which is complementary to that obtained from the commonly used COSY (see p 53). Optimized experimental procedures to minimize spectral artifacts and maximize the signal-to-noise ratio (see SNR) have been described and applied to the protein basic pancreatic trypsin inhibitor [4].

References
1. Piantini U, Sørensen OW, Ernst RR (1982) J. Am. Chem. Soc. 104: 6800
2. Shaka AJ, Freeman R (1983) J. Magn. Reson 51: 169
3. Rance M, Sørensen OW, Bodenhausen G, Wagner G, Ernst RR, Wüthrich K (1983) Biochem. Biophys. Res. Commun. 117: 479
4. Müller N, Ernst RR, Wüthrich K (1986) J. Am. Chem. 108: 6482

MQ-NMR
Acronym for **M**ultiple-**Q**uantum-**N**uclear **M**agnetic **R**esonance

Following the pioneering work of three groups in the last 10 years [1–5], multiple-quantum NMR is now a well-established technique. In particular, FOURIER transform double-quantum ($\Delta m = \pm 2$), triple-quantum ($\Delta m = \pm 3$), etc., methods can be used to (i) reveal small chemical shifts, otherwise masked by broad quadrupole interactions, and (ii) to simplify NMR spectra by successive elimination of single-quantum, double-quantum, etc., transitions.

For a review of both the development of MQ-NMR and its applications, the interested reader is referred to [6].

But, the mathematical treatment of MQ-NMR will be discussed here in more detail. In much of the early work, the mathematical treatment of MQ-NMR was based either on the fictitious spin-1/2 formalism [3,5], or matrix transformations of the density matrix following radio-frequency pulses or evolution periods [7].

Both approaches have recently been criticized [8] from different points of view: (1). the use of fictitious spin-1/2 operators may be helpful because of its very simple commutation relationships, however, the method suffers from three drawbacks: (a) the fictititous spin-1/2 operators do not form a basis set, (b) the operators themselves are mixtures of tensors of different rank, and (c) contact with the physical reality can be lost when manipulating these operators; (2). the alternative procedure based on matrix transformations of the density matrix [7] leads to lengthy equations even for the case of $I = 1$.

By way of contrast, however, Sanctuary and co-workers [9] have based their own mathematical treatment of MQ-NMR on the application of irreducible tensor operators. But, Bowden and Hutchison [8] have developed an alternative approach to that of Sanctuary [9], which follows more the spirit of Slichter [10]. They started their tensor operator formalism for MQ-NMR in the case of spin-1 nuclei [8] and extended them for spin-3/2, spin-2, spin-5/2 and general I [11].

J. Baum and A. Pines [12] have recently presented a time-resolved MQ-NMR experiment which enables us to determine the spatial distribution of atoms in materials lacking long-range order. In particular, they have studied the size and extent of atomic clustering in those materials. Model experiments have been demonstrated on the hydrogen distribution in selectively deuterated organic solids such as p-hexyl-p'-cyanobiphenyl,1,8-dimethyl-naphthalene-d_6 in 1,8-di-methyl-naphthalene-d_{12}, or hexamethylbenzene and in hydrogenated amorphous silicon, too [12].

References

1. Hatanaka H, Terao T, Hashi T (1975) J. Phys. Soc. Japan 39: 835
2. Hatanaka H, Hashi T (1975) J. Phys. Soc. Japan 39: 1139
3. Vega S, Pines A (1977) J. Chem. Phys. 66: 5624
4. Vega S (1978) J. Chem. Phys. 68: 5518
5. Wokaun A, Ernst RR (1977) J. Chem. Phys. 67: 1752
6. Bodenhausen G (1981) Progr. NMR Spectrosc. 14: 137
7. Hoatson GL, Packer KJ (1980) Mol. Phys. 40: 1153
8. Bowden GJ, Hutchison WD (1986) J. Magn. Reson. 67: 403
9. Sanctuary BC, Halstead TK, Osment PA (1983) Mol. Phys. 49: 753
10. Slichter CP (1963) Principles of magnetic resonance, Harper and Row, New York
11. Bowden GJ, Hutchison WD, Khachan J (1986) J. Magn. Reson. 67: 415
12. Baum J, Pines A (1986) J. Am. Chem. Soc. 108: 7447

M

MQT

Abbreviation of **M**ultiple-**Q**uantum **T**ransition, Methods used in Nuclear Magnetic Resonance (see NMR)

In NMR, multiple-quantum transition methods [1–5] have become very powerful tools in structure elucidation in the fields of organic chemistry, biochemistry, and life sciences. A lot of these new procedures have been given acronyms or abbreviated specifically. Most of them are explained under their specific acronyms in this dictionary.

References

1. Wokaun A, Ernst RR (1978) Mol. Phys. 36: 317
2. Wokaun A, Ernst RR (1979) Mol. Phys. 38: 1579
3. Bodenhausen G (1981) Prog. Nucl. Magn. Reson. Spectrosc. 14: 137
4. Thomas MA, Kumar A (1983) J. Magn. Reson. 54: 319
5. Wagner G, Zuiderweg ERP (1983) Biochem. Biophys. Res. Commun. 113: 854

MREV-8
A pulse sequence standing for **M**ansfield, **R**him, **E**lleman, and **V**aughan used in NMR

A typical line-narrowing sequence [1] to reduce the effective size of dipolar coupling during evolution (t_1) and detection (t_2) periods. The echoes are created in the t_1 and t_2 periods so that a simple magnitude calculation will provide a suitably phased image [2]. This method is useful for NMR images of solids [3] and is adaptable to any fast recovery solid-state NMR spectrometer having 2D FT NMR software and options for applying magnetic gradients methods.

References
1. a) Mansfield R (1971) J. Phys. C4: 1444
   b) Rhim W-K, Elleman DD, Vaughan RW (1973) J. Chem. Phys. 58: 1772
   c) Rhim W-K, Elleman DD, Vaughan RW (1973) J. Chem. Phys. 59: 3740
2. a) Bax A, Mehlkopf AF, Smidt J (1979) J. Magn. Reson. 35: 373;
   b) Bax A, Freeman R, Morris GA (1981) J. Magn. Reson. 43: 333
3. Ref. 2b

MRIE
Acronym for **M**agnetic **R**esonance **I**maging by the **E**arth's Magnet Field

M

Recently [1], an attempt has been made to employ the earth's magnetic field for nuclear magnetic resonance imaging (see NMRI) magnetic field. MRIE has been performed primarily on a home-made setup using the so-called VARIAN-PACKARD method [2,3]. The construction scheme was similar to that described by Bené [4], except that at the field switching the polarizing field decays like a damping oscillator with a time constant of 3 ms.

In spite of the fact that the earth's magnetic field is about ten thousand times weaker than the magnetic fields commonly used in NMRI, the estimated [5] signal-to-noise ratio (see SNR) is only a few times lower than those in high fields under the assumption that the volume of interest (see VOI) is about or more than 1 dm³, and that the prepolarization is nearly the same as the magnetization in high magnetic fields.

A so-called *selective polarization* has been realized for MRIE [6] and the corresponding images of phantoms have been obtained via the *projection/reconstruction* techniques [7].

It will be of interest to hear about further developments in MRIE, because the generation of the required magnetic fields for NMRI has become "not only a technological challenge but also a decisive economic factor" [1].

References
1. Stepišnik J, Eržen V, Mihajlovič D: poster presented at the 8th European Experimental NMR Conference June 3–6, 1986, Spa/Belgium
2. Packard ME, Varian R (1954) Phys. Rev. A94: 941
3. Waters GS, Francis PD (1958) Rev. Scientific Instr. 35: 88
4. Bené GJ (1980) Physics Reports 58: 213
5. Hoult DF, Lauterbur PC (1979) J. Magn. Reson. 34: 425
6. Stepišnik J, Eržen V, Koš M (to be published) see Ref. [1]

MT
Abbreviation of **M**agnetization **T**ransfer

Magnetization-transfer is an effective experiment in nuclear magnetic resonance [1–5] in general, and has been established as powerful tool for determining chemical exchange rates [6–13], especially in the case of two-site exchange [13]. In this particular case it was shown that, in general, the rate constants and the corresponding individual longitudinal relaxation rates can be determined simultaneously only from a complementary set of magnetization-transfer experiments [13]. In the meantime it has been demonstrated that a magnetization-transfer experiment can provide reliable rate constants from a spin system consisting of n exchanging sites only if the experiment is separated into n individual inversion procedures containing complementary information [14]. In such a case, however, it is possible to determine the $n(n-1)/2$ exchange rates as well as the n longitudinal relaxation rate even if the inverting rf pulses are not completely selective. The authors [14] have used the well-known DANTE pulse sequence (see p 65) consisting of 36 consecutive pulses.

References
1. Forśen S, Hoffman RA (1963) J. Chem. Phys. 39: 2892
2. Alger JR, Prestegard JH (1977) J. Magn. Reson. 27: 137
3. Campbell ID, Dobson CM, Ratcliffe RG (1977) J. Magn. Res. 27: 455
4. Dahlquist FW, Longmuir KJ, Du Vernet RB (1975) J. Magn. Res. 17: 406
5. Campbell ID, Dobson CM, Ratcliffe RG, Williams RJP (1978) J. Magn. Res. 29: 397
6. Waelder SF, Redfield AG (1977) Biopolymers 16: 623; Johnston PD, Redfield AG (1977) Nucleic Acid Res. 4: 3599; Johnston PD, Redfield AG (1981) Biochemistry 20: 1147, 3996
7. Krishna NR, Huang DH, Glickson JD, Rowan R III, Walther R (1979) Biophys. J. 26: 345
8. Ugurbil K, Schulman RG, Brown TR (1979) In: Shulman RG (ed) Biological application of magnetic resonance, Academic, New York, p 537
9. Perrin CL, Johnston ER, Collo CP, Kobrin PA (1981) J. Am. Chem. Soc. 103: 4691
10. Hvidt AA, Gesmar H, Led JJ (1983) Acta Chem. Scand. B 37: 227
11. Alger JR, Shulman RG (1984) Q. Rev. Biophys. 17: 83
12. Knight CTG, Merbach AE (1984) J. Am. Chem. Soc. 106: 804
13. Led JJ, Gesmar H (1982) J. Magn. Reson. 49: 444
14. Gesmar H, Led JJ (1986) J. Magn. Reson. 68: 95

M

MUDISM
Acronym for **M**ulti**di**mensional **S**tochastic **M**ethod or **M**ulti**di**mensional Stochastic **M**agnetic Resonance

In the field of the stochastic nuclear magnetic resonance (see STNMR) [1] MUDISM became an interesting approach [2,3] to this non-linear system analysis in NMR spectroscopy (4–6).

Basic attempts involving a general perturbation concept [2] have been extended for the analysis of three-dimensional lineshapes from stochastic NMR (see STNMR) of two-level systems [7]. Two most important results have been obtained: Firstly, as one-dimensional sub-diagonal cross-section, which is similar to a continuous wave (see CW) or FOURIER-transformed (see FT or FFT) NMR spectrum, can exhibit a narrower lineshape than obtained in 1D-FT-NMR spectroscopy. Secondly, the relaxation times T_1 (spin-lattice) and T_2 (spin-spin) can be determined from the lineshapes of selected cross-sections, independent of magnetic field inhomogeneity.

References

1. Blümich B (1987) Prog. NMR Spectrosc. 19: 331
2. Blümich B, Ziessow D (1983) Mol. Phys. 48: 955
3. Blümich B (1984) Mol. Phys. 51: 1283
4. Blümich B (1981) Dissertation, Technical University, Berlin
5. Blümich B, Ziessow D (1981) Ber. Bunsenges. phys. Chem. 84: 1090; (1981) Bull. Magn. Reson. 2: 299; (1981) Nachr. Chem. Tech. Lab. 29: 291
6. Blümich B, Ziessow D (1983) J. Chem. Phys. 78: 1059
7. Blümich B (1983) Mol. Phys. 48: 969

MVA
Abbreviation of Multi-Variate Analysis

MVA has been used as a new concept for *pattern recognition* [1] *in two-dimensional nuclear magnetic resonance* (see 2D-NMR) or as an alternative to other different approaches in this field [2–7]. A data representation of 2D-NMR spectra (see p 10) suitable for MVA has been introduced [1]. The obtained results show that MVA can be used to separate mixtures of different spin systems, regardless of coupling approximation, and to classify them. *Principal component analysis* (see PCA) for 2D-NMR has been described, too. 2D simulations in the weak coupling approximation, of the six different coupling topologies that a four-spin system can assume, have been generated and PCA has been applied to reduce each entire spectrum to a simple pattern. Extensions of this MVA approach to an automated spectral analysis of weakly and strongly coupled spin systems as found in biopolymers have been outlined, too [1].

References

1. Grahn H, Delagio F, Delsuc MA, Levy GC (1988) J. Magn. Reson. 77: 294
2. Meier BU, Bodenhausen G, Ernst RR (1984) J. Magn. Reson. 60: 161
3. Pfändler P, Bodenhausen G, Meier BU, Ernst RR (1985) Anal. Chem. 57: 2510
4. Bolton PH (1986) J. Magn. Reson. 70: 344
5. Mádi Z, Meier BU, Ernst RR (1987) J. Magn. Reson. 72: 584
6. Pfändler P, Bodenhausen G (1986) J. Magn. Reson. 70: 71
7. Neidig KP, Bodenmüller H, Kalbitzer HR (1984) Biochem. Biophys. Res. Commun. 125: 1143

MWFT
Abbreviation of Microwave FOURIER Transform (Spectroscopy)

In more than 40 years, microwave studies of the rotational spectra of free molecules in the gaseous phase have led to very important information about molecular parameters and data. In common experiments one uses absorption of monochromatic microwave radiation (in the frequency range from 4 to 40 GHz) through molecular gases in frequency dependence. Such absorption spectroscopy in the "frequency domain" contains the specific molecular information in measured isolated or overlapping spectral lines. The center frequencies are characteristics of the molecular rotational energy terms.

On the other hand we can define MWFT as a spectroscopy in the "time domain" because of its time response after excitation by microwave pulses (see (Fig. 17). FOURIER transformation (see FT) of a transient emission signal leads to a rotational spectrum in the power mode.

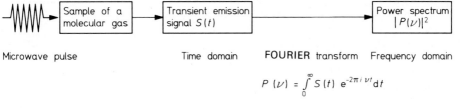

$$P\,(\nu) = \int\limits_{0}^{\infty} S\,(t)\;e^{-2\pi i\,\nu t}\mathrm{d}t$$

Fig. 17

So far only a few MWFT spectrometers have been built up in some laboratories and used in studies of the rotational spectra of gases in thermodynamic equilibrium [1–7]. Advantages of the FT mode of microwave spectroscopy and new application possibilities have recently been reviewed [8].

In the last few years the MWFT spectroscopy has been used to clear up traditional questions of rotational spectroscopy dealing with structures and molecular dynamics [9]. But, the special features of MWFT and its high sensitivity and resolution will lead to substantial new insights in the near future [10].

M

References

1. Ekkers J, Flygare WH (1976) Rev. Sci. Instrum. 47: 448
2. Bestmann G, Dreizler H, Mäder H, Andresen U (1980) Z. Naturforsch. 35a: 392
3. Bestmann G, Dreizler H (1982) Z. Naturforsch. 37a: 58
4. Bestmann G, Dreizler H, Fliege E, Stahl W (1983) J. Mol. Struct. 97: 215
5. Stahl W, Bestmann G, Dreizler H, Andresen U, Schwarz R (1985) Rev. Sci. Instrum. 56: 1759
6. Fliege E, Dreizler H (1986) Z. Naturforsch. 42a: 72
7. Oldani M (1985) Dissertation, ETH Zürich/Switzerland
8. Dreizler H (1986) Mol. Phys. 59: 1
9. Gordy W, Cook RL (1984) Microwave molecular spectra, Wiley, New York
10. Mäder H, Dreizler H (1987) Nachr. Chem. Tech. Lab. 35: 468

NEP
Abbreviation of **N**oise **E**quivalent **P**ower

The sensitivity of infrared detectors (see IR) is commonly expressed in terms of the noise equivalent power (NEP) of the detector, which is the ratio of the root mean square (rms) detector noise voltage, V_n, in volts $Hz^{-1/2}$, to the voltage responsivity, R_v, measured in volts per watt, that is,

$$NEP = \frac{V_n}{R_v} \quad [W\,Hz^{-1/2}] \tag{1}$$

NEP is important for the determination of the signal-to-noise ratio (see SNR) obtainable in any IR measurement [1].

Reference
1. Griffiths PR, de Haseth JA (1986) Fourier transform infrared spectrometry, Wiley, New York, chap 7 (Chemical analysis, vol 83)

NERO-1
Acronym for **N**on-linear **E**xcitation with **R**ejection **o**n-Resonance

NERO-1 is one of the new pulse sequences used for solvent signal suppression (see SST) without phase distortion in high-resolution nuclear magnetic resonance (see NMR) [1].
The pulse sequence may be written as

$$120° - \tau_1 - \overline{115°} - \tau_2 - \overline{115°} - 2\tau_3 - 115° - \tau_2 - 115° - \tau_1 - \overline{120°} - \tau_r \ ,$$

where τ_i are interpulse delays, pulses 180° out of phase having an overbar. Note, that the pulses do not simply alternate in phase. The delays between the pulses must be calculated according to the resonance offset of the center of the desired spectral region to be excited. NERO-1 is superficially similar to the earlier binomial sequences, being an antisymmetric set of short pulses alternating with delays. However, the presence of large flip angles (115 to 120°) betrays its radically different construction and properties.

Reference
1. Levitt MH, Roberts MF (1987) J. Magn. Reson. 71: 576

NIR
Abbreviation of **N**ear **I**nfrared (Spectroscopy)

The NIR region or sometimes called *overtone region* of infrared spectroscopy (see IR and FT-IR) ranges from 5000 to 12500 cm^{-1} in practice.
For instrumental and theoretical reasons Kaye [1] originally assigned the NIR region from 0.7 to 3.5 μm (14285–2860 cm^{-1}) and reviewed this region in two papers [1,2] including a correlation chart.

Later on other correlation charts were published by Goddu and Delker [3] and the Anderson Physical Laboratory. In 1960 a bibliography dealing with NIR was published by Kaye [4]. A review article was presented in 1968 by Whetsel [5].

In 1971, Wendisch described the special structural identification possibilities of NIR in the field of cyclopropane derivatives [6]. Some years later almost new correlation charts for the NIR region were presented by Colthup, Daly, and Wiberly [7].

In 1983, NIR analysis was called a "sleeping technique" [8]. Four years later, Davies wrote an exciting article (NIR Spectroscopy: Time for the Giant to Wake UP!) about the theory of NIR spectroscopy and the basic steps of its development, scope and limitations [9].

He presented applications of NIR analysis in the areas of agriculture and food [10,11] and discussed clinical applications, introduced by Norris and co-workers [12] which showed that fat in the human body can be determined rapidly and non-invasively by NIR using a fibre-optic probe. Callis et al. demonstrated the location of a breast cancer tumor by using a combination of visible and NIR spectroscopy [13], but today they have to compete with nuclear magnetic resonance imaging (see NMRI) and in-vivo-NMR spectroscopy (see p 145). Other examples of NIR analytical work range from mineralogy, geochemistry, and polymer chemistry to pharmaceutical disciplines [9].

On the other hand, Davies [9] has pointed out that improved methods of data analysis were important stimulants in the field of NIR. In to this category fall papers of new analysis of NIR data including GAUSS-JORDAN linear algebra [14], principal component analysis (abbreviated as PCA) [15], MAHALANOBIS distance [16], FOURIER transformation (see FT and FFT) [17], and partial least-squares methods [18].

Much more recently Davies [19] has discussed the possibilities of NIR spectroscopy in the field of process control.

N

References

1. Kaye W (1954) Spectrochim. Acta 6: 257
2. Kaye W (1955) Spectrochim, Acta 7: 181
3. Goddu RF, Delker DA (1960) Anal. Chem. 32: 140
4. Kaye W (1960) In: Clark GL (ed) The encyclopedia of spectroscopy, Van Norstrand-Reinhold, Princeton, p 409
5. Whetsel KB (1968) Appl. Spectrosc. 2: 1
6. Wendisch D (1971) In: HOUBEN-WEYL (eds) Carbocyclische Dreiring-Verbindungen, Georg Thieme, Stuttgart, p 31 (Methoden der Organischen Chemie, vol IV/3)
7. Colthup NB, Daly LH, Wiberly SE (1975) In: Introduction to infrared and Raman spectroscopy, 2nd edn, Academic, New York, p 371
8. Wetzel DL (1983) Anal. Chem. 55: 1165A, 1166A, 1170A, 1172A, 1174A, and 1176A
9. Davies AMC (1987) European Spectroscopy News 73: 10
10. Starr C, Morgan AG, Smith DB (1981) J. Agric. Sci., Camb. 97: 107
11. Osborn BG, Feam T (1986) In: Near infrared spectroscopy in food analysis, Longman Scientific and Technical, Harlow
12. Conway JM, Norris KH, Bodwell CE (1984) Amer. J. Clinical Nutrition 40: 1123
13. Anon. (1986) Anal. Chem. 58: 874A
14. Honigs DE, Freeling JM, Hieftje GM, Hirschfeld TB (1983) Appl. Spectrosc. 37: 491
15. Cowe IA, McNicol JW (1985) Appl. Spectrosc. 39: 257
16. Mark H (1986) Anal. Chem. 58: 379
17. Giesbrecht FG, McClure WF, Hamid A (1981) Appl. Spectrosc. 35: 210
18. Martens M, Martens H (1986) Appl. Spectrosc. 40: 303
19. Davies AMC (1988) European Spectroscopy News 79: 18

NLLSQ
Abbreviation of **N**on-**L**inear **L**east-**Sq**uares Calculations Used in different spectroscopy disciplines for spectral analysis.

NMR
Acronym for **N**uclear **M**agnetic **R**esonance

NMR was originally detected, in 1945, independently by 2 groups: under BLOCH at Stanford University [1] and PURCELL at Harvard University [2]. It became, in an extraordinarily short time, the most powerful spectroscopic method in structure elucidation of chemical compounds.

The phenomenon of NMR can be explained by using classic equations, the so-called BLOCH's equations (see p 29), or by quantum-mechanical treatment using the HAMILTONian operator (see HAMILTONian).

The physical foundation of NMR spectroscopy lies in the magnetic properties of atomic nuclei. The interaction of the nuclear magnetic moment of a given nucleus, μ, with an external magnetic field, H_o or B_o, leads in accordance to the results of quantum-mechanics, to a nuclear energy level diagram, because the magnetic energy of the nucleus is restricted to certain discrete values E_i, the so-called *eigenvalues*. Associated with these eigenvalues are the *eigenstates*, which are the only states in which an elementary particle can exist. They are also called *stationary states*. Through a high-frequency transmitter, transitions between eigenstates can be stimulated. The absorption of energy can be detected, amplified and registrated as a spectral line, the so-called *resonance signal*. In such a schematically described way one obtains a *spectrum* from a chemical compound containing atoms whose nuclei have non-zero magnetic moments. All other nuclei with *even* mass and atomic number are therefore NMR-*inactive*.

The following statements about the NMR *spectrum* should be noted here:

a) Several *resonance signals* correspond to different nuclei (mostly protons) in a given molecule. They arise because the nuclei reside in different chemical environments and are separated by the so-called *chemical shift* (see CS)

b) The area under a resonance signal is strictly proportional to the number of nuclei (=protons) that give rise to the signal and can be measured via integration. This fact explains directly the application possibility of NMR spectroscopy for the *quantitative analysis* of mixtures.

c) Normally not all spectral lines are simple singlets. Splittings of the NMR signals are caused by the *spin-spin coupling constants*, magnetic interactions of one nucleus with others, abbreviated as CC (see p 38) and symbolized with J. The magnitude of J is measured in Hertz (Hz) and is field-independent in contrast to the chemical shift.

d) The *lineshape* of an NMR line is determined or governed by different *relaxation mechanisms* and *times*. T_2 or T_2^* are symbols for the spin-spin or transverse relaxation time, whereas T_1 stands for the spin-lattice or longitudinal relaxation time.

e) The *temperature dependence* of a given NMR spectrum often indicates— besides decomposition of a compound—the *dynamic behaviour* of a molecule under investigation. NMR spectroscopy can be used therefore for dynamic studies (see DNMR).

f) All the features mentioned above demonstrate the potentials of NMR spectroscopy in the field of structure elucidation of chemical compounds in a relatively simple manner without discussions of its problems, scope and limitations. But, if, for example, the splittings or the fine structures of the *patterns* are much more complicated than anticipated by the so-called *rules of 1st-order analysis* the determination of chemical shifts and coupling constants from a given proton NMR spectrum demands other methods such as utilizing decoupling techniques, higher magnetic fields, lanthanide shifts-reagents (see LSR), aromatic solvents (see ASIS), and computer-assisted analysis of higher-order spin systems based on the quantum-mechanical treatment of NMR (see LAOCOON and other programs such as DAVINS, etc.).

In addition, the most important deficiency of NMR spectroscopy is its low sensitivity in comparison to infrared spectroscopy (see IR and FT-IR) or to electron spin or paramagnetic resonance (see ESR and EPR).

After the successful application of NMR in chemistry, biochemistry, molecular biology and material sciences, NMR spectroscopy is now starting to play an important rule in the field of clinical medicine. Many hopeful applications have emerged so far, such as studies on whole tissues in vitro and in vivo. But, the almost promising application of NMR in medicine will be the non-invasive imaging of entire biological organisms (see NMRI, MRI, and TMR).

According to the wide-spread application spectrum of NMR today, the interested reader will find under the items of this dictionary a plethora of them dealing with NMR. On the other hand, different NMR techniques themselves have produced a lot of acronyms, abbreviations, and symbols which are explained in this dictionary.

N

References

1. Bloch F, Hansen WW, Packard M (1946) Phys. Rev. 69: 127
2. Purcell EM, Torrey HC, Pound RV (1946) Phys. Rev. 69: 37

NMRD
Acronym for **N**uclear **M**agnetic **R**elaxation **D**ispersion

The dependence of relaxation rates (or times) on magnetic field has been termed *nuclear magnetic relaxation dispersion*, or NMRD, and the study of all aspects of relaxation phenomena, *relaxometry* [1].

Measurement of relaxation rates, particularly $1/T_1$ (T_1 stands for the spin-lattice relaxation time), over a wide range is facilitated by a field-cycling method first developed by Redfield [2].

Recently [3] an automated (pc controlled) NMRD *field-cycling relaxometer* has been described which can measure $1/T_1$ of solvent or tissue protons at fields corresponding to 0.01 up to 60 MHz, proton LARMOR frequency, or 2.5×10^{-4} to 1.5 T, for rates between 0.1 and $40\,s^{-1}$, with accuracies on the order 0.5% for sample volumes of 0.5 ml. The sample temperature can be maintained within 0.1 °C between -10 and $+40$ °C. This system has been used for a variety of proton studies on proteins, tissues and contrast agents important in the field of *nuclear magnetic resonance imaging* (see NMRI).

Selected applications of multinuclear field-cycling relaxometry on the same system have been reported [4]: Proton relaxation rates of aqueous gels doped with paramagnetic species have been measured since these materials are often used as standard test samples for NMRI. Dispersion profiles show that, in contrast to the known effect of manganese, the paramagnetic contribution of the gadolinium ion is rather complex; for instance, the relaxation rates of water in agar and agarose gels exhibit a high sensitivity to the applied magnetic field—especially between 0.12 and 1.2 T—leading to a maximum in the high field region. Perfluorocarbons and their aqueous emulsions have been studied via fluorine-19 NMRD because of their potential application as blood substitutes. Effects of some paramagnetic compounds have been observed and discussed in terms of their possible use in the field of fluorine magnetic resonance imaging.

The importance of relaxometric studies (T_1 and T_2, the spin-spin relaxation time) for biochemistry and life sciences may be illustrated here with the following contributions: (a) diffusion of water molecules in the presence of magnetic dipoles [5], (b) nuclear magnetic relaxation in collagen fibers and the cellular water dynamics [6-8], (c) proton relaxation rates in human plasma, serum and protein solutions [9,10], and (d) proton NMR studies of the different water systems in moistened spores of *Phytomyces Blakesleanus* [11].

References

1. See Ref. [3]
2. Redfield AG, Fite W, Bleich HE (1968) Rev. Sci. Instrum. 39: 710
3. Brown RD III, Koenig SH (1986) poster presented at the 8th European Experimental NMR Conference, Spa/Belgium, June 3–6, 1986
4. Müller RN, van der Elst L, van Haverbeke Y, Brown RD III, Koenig SH (1986) poster presented at the 8th European Experimental NMR Conference (abbreviated EENC), Spa/Belgium, June 3–6, 1986
5. Gillis P (1986) poster presented at the 8th EENC, Spa/Belgium, June 3–6, 1986
6. Hazlewood CF, Chang DC, et al. (1974) J. Biophys. 14: 583
7. Hazlewood CF, Chang DC, et al. (1976) J. Biophys. 16: 1043
8. Peto S, Gillis P (1986) poster presented at the 8th EENC, Spa/Belgium, June 3–6, 1986
9. Hallenga K, Koenig SH (1976) Biochemistry 15: 4255
10. Raeymaekers HH, Borghys D, Eisendrath H, van Craen J (1986) poster presented at the 8th EENC, Spa/Belgium, June 3–6, 1986
11. Francois A, Overloop K, van Gerven L (1986) poster presented at the 8th EENC, Spa/Belgium, June 3–6, 1986

N

NMRI or MRI

Acronym for **N**uclear **M**agnetic **R**esonance **I**maging or **M**agnetic **R**esonance **I**maging

NMRI is the most promising application of NMR in medicine as a non-invasive imaging technique for entire living organisms.

The first experimental demonstration of feasibility of macroscopic imaging by NMR was given in 1972 at Standford University, CA by P.C. Lauterbur [1,2]. The application of NMR studies of the human body, especially for detection of different types of cancer, was first advocated by R. Damadian [3,4].

The crucial idea of Lauterbur [1,2], was to utilize a magnetic field *gradient* to disperse the NMR frequencies of the different volume elements. The NMR spectrum of an object or phantom recorded in the presence of a strong linear magnetic gradient may be considered as a one-dimensional projection of the three-dimensional proton density onto the direction of this field gradient. All nuclei (= protons) in a plane perpendicular to the field gradient will experience the same magnetic field and contribute to the signal amplitude at quite the same frequency. For a complete representation—or imaging—of an object or phantom, each volume element in the space must have its correspondence in the frequency-domain. A full NMR image therefore has the form of a three-dimensional spectrum were the signal intensities represent the local spin density. Much technology is common to both NMRI and conventional two-dimensional nuclear magnetic resonance (see 2D-NMR) spectroscopy.

The interested reader may remember that Paul C. Lauterbur [1,2] at first named his proposed technique *zeugmatography*. Later on, NMRI, MRI, *topical magnetic resonance* (see TMR), *NMR tomography* (in analogy to X-ray computer tomography) [5–8] or *Kernspintomographie* (used in german speaking countries by their radiologists) have been used as acronyms, names, or abbreviations in this field.

In the following part we will only classify the different principal techniques used for NMRI purposes, because we have always to bear in mind that the major handicap of NMR is its low inherent sensitivity [9,10].

There are four types of experiment in the field of NMRI.

1. *Sequential point* techniques. A single volume element is selectively excited and observed at a time [11–15]. These techniques produce a direct image without any intermediate image processing. Dependent on the low sensitivity the measurement time is relatively long. But, sequential point techniques are of particular merit if a localized area has to be studied in more detail.
2. *Sequential line* procedures. To increase the signal-to-noise ratio (see SNR), an entire line of the volume elements is excited and measured simultaneously [16–35]. Normally these techniques utilize a one-dimensional FOURIER experiment for excitation and observation, and therefore for reconstruction of the line image only a simple FOURIER transformation (see FT) is necessary.
3. *Sequential plane* methods. The sensitivity can further be increased when an enitre plane of volume elements is excited and observed at once [36–47]. Two-dimensional reconstruction of the image can, only in some procedures, be achieved by a 2D-FT process [45,46]. In other cases, alternative reconstruction approaches, such as filtered back projection, are required.
4. *Simultaneous* techniques. The almost sensitive methods involve simultaneous excitation and registration of the entire three-dimensional phantom or object. But, such a type of experiment is quite demanding with regard to the amount of data to be acquired and the long performance time [48].

In just a few years a plethora of acronyms in the field of NMRI has been published. Thus it is not possible to explain all these acronyms or abbreviations in detail in this dictionary. But, the interested reader will find in Table 1 a list of those items which are *not* explained in detail in this dictionary.

Furthermore the interested reader is referred to five monographies in the field of NMRI written since 1980 [63–67].

Pars pro toto, we will discuss some forthcoming papers in the field of NMRI.

Table 1. A list of some acronyms used in NMRI after 1984

Acronym	Explanation	Reference
DRESS	Depth-Resolved Surface-Coil Spectroscopy	49
HASP	Harmonically Analyzed Sensitivity Profile	50
ISIS	Image-Selected In Vivo Spectroscopy	51, 52
SEU	Selective Excitation Unit	—
SPACE	Spatial and Chemical Shift-Encoded Excitation	53
SPARS	Spatially Resolved Spectroscopy	54
STE	Stimulated Spin-Echo	—
STEAM	Stimulated Echo Acquistion Mode	55
TART	Tip-Angle Reduced T_1 Imaging	56
VOI	Volume of Interest (sometimes ROI = Region of Interested, is used)	—
VSE	Volume-Selective Excitation	57–60
WIMP	Wireless Implanted Magnetic Resonance Probes	61, 62

References

1. Lauterbur PC (1972) Bull. Am. phys. Soc. 18: 86
2. Lauterbur PC (1973) Nature 242: 190
3. Damadian R (1971) Science 171: 1151
4. Damadian R (1974) U.S. Patent 3.789.832
5. Ledley RS, Di Chiro G, Luessenhop AJ, Twigg HL (1974) Science 186: 207
6. Robinson AL (1975) Science 190: 542, 647
7. Brooks RA, Di Chiro G (1976) Phys. Med. Biol. 21: 689
8. Ter-Pogossian MM (1977) Sem. Nucl. Med. 7: 109
9. Brunner P, Ernst RR (1979) J. Magn. Reson. 33: 83
10. Hoult DI, Lauterbur PC (1979) J. Magn. Reson. 34: 425
11. Hinshaw WS (1974) Phys. Lett. A48: 87
12. Hinshaw WS (1974) Proc. 18th Ampère Congress, p. 433, Nottingham/UK
13. Hinshaw WS (1976) J. Appl. Phys. 47: 3709
14. Damadian R, Goldsmith M, Minkhoff L (1977) Physiol. Chem. Phys. 9: 97
15. Damadian R, Goldsmith M, Koutcher JA (1978) Naturwissenschaften 65: 250
16. Hinshaw WS, Bottomley PA, Holland GN (1977) Nature 270: 722
17. Hinshaw WS, Andrew ER, Bottomley PA, Holland GN, Moore WS, Worthington BS (1978) British J. Radiol. 51: 273
18. Holland GN, Bottomley PA, Hinshaw WS (1977) J. Magn. Reson. 28: 133
19. Brooker HR, Hinshaw WS (1978) J. Magn. Reson. 30: 129
20. Hinshaw WS (1983) Proc. IEEE 71: 338
21. Andrew ER, Bottomley PA, Hinshaw WS, Holland GN, Moore WS, Simaroj C (1977) Phys. Med. Biol. 22: 971
22. Andrew ER, Bottomley PA, Hinshaw WS, Holland GN, Moore WS, Simaroj C (1978) Proc. 20th Ampère Congress, Tallinn/USSR
23. Mansfield P, Grannell PK (1973) J. Phys. C6: L422
24. Mansfield P, Grannell PK (1975) Phys. Rev. B12: 3618
25. Mansfield P, Grannell PK, Maudsley AA (1974) Proc. 18th Ampère Congress, p 431, Nottingham/UK
26. Mansfield P, Maudsley AA, Baines T (1976) J. Phys. E9: 271
27. Mansfield P, Maudsley AA (1976) Proc. 19th Ampère Congress, p 247, Heidelberg/Germany
28. Gerroway AN, Grannell PK, Mansfield P (1974) J. Phys. C: Solid State Phys. 7: L457
29. Mansfield P, Maudsley AA (1976) Phys. Med. Biol. 21: 847
30. Mansfield P, Maudsley AA (1977) British. J. Radiol. 50: 188
31. Mansfield P (1976) Contemp. Phys. 17: 553
32. Mansfield P, Pykett IL (1978) J. Magn. Reson. 29: 355

33. Hutchinson JMS, Goll CC, Maillard JR (1974) Proc. 18th Ampère Congress, p 283, Nottingham/UK
34. Sutherland RJ, Hutchinson JMS (1978) J. Phys. E: Sci. Instrum. 11: 79
35. Hutchinson JMS, Sutherland RJ, Maillard JR (1978) J. Phys. E: Sci. Instrum. 11: 217
36. Lauterbur PC (1973) Proc. 1st Int. Conf. on Stable Isotopes in Chem., Biol., and Medicine
37. Lauterbur PC (1974) Pure Appl. Chem. 40: 149
38. Lauterbur PC (1974) Proc. 18th Ampère Congress, p 27, Nottingham/UK
39. Lauterbur PC (1977) In: Dwek RA, Campbell ID, Richards RE, Williams RJP (eds) NMR in biology, Academic, London, p 323
40. Lauterbur PC, Kramer DM, House WV, Chen C-N (1975) J. Am. Chem. Soc. 97: 6866
41. Lauterbur PC (1979) IEEE Trans. Nucl. Sci. NS26: 2808
42. Mansfield P, Maudsley AA (1976) J. Phys. C: Solid State Phys. 9: L409
43. Mansfield P, Maudsley AA (1977) J. Magn. Reson 27: 101
44. Mansfield P (1977) J. Phys. C: Solid State Phys. 10: L55
45. Kumar A, Welti D, Ernst RR (1975) Naturwissenschaften 62: 34
46. Kumar A, Welti D, Ernst RR (1975) J. Magn. Reson. 18: 69
47. Hoult DI (1978) J. Magn. Reson. 33: 183
48. For further details see: Ernst RR, Bodenhausen G, Wokaun A (1987) Nuclear magnetic resonance in one and two dimensions, Clarendon, Oxford, chap 10
49. Bottomley PA, Foster TH, Dorrow RD (1984) J. Magn. Reson. 59: 338
50. Pekar J, Leigh JS Jr, Chance B (1985) J. Magn. Reson. 64: 115
51. Ordidge RJ, Connelly A, Lohman JA (1986) J. Magn. Reson. 66: 283
52. Tychko R, Pines A (1984) J. Magn. Reson. 60: 156
53. Doddrell DM, Brooks WM, Bulsing JM, Field J, Irving MG, Baddeley H (1986) J. Magn. Reson. 68: 367
54. Luyten PR, Marien AJH, Sijtsma B, Den Hollander JA (1986) J. Magn. Reson. 67: 148
55. Frahm J, Merboldt K-D, Hänicke W (1987) J. Magn. Reson. 72: 502
56. Mareci TH, Sattin W, Scott KN, Bax A (1986) J. Magn. Reson. 67: 55
57. Post H, Ratzel D, Brunner P (1982) German patent, No. P 3209263.6, 13 March 1982
58. Aue WP, Müller S, Cross TA, Seelig J (1984) J. Magn. Reson. 56: 350
59. Müller S, Aue WP, Seelig J (1985) J. Magn. Reson. 63: 530
60. Müller S, Aue WP, Seelig J (1985) J. Magn. Reson. 65: 332
61. Harihara Subramanian V, Schnall MD, Leigh JS Jr (1985) Proceedings of the Society of Magnetic Resonance in Medicine, London, Aug. 18–23
62. Schnall MD, Barlow C, Harihara Subramanian V, Leigh JS Jr (1986) J. Magn. Reson. 68: 161
63. Kaufmann L, Crooks LE, Margulis AR (1981) Nuclear magnetic resonance imaging in medicine, Igakun-Shoin, Tokyo
64. Jaklovsky J (1983) NMR imaging, a comprehensive bibliography, Addison Wesley, Reading/USA
65. Wende S, Thelen M (eds) (1983) Kernresonanz-Tomographie in der Medizin, Springer, Berlin, Heidelberg, New York
66. Roth K (1984) NMR-Tomographie und Spektroskopie in der Medizin, Springer, Berlin, Heidelberg, New York
67. Morris PG (1986) Nuclear magnetic resonance imaging in medicine and biology, Clarendon, Oxford
68. Johnson GA, Thompson MB, Gewalt SL, Hayes CE (1986) J. Magn. Reson. 68: 129
69. Callaghan PT, Eccles CD (1987) J. Magn. Reson. 71: 426
70. Murphy-Boesch J, So GJ, James TL (1987) J. Magn. Reson. 73: 293
71. Frahm J, Hänicke W, Merboldt K-D (1987) J. Magn. Reson. 72: 307
72. Blümich B (1981) Dissertation, TU Berlin/Germany
73. Blümich B (1984) J. Magn. Reson. 60: 37
74. Blümich B, Spiess HW (1986) J. Magn. Reson. 66: 66
75. Liu X-R, Lauterbur PC, Marr RB (1985) Soc. Magn. Reson. Med., 4th Ann. Meeting, London
76. Liu X-R (1985) Dissertation, University of New York, Stony Broock
77. Bernardo M Jr, Chaudhuri D, Liu X-R, Lauterbur PC (1985) Soc. Magn. Reson. Med., 4th Ann. Meeting, London
78. Blümich B (1987) Prog. NMR Spectrosc. 19: 331
79. Ogan MD, Brasch RC (1985) Ann. Rep. on Med. Chem. 20: 277
80. Suits BH, White D (1984) Solid State Comm. 50: 294
81. Soroko LM (1983) Fortschr. d. Phys. 31: 419

82. Luca F, Maraviglia B (1986) J. Magn. Reson. 67: 7
83. Kucera LJ, Brunner P, Boesch C: Bruker Report 1/1986: 30
84. Samoilenko AA, Artemor DY, Sibeldina LA: Bruker Report 2/1987: 30
85. Maudsley AA (1986) J. Magn. Reson. 69: 488
86. Karis JP, Johnson GA, Glover GH (1987) J. Magn. Reson. 71: 24

Recently [68] the problems of NMRI at microscopic resolution have been studied in detail. Resolution limits in NMRI are imposed by bandwith considerations, available magnetic gradients for spatial encoding, and signal-to-noise ratios (see SNR).

A modification of a commercially available clinical NMRI device with picture elements of $500 \times 500 \times 5000$ μm to yield picture elements of $50 \times 50 \times 1000$ μm has been reported [68]. The resolution enhancement has been achieved by using smaller gradient coils permitting gradient fields smaller than 0.4 mT/cm. Significant improvements in SNR have been achieved with smaller radio-frequency coils, close attention to the choice of bandwidth and signal averaging, too.

The improvements will permit visualization of anatomical structures in the rat brain with an effective diameter of 1 cm with the same definition in human imaging. These techniques and modifications should open a number of basic sciences such as embryology, plant sciences, and teratology to the potentials of NMRI.

Much more recently [69] the limits to resoltuion in NMRI, arising from available signal-to-noise ratios (see SNR), have been examined. It has been shown that NMR microscopy can achieve a transversal resolution comparable with the optical limit of order of 1 μm but with an imaging time of around 1000 seconds. The specific imaging techniques of FOURIER zeugmatography (FZ) and filterd back projection (FBP) have been considered in detail. It has been demonstrated particularly that these techniques will treat the signal and the noise quite differently. Significant differences arise when smoothing filters are used and exact expressions have been given for image signal to noise and resolution for both FZ and FBP under conditions of optimal and asymptotic broadening [69].

Because imaging techniques tend to suffer distortions from variations in nutation angle (see NNMR) and sample concentration when used to profile the used B_1 field, a group of scientists [70] has examined the use of the nutation rate, ω_1, as a means of measuring B_1. Combining the rotating-frame experiment with the spin-echo imaging technique, they have developed a two-dimensional procedure for imaging of ω_1 in only one dimension. Under using this technique, they have measured the profile of a 2 cm single loop coil and compared it with a fit of a computed profile. The agreement was within $\pm 1\%$ over a range in which B_1 varied nearly fivefold in magnitude. Measurements with perturbing elements placed in the direct neighborhood of the coil indicated that the chosen method will be useful in detecting small differences in B_1 and unaffected by small local inhomogeneities in the static field used in the experiment.

Recently J. Frahm and his group have studied the transverse coherence in rapid FLASH nuclear magnetic resonance imaging [71]. They have proposed two modifications of the original FLASH sequence that either eliminate ("spoil") or include ("refocus") the effects of transverse coherences in rapid images.

Stochastic radio-frequency excitation has also been proposed in NMRI [72,73] because of reduced radio-frequency exposure of the patient and less stringent transmitter requirements. Stochastic NMRI is presently being studied by two groups [73–77] and has been reviewed recently [78].

Maudsley reported recently (85) that the well-known Carr-Purcell-Meiboom-Gill (see CPMG) pulse sequence is no longer satisfactory for imaging. He proposed a modified pulse sequence which provides a significant improvement in the ability to compensate for non-linear refocusing pulses.

SNR improvements in three-dimensional NMR microscopy have been obtained using limited-angle excitation [86]. This limited-angle approach has been verified experimentally in small animal studies of the brain of a 200 g rat.

The role of contrast enahncing agents in NMRI has been discussed by Ogan and Brasch [79].

The combination of NMRI and *in vivo*-NMR spectroscopy (see p 145) will become a most powerful tool in biology, biochemistry, and medicine in the very near future.

Some new publications have demonstrated [80–84] the application of NMRI for the investigation of solid state samples in the field of material sciences.

NMRIT/NMREN
Abbreviations of a computer program for the analysis of NMR (see NMR) spectra, where IT stands for Iteration and EN for Energy levels

This program or approach of Reilly and Swalen [1] has been used for the analysis of high-resolution proton magnetic resonance spectra in the past.

Like LAOCOON (see p 155) calculation starts with an estimated parameter set ($v_i^{(0)}$ and $J_{ij}^{(0)}$). In the first calculation process, i.e. by the diagonalization of the HAMILTONian (see p 122) matrix set up according to special rules, there is obtained in addition to the eigenvalues the unitary matrix U of the corresponding eigenvectors. In the next step one has to assign the lines of the observed NMR spectrum to those of the trial spectrum. The computer will get information like Eq. (1)

$$E_p - E_q = f_{pq} \tag{1}$$

where E_p and E_q are the calculated eigenvalues and f_{pq} is the experimentally measured energy difference. The NMRIT/NMREN program then calculates with the aid of this assignment a new set of eigenvalues.

Since generally more transitions than eigenvalues exist, the system of equations such as Eq. (1) is overdetermined, but only of the order $n-1$. A solution can only be found if an additional equation involving the sum of two or more eigenvalues exists. Relation (2) satisfies this condition:

$$\sum_{i=1}^{n} E_i = 0 \tag{2}$$

In the program, the next step involves a transformation with the unitary matrix $\hat{U}$ in order to generate new diagonal elements $H_{pp} = \sum_q \hat{U}_{pq}^2 E_q$ of the HAMILTONian matrix. From these a new and better set of parameters $v_i^{(1)}$, $J_{ij}^{(1)}$ can be calculated. The uncertainty only lies in the use of the approximate matrix $\hat{U}$ that is still based on the first parameter set $v_i^{(0)}$, $J_{ij}^{(0)}$. If after the nth cycle no further improvement in the parameters is achieved analysis may be considered as solved.

References

1. Swalen JD, Reilly CA (1962) J. Chem. Phys. 37: 21, for a variation of the Swalen/Reilly program see: Ferguson RC, Marquardt DW (1964) J. Chem. Phys. 41: 2087

NMR-thermometer
Synonym for an Internal Thermometer used in Nuclear Magnetic Resonance (see NMR)

The usual way of measuring and controlling the sample temperature in a high-resolution NMR experiment is to place a thermocouple in the heating or cooling gas stream close to the bottom of the sample tube. But, this technique has the principal disadvantage that the measured temperature is not identical to the temperature within the sample tube itself. This difference in temperature can be considerable, especially under the effects of strong decoupling fields [1].

Therefore, various indirect procedure using a so-called "internal NMR thermometer" have been designed [2]. For proton magnetic resonance the temperature dependence of the chemical shift difference between the oxygen- and carbon-bonded protons in methanol (for low temperatures) and ethylene glycol (for high temperatures) has been used so far.

Quite similar temperature calibrations for heteronuclei such as carbon-13 [3], phosphorus [4], fluorine [5], cobalt [6], thallium [7] have been proposed using suitable compounds as standards. The disadvantage of most of these standard samples is their restriction to only one nucleus.

Recently K. Roth has demonstrated a universal NMR thermometer for all nuclei of NMR interest [8]. This thermometer is an internal deuterium thermometer. The temperature can be indirectly determined by using a suitable deuterated reference compound and the lock channel for observing the deuterium rapid-passage NMR spectrum. By calculating the autocorrelation spectrum by two forward FOURIER transformations (see FT), the relative shift differences can easily be determined. The proposed procedure can be performed on most commercial NMR spectrometers and is completely independent of the frequency of the observed nucleus and/or any high-power proton decoupling of the sample. In practice, the very small shift differences between the deuterium and proton signals of the compounds used [9] can be neglected.

Recently [10] so-called NMR lineshape thermometry at low (millikelvin) temperatures has been proposed. The method is self-calibrating, and measures the actual spin temperature of nuclei within the sample, so that any problems arising from thermal contact between spins and lattice automatically become evident.

The fact that the lineshape is temperature dependent is well known and has been discussed for several years in the context of electron spin resonance (see ESR and EPR where the necessary conditions can be reached at relatively moderate temperatures [11–13]. Later on a detailed discussion for the NMR case given by Abragam et al. [14].

Because of practical interest in NMR for studying dynamic processes in biological systems, where electrolyte concentrations are intrinsically high, recently an Australian group [15] has investigated a number of different approaches to obtaining the true sample temperature where radio-frequency heating may be significant. Ethylene glycol as a thermometer for X-nucleus NMR spectroscopy in biological samples has been found useful [15].

References

1. Led JJ, Petersen SB (1978) J. Magn. Reson. 32: 1; McNair DS (1981) J. Magn. Reson. 45: 490
2. Van Geet AL (1968) Anal. Chem. 40: 2227; (1970) 42: 679; Raford S, Fisk CL, Becker ED (1979) Anal. Chem. 51: 2050; Hansen EW (1985) Anal. Chem. 57: 2993
3. Vidrine DR, Petersen PE (1985) Anal. Chem. 48: 1301; Combrisson S, Prange T (1975) J. Magn. Reson. 19: 108; Schneider HJ, Freitag W, Schommer M (1975) J. Magn. Reson. 18: 393
4. Gupta RK, Gupta P (1980) J. Magn. Reson. 40: 587
5. Bornais J, Brownstein S (1978) J. Magn. Reson 29: 207
6. Bailey JT, Levy GC, Wright DA (1980) J. Magn. Reson. 37: 353
7. Foster MJ, Gillies DG (1985) J. Magn. Reson. 65: 497
8. Roth K (1987) Magn. Reson. Chem. 25: 429
9. Evans DE: Chem. Commun. 1982: 1226
10. Kuhns P, Gonen O, Waugh J (1987) J. Magn. Reson. 72: 548
11. McMillan M, Opechowski W (1960) Can. J. Phys. 38: 1168
12. Hornig AW, Hyde JS (1963) J. Mol. Phys. 6: 33
13. Schmugge TJ, Jeffries CD (1965) Phys. Rev. A138: 1785
14. Abragam A, Chapellier M, Jacquinot JF, Goldman M (1973) J. Magn. Reson. 10: 322
15. Bubb WA, Kirk K, Kuchel PW (1988) J. Magn. Reson. 77: 363

NNMR
Acronym for **N**utation **N**uclear **M**agnetic **R**esonance

NNMR is a method used for studies of quadrupole nuclei in solids.

The interaction between the nuclear electric quadrupole moment and the electric field gradient allows one to get information about the microstructure around the nucleus in question. The quadrupole interaction parameters e^2qQ/h and η can be studied with nuclear quadrupole resonance (see NQR) where no or only small magnetic fields are present. NQR, however, often lacks sensitivity and in addition one has to search over a wide range of frequencies. In an NNMR experiment, the quadrupole interaction can be treated as a perturbation with respect to the ZEEMAN interaction. In that case all transitions m $\leftrightarrow$ m' between eigenstates of the ZEEMAN HAMILTONian shift in first order, except for the $1/2, -1/2$ which only shifts in second order. As a result one obtains characteristic powder patterns for $1/2, -1/2$ transition whereas all other transitions are usually broadened beyond detection. So, in principle, it must be possible to determine the parameters e^2qQ/h and η in a single NMR experiment [1]. Magic angle spinning (see MAS, MASS, or MAR) yields powder patterns for the $1/2, -1/2$ transition which are nearly 4 times smaller than for static samples, and thus is useful for enhancing the spectral resolution while retaining the information about the interesting quadrupole parameters intact [2,3]. However, in many cases the characteristic lineshapes are blurred by a spread in chemical shift (see CS) and /or by the presence of more than one quadrupole interaction.

These problems are overcome by the application of a so-called two-dimensional nutation nuclear magnetic resonance (2D-NNMR experiment) [4–6] by which the quadrupole interaction is separated from the chemical shift.

This experiment is divided into two time domains: In the evolution period (t_1) the spin system is irradiated with a radio-frequency (rf) magnetic field, B_1, and studied in the rotating frame. This evolution period depends on the ratio of the secular part of the quadrupole interaction and the rf field, whereas CS interactions are negligible. During the detection period (t_2) the free induction decay (see FID) due to both chemical shift interaction and second order quadrupole interaction is collected. In the f_1-dimension the spectra are specific for a certain ratio of e^2qQ/h and γB_1, so with γB_1 known one can determine the quadrupole interactions parameters e^2qQ/h and η. In addition, spins with overlapping resonances in the f_2-dimension, but with different quadrupole interaction parameters can be separate in the f_1-dimension.

Recently this new method has been used to investigate the hydration of zeolite NaA [7]. 2D-^{23}Na-NNMR measurements have been applied to observe the change of quadrupole parameters of sodium ions in zeolite NaA when the zeolite is loaded in steps with water: Relaxation effects in the rotating frame have studied here by the combination of rotatory echoes with NNMR.

References

1. Baugher JF, Taylor PC, Oya T, Bray PJ (1969) J. Chem. Phys. 50: 4914
2. Samoson A, Kundla E, Lippmaa E (1982) J. Magn. Reson. 49: 350
3. Kentgens APM, Scholle KFMGJ, Veeman WS (1983) J. Phys. Chem. 87: 4357
4. Samoson A, Lippmaa E (1983) Chem. Phys. Lett. 100: 205.
5. Samoson A, Lippmaa E (1983) Phys. Rev. B 28: 6567
6. Kentgens APM, Lemmens JJM, Geurts FMM, Veeman WS (1987) J. Magn. Reson. 71: 62
7. Tijink GAH, Janssen R, Veeman WS (1987) J. Am. Chem. Soc. 109: 7301.

N

NOECOSS
Acronym for **N**uclear **OVERHAUSER** **E**nhancement **Cor**relation with **S**hift **S**caling

NOECOSS is a new pulse sequence used to enhance resolution by frequency scaling along the ω_1 (or f_1) axis [1] in the conventional NOESY experiment (see NOE, 1D-NOE, 2D-NOE, and NOESY) in nuclear magnetic resonance (see NMR). With Fig. 18 the pulse sequence used [1] is characterized.

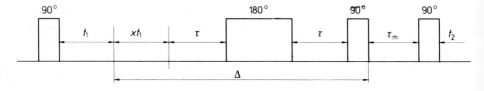

Fig. 18

In Fig. 18, Δ is a constant time delay that helps in J-scaling suppression, whereas τ is a delay which changes synchronously with the evolution time period t_1, and τ_m represents the so-called mixing time. For comparison (see SECOSS) in this dictionary.

In the meantime general criteria have been described [2] for evaluating both resolution and sensitivity enhancements in two-dimensional scaling experiments.

Much more recently [3] the combined use of COSS (see p 53) and S. COSY (see p 242) has been suggested for the recording of *pure-phase* shift-scaled NMR spectra.

References

1. Hosur RV, Ravikumar M, Sheth A (1985) J. Magn. Reson. 65: 375
2. Ravikumar M, Sheth A, Hosur RV (1986) J. Magn. Reson. 69: 418
3. Hosur RV, Sheth A, Majumdar A (1988) J. Magn. Reson. 76: 218

NOE-DIFF
Acronym for Nuclear **OVERHAUSER E**nhancement (or Effect) **Diff**erence Spectroscopy

The proton/proton NOE is today perhaps the single most powerful tool for elucidating details of the solution-state conformations of biologically interesting molecules. With modern high-field nuclear magnetic resonance (see NMR) spectrometers, fractional changes as small as 0.2 per cent can be quantitatively measured, especially when direct coherent accumulation of a *"difference"* free induction decay (see FID) is employed [1]. To achieve high precision in the case of dilute solutions (2 to 20 mM), for which intermolecular terms can be ignored, it is necessary to collect thousand or more transients and to spend six or more hours of accumulation time if a full relaxation recovery period (up to $5T_1$) is employed. Access to instrument time thus stands as the primary deterrent to the widespread use of NOE data in geometry determination.

The effect of substantial reduction in the preparatory waiting period used in the collection of NOE spectra has been examined experimentally for systems near the extreme narrowing limit with particular emphasis on the truncated driven and selective inversion-recovery techniques which do not require a long evolution [2]. Over short evolution periods the NOE buildup rate should be directly proportional to the cross-relaxation rate by either method. The transient NOE experiment still provides precise ratios of cross-relaxation rates even when the waiting time is less than T_1. Comparisons between standard (long waiting time) and rapid acquisition protocols indicate that the rapid acquisition method [2] can, with the proper use of integration standards, be applied for quantitative NOE-DIFF studies.

Recently [3] computer simulation of transient NOE experiments with short preparatory delays has been reported.

NOE-DIFF spectroscopy has been used for the editing of in vivo carbon-13 NMR spectra [4]. By this technique it was possible to separate trehalose and poly-saccharide responses in *Methylobacterium extorquens* [4].

References

1. Chapman GE, Abercrombie BD, Cary PD, Bradbury EM (1978) J. Magn. Reson. 31: 459
2. Anderson NH, Nguyen KT, Hartzell CJ, Eaton HL (1987) J. Magn. Reson. 74: 195
3. Eaton HL, Anderson NH (1987) J. Magn. Reson. 74: 212
4. Barnard GN, Sanders JKM (1987) J. Magn. Reson. 74: 372

NOESY
Acronym for two-dimensional **N**uclear **O**verhauser **E**nhanced **S**pectroscopy

This procedure in proton nuclear magnetic resonance uses the NOE as a powerful tool in structure elucidation in a two-dimensional method [1,2] and has become indispensible for resonance assignments in complex molecules such as proteins [3–6], and the spatial structures in solution of some macromolecules have already been determined on the basis of NOE-data alone [7–10]. In the meantime the so-called "NOE cross peaks" [11] are under more quantitative investigations [12].

Recently a NOESY-type experiment with multiple mixing periods and acquisitions within a single-pulse sequence was used to obtain a complete NOE correlation map of a small organic molecule [13]. Incorporating multiple mixing periods and acquisitions can eliminate the need for repeating NOESY experiments with different mixing periods to observe dipolar exchange of rather different rates or to determine exchange rates.

Pure absorption phase two-dimensional nuclear OVERHAUSER spectroscopy (NOESY), using a short cycle time protocol, has been used recently [14] for studies of the cross-relaxation between protons of ligands bound to proteins.

The interested reader is also referred to the acronym *2D-NOE* in this dictionary.

References

1. Jeener J, Meier BH, Bachmann P, Ernst RR (1979) J. Chem. Phys. 71: 4546
2. Kumar A, Ernst RR, Wüthrich K (1980) Biochem. Biophys. Res. Commun. 95: 1
3. Wagner G, Wüthrich K (1982) J. Mol. Biol. 155: 347
4. Wider G, Lee KH, Wüthrich K (1982) J. Mol. Biol. 155: 367
5. Williamson MP, Marion D, Wüthrich K (1984) J. Mol. Biol. 173: 341
6. Wemmer D, Kallenbach NR (1983) Biochemistry 22: 1901
7. Braun W, Wider G, Lee KH, Wüthrich K (1983) J. Mol. Biol. 169: 921
8. Arseniev AS, Kondakov VI, Maiorov VN, Bystrov VF (1984) FEBS Lett. 165: 57
9. Williamson MP, Havel TF, Wüthrich K (1985) J. Mol. Biol. 182: 295
10. Zuiderweg ERP, Billeter M, Boelens R, Scheek RM, Wüthrich K, Kaptein R (1984) FEBS Lett. 174: 243
11. Havel TF, Wüthrich K (1985) J. Mol. Biol. 182: 281
12. Denk W, Baumann R, Wagner G (1986) J. Magn. Reson. 67: 386
13. Meyerhoff DJ, Nunlist R, O'Connell JF (1987) Magn. Reson. Chem. 25: 843
14. Andersen NH, Eaton HL, Nguyen KT (1987) Magn. Reson. Chem. 25: 1025

N

NORD
Acronym for **N**oise-**M**odulated **O**ff-**R**esonance **D**ecoupling

K. Roth [1–4] has shown that in addition to the well-established double resonance methods [5] the noise-modulated off-resonance decoupling (NORD) technique results in further spectral information for assignments of individual signals and structure elucidation of chemical compounds. NORD-spectra of secondary carbons with two chemically non-equivalent protons have been investigated theoretically and experimentally too. According to K. Roth [4], there exists a linear relationship between linewidth of secondary carbon signals and the chemical shift difference of the two directly bonded protons. Analysis of such linewidths allows an easy determination of symmetry properties of spin systems and normally leads to an unambiguous assignment of both carbon and proton signals.

References

1. Roth K (1977) Org. Magn. Reson. 9: 414
2. Roth K (1977) Org. Magn. Reson. 10: 56
3. Roth K (1980) J. Magn. Reson. 40: 489
4. Roth K (1981) J. Magn. Reson. 42: 132
5. Philipsborn W von (1971) Angew. Chemie 83: 470

NOVEL
Acronym for **N**uclear Spin **O**rientation **V**ia **E**lectron Spin **L**ocking

NOVEL seems to be the first observation of resonant transfer of electron spin polarization to a nuclear spin system [1] and is, in some aspects, similar to the technique of dynamic nuclear polarization (see DNP) [2]. The nuclear spin system is present in a solid doped with paramagnetic centres and pulsed microwave irradiation is applied to transfer the electron spin polarization of these paramagnetic centres to the nuclear spins.

NOVEL can be seen to be closely related to the method of Hartmann and Hahn (see HAHA) [3] for transferring spin polarization from one nuclear spin species to another. With NOVEL one brings electron spins into a so-called *spin-locked state in the rotating frame* where they will have the same resonance frequency as the nuclear spins in the laboratory frame. The presence of *dipolar interaction* between the two spin species then allows for the resonant transfer of the electron spin polarization to the nuclear spins.

Using NOVEL, the observation of enhanced nuclear polarization due to *flip-flop* transitions between locked electron and nuclear spins has been demonstrated [1]. It appears that the polarization rate in the NOVEL scheme is very rapid in comparison to conventional DNP making it an attractive technique for polarizing nuclear spins.

References

1. Henstra A, Dirksen P, Schmidt J, Wenkebach WT (1988) J. Magn. Reson. 77: 389
2. Abragam A, Goldman M (1982) Nuclear magnetism: Order and disorder, Clarendon, Oxford
3. Hartmann SR, Hahn EL (1962) Phys. Rev. 128: 2042

NQCC
Abbreviation of **N**uclear **Q**uadrupole **C**oupling **C**onstants

As in nuclear magnetic resonance (see NMR), coupling constants (see CC) are the relationships between different nuclei in nuclear quadrupole resonance (see NQR) and have been called NQCC.

NQR
Abbreviation of **N**uclear **Q**uadrupole **R**esonance Spectroscopy

Since its introduction by Dehmelt and Krüger [1,2] in 1950, nuclear quadrupole resonance spectroscopy has been increasingly used for the investigation of chemical bonds in crystalline solids. NQR is applicable to compounds of 130 isotopes whose nuclear spin is greater than 1/2 [3]. These nuclei have a nuclear charge distribution in the form of an ellipsoid of rotation, and consequently possess a nuclear quadrupole moment, whose magnitude and sign depends on the nature and magnitude of the ellipticity. Definition of the scalar electric nuclear quadrupole moment is given by Eq. (1) and symbolized by Fig. 19.

$$eQ = \int \rho \, r^2 (3 \cos^2 \vartheta - 1) \, d\tau \tag{1}$$

Here, e corresponds to the nuclear charge, and I is the nuclear spin quantum number.

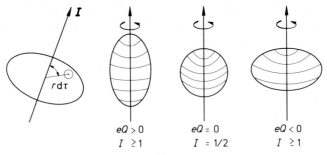

Fig. 19

Primarily from the pioneering work of Townes and Dailey [4], it was possible to use the NQR spectroscopic data to establish the electronic structure of the chemical bond. The nuclear quadrupole can take positions that differ in energy in relation to the direction of I in the inhomogeneous internal electric field of the molecule. The energies are determined by the magnitude of the quadrupole moment eQ and the field inhomogeneity at the position of the nucleus q. The possible orientations of the nucleus, i.e. the number of terms, are governed by the spin quantum number I. The field gradient provides the chemical information. Its magnitude is determined exclusively by the bonding electrons in incomplete shells, mainly p electrons. The influence of d electrons is weaker than that of the p electrons by a factor of about 10. The NQR frequencies are proportional to the field gradients.

The ionic character of a bond, the degree of s-hybridization, and the multiple bond component (conjugation effect) have been studied by NQR. Besides these bond studies, NQR have been used in the past for structure elucidation of chlorine [5,6] and nitrogen [14] containing chemical compounds. The very frequent predominance of the ionic contribution in chlorine compounds was the basis for the manifestation of so-called chemical shift tables (in analogy to NMR) in chlorine-35 NQR [7–12]. Similar classifications are based on the electronic properties of atoms or groupings, which are usually described by HAMMETT's constant and TAFT's (inductive effects).

In nitrogen-14 NQR there were two problems in constructing analogous "shift" tables: (1), for the nitrogen atom, because of its ^4S ground state, $e^2 q_{at} Q = 0$, so that a value of about 8.4 MHz must be assumed for an unpaired p electron [13], and (2), the multiple coordination of the bonded atoms leads to problems in establishing the principal axis of the field gradients, which make a general representation difficult. However, the method of Townes and Dailey [4] can be used here again [13], and within certain classes of substances, the same concept as for chlorine compounds can be applied. In the past there was a common belief [14] that NQR may complement nuclear magnetic resonance (see NMR) in the case of quadrupolar nuclei. But, the amount of substances required, their purity, and their good crystallizability were the almost limiting factors in the filed of NQR for structure elucidation. On the other hand, NMR of quadrupolar nuclei has received much more support in the last years from new methods and techniques.

The present state and further developments of the NQR technique was recently discussed at the "8th International Symposium on NQR Spectroscopy" at Darmstadt/Germany in 1985 [15]. Problem solutions in the fields of electronic structure and molecular dynamics have to be present in the next future. New techniques using SQUID magnetometers (acronym standing for *s*uperconducting *qu*antum *i*nterference *d*evice) or FOURIER transform NQR seem to be helpful in new applications, but they have to compete with all the fields of nuclear magnetic resonance (see NMR) spectroscopy.

Much more recently [16] we had to learn that corrections to the theory of the temperature dependence of NQR frequencies are necessary:

The strict expression for the statistical mean-square of libration $\langle \sin^2\phi \rangle$ for the harmonic torsional oscillator has been derived [16]. The influence of replacement of the approximate expression, $\langle \phi^2 \rangle$, used in BAYER's theory, by the strict expression $\langle \sin^2\phi \rangle$ on the results of the direct numerical analysis of the temperature dependence of NQR frequencies has been studied in detail [16].

The problem arising from the influence of anharmonicity of the torsional potential on the temperature dependence of $\langle \sin^2\phi \rangle$ has been discussed, too [16].

N

References

1. Dehmelt HG, Krüger H (1950) Naturwissenschaften 37: 111, 398; (1951) Z. Physik 129: 401
2. Dehmelt HG (1951) Z. Physik 130: 356
3. Lucken EAC (1969) Nuclear quadrupole coupling constants, Academic, London
4. Townes CH, Dailey BP (1949) J. Chem. Phys. 17: 782; (1950) Phys. Rev. 78: 346; (1952) J. Chem. Phys. 20: 35; (1955) J. Chem. Phys. 23: 118
5. Brame EG Jr (1971) Anal. Chem. 43: 35
6. Fitzky HG (1971) GIT-Fachz. Lab. 15: 922, 1100
7. Bray PJ, Barnes RG (1957) J. Chem. Phys. 27: 551
8. Fedin EI, Semin GK (1960) Zh. Strukt. Khim. 1: 464
9. Hooper HO, Bray PJ (1960) J. Chem. Phys. 33: 334
10. Brame EG Jr (1967) Anal. Chem. 39: 517
11. Roll DB, Biros FJ (1969) Anal. Chem. 41: 407
12. Fitzky HG (1971) paper presented at the XVIth Colloquium Spectroscopicum Internationale, Heidelberg 1971, Adam Hilger, London, vol 1, p 64
13. Schemp E, Bray PJ (1968) J. Chem. Phys. 49: 3450
14. Fitzky HG, Wendisch D, Holm R (1972) Angew. Chem. internat. Edit. 11: 979
15. Smith JA (1986) Z. Naturforsch. 41a: 453
16. Kalenik J, Bereszyuski Z (1987) J. Magn. Reson. 74: 105

NSE
Acronym for **N**uclear **S**olid **E**ffect

NSE is a method for polarization transfer which occurs via the so-called forbidden transitions. The NSE is analogous to the solid effect associated with dynamic nuclear polarization (see DNP). In DNP, the nuclei are polarized by saturation of a forbidden transition involving simultaneous flips or flip-flops of an electron and nuclear spin. This transition takes place at the sum or difference of the electron and nuclear LARMOR frequencies. In NSE, however, the abundant nuclear spin assume the role of the electrons, so that the forbidden transition is at the sum or difference of the LARMOR frequencies of the abundant and rare nuclei. The NSE technique was first applied in 1958 by Abragam and Procter [1] to lithium fluoride at low temperatures, but, to the best of our knowledge the possibility of combining this technique with modern nuclear magnetic resonance (see NMR) methods was never explored before 1986. Nevertheless, the NSE method offers a lot of advantages, and it has been shown [2] that for polyethylene at 80 K the carbon-13 spectrum obtained via NSE is superior to that obtained via cross polarization (see CP). Recently, the NSE has been investigated much more fundamentally [3].

References

1. Abragam A, Procter WR (1958) C.R. Acad. Sci. 246: 2253
2. Wind RA, Yannoni CS (1986) J. Magn. Reson. 68: 373
3. Wind RA, Yannoni CS (1987) J. Magn. Reson. 72: 108

NYQUIST's CRITERION or FREQUENCY

The name of an important condition, necessarily used both in Nuclear Magnetic Resonance (see NMR) and Infrared (see IR) Spectroscopy in the FOURIER Transform (see FT) Mode

N

Any waveform that is a sinusoidal function of time or distance can be sampled unambiguously using a sampling frequency greater than or equal to twice the bandwidth of the system [1], and is known as NYQUIST's criterion or frequency [see Eq. (1)]. When holding this condition, the signal of interest may then be effectively recorded without any loss of information. For further details see the literature in the field of NMR and IR spectroscopy.

$$N = 2 \cdot \Delta F \cdot T \tag{1}$$

Reference

1. Woodward M (1955) Probability and information theory, Pergamon, New York

ODMR
Acronym for **O**ptically **D**etected **M**agnetic **R**esonance

The effect of ODMR itself and its application possibilities have been reviewed recently [1].

Reference
1. Lynch WB, Pratt DW (1985) Magn. Reson. Rev. 10: 111

ODPAS
Acronym for **O**ptically **D**etected **P**hoto**a**coustic **S**pectroscopy

Optically detected photoacoustic spectroscopy (PAS) is an alternative technique to the commonly used piezoelectric detection in PAS. The optical approach to detect the photoacoustic pulse within the sample can overcome several of the known‚problems associated with the piezoelectric measurements.

The experiment uses a probe laser beam (see LASER) in the sample, parallel to the excitation beam; the probe beam is deflected as the ultrasonic density wave passes through it following the excitation pulse.

This approach to PAS detection has been demonstrated by Tam and his co-workers [1] in its application for the determination of ultrasound velocities in liquids, vapors, gases, and flames.

Recently [2] the application of ODPAS to trace-level absorption measurements in condensed phase samples has been considered. The detection capabilities of operating in the adiabatic (=photoacoustic) versus the isobaric (=photo-thermal) time regimes have been compared in detail [2]. The rise-time of the beam deflection depends on the beam spot size and the ultrasound velocity and may be in practice as short as 50 ns. This time-resolution allows effective stray light rejection. Acoustic coupling problems are also eliminated since the acoustic wave is detected within the sample. It has been pointed out [2] that the ODPAS technique is a non-contact, remote sensing method which allows its application to cryogenic samples and for monitoring situations.

References
1. See Ref. [2]
2. Poston PE, Harris JM (1987) paper presented at the Pittsburgh Conference, Atlantic City, NJ, March 9–13, 1987

OFF-Resonance
Abbreviation of "off-resonance decoupling"

Coherent proton decoupling off-resonance [1,2] has the effect of reducing, in carbon-13 NMR spectra, the one-bond ^{13}C–H couplings to a fraction of thier actual value while generally removing long-range coupling constants. The resulting so-called reduced splitting J^r was found to be related to the frequency offset (Δv), of the decoupler from resonance and the decoupler power (H_2) (normally expressed in hertz) ($\gamma H_2/2\pi$), where γ stands for the magnetogyric ratio of the involved proton. The phenomenon is most easily understood in terms of a simple sector diagram [3] based on the rotating frame of reference. More details may be written in terms of quantum-mechanical treatment (see HAMILTONion operator). In the meantime, other interpretation aids or assignment techniques have become more and more important in carbon-13 NMR.

References
1. Ernst RR (1966) J. Chem. Phys. 45: 3845
2. Weigert FJ, Jautelat M, Roberts JD (1968) Proc. Nat. Acad. Sci. U.S. 60: 1152
3. See Hoffman RA, Forsén S (1966) In: Emsley JW, Feeney J, Sutcliffe LH (eds) High resolution nuclear magnetic double and multiple resonance, Pergamon, Oxford

ORD

Abbreviation of **O**ptical **R**otatory **D**ispersion. ORD is like CD (see p 41) one of the most important chiroptical techniques and will be explained together with CD in this dictionary

Here, we will only present Fig. 20 for the definition of *optical rotation* and *ellipticity*. Please, note that the elliptic transmitted light is seen from an observer facing the direction of the propagated light.

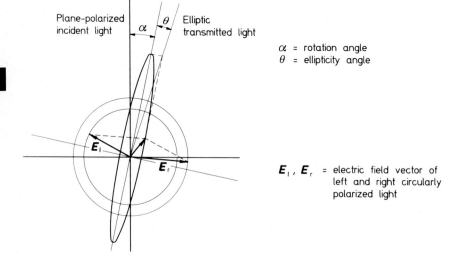

α = rotation angle
θ = ellipticity angle

E_l, E_r = electric field vector of left and right circularly polarized light

Fig. 20

OS

Abbreviation of **OVERHAUSER** Shift

The OS is the shift of the conduction electron LARMOR frequency induced by the polarization of magnetic nuclei [1,2]. The OS is an "inverse KNIGHT shift" and it is an extremely sensitive way of measuring the KNIGHT shift and the NMR relaxation times, since the nuclei are dynamically polarized in these substances and the enhancement factor reaches values of 550 for protons and 2000 for carbon-13. A conventional electron spin resonance (see ESR) spectrometer equipped with ENDOR (see p 87) facilities can be used to measure the OS.

References
1. Denninger G, Stöcklein W, Dormann E, Schwoerer M (1984) Chem. Phys. Lett. 107: 222
2. Stöcklein W, Denninger G (1986) Mol. Cryst. Liq. Cryst 136: 335

O

PCA
Acronym for **P**rincipal **C**omponent **A**nalysis

PCA is a general concept in solving problems arising from natural sciences [1,2]. PCA has been used in connection with a multi-variate analysis (see MVA) in two-dimensional nuclear magnetic resonance (see 2D-NMR and 2D-NMR spectra) recently (3) as an ideal solution in pattern recognition analysis.

PCA can be thought of as a graphical procedure which simplifies multi-dimensional data by projecting them down to fewer dimensions. The PCA method orients the projections according to the directions of maximum variance in the data. Thus, PCA can be used to find out consistent ways of viewing each class, and also as a powerful tool to reduce complicated multi-dimensional representations to more and less simple patterns in two or three dimensions.

References
1. Joliffe IT (1986) Principal component analysis, Springer, Berlin Heidelberg New York
2. Wold S, Albano C, Dunn WJ III, Edlund U, Esbensen K, Geladi P, Hellberg S, Johansson E, Lindberg W, Sjöström M (1984) In: Kowalski BR (ed) Chemometrics mathematics and statistics in chemistry, NATO ASI Series No. 138, Reidel, Dordrecht
3. Grahn H, Delaglio F, Delsuc M, Levy GC (1988) J. Magn. Reson. 77: 294

P-ENDOR
Acronym for **P**ulsed **E**lectron **N**uclear **Do**uble **R**esonance

In comparison to the so-called continuous wave endor spectroscopy (see CW-ENDOR) this new method in electron spin or paramagnetic resonance (see ESR or EPR) uses the advantages of pulsed procedures in magnetic resonance [1] over common continuous wave methods in this field. In such a pulsed ENDOR procedure the radio-frequency field will be switched on between the second and third microwave pulse of a three-pulse sequence. The amplitude of the stimulated echo will be measured as a function of the radio frequency [2]. The registered frequency spectrum may be equivalent to the conventional ENDOR spectrum. But, in practice, there is a remarkable difference: intensities of a pulsed ENDOR spectrum are not under control of the different relaxation rates.

References
1. Schweiger A (1986) Chimia 40: 111
2. Mims WB (1965) Proc. Soc. (London) 283: 452

P.E. COSY
Acronym for **P**rimitive **E**xclusive **Cor**relation **S**pectroscopy

Recently, L. Müller [1] has introduced an alternative technique to the elegant two-dimensional nuclear magnetic resonance (see 2D-NMR) experiment called "exclusive correlation spectroscopy" by R.R. Ernst and co-workers [2] (see E. COSY). The new technique has been named primitive E. COSY or P.E. COSY because of its simplicity. This simplification will be helpful in the computer-assisted analysis of COSY spectra when using pattern recognition procedures [3,4] as well as in the determination of sign and magnitude of J couplings.

References

1. Müller L (1987) J. Magn. Reson. 72: 191
2. Griesinger C, Sørensen OW, Ernst RR (1985) J. Am. Chem. Soc. 107: 6394
3. Meier BU, Bodenhausen G, Ernst RR (1984) J. Magn. Reson. 60: 161
4. Pfändler P, Bodenhausen G, Meier BU, Ernst RR (1985) Anal. Chem. 57: 2510

P(A)NIC

French acronym for Polarisation Nucléaire (Dynamique) Induite Chimiquement (see CIDNP), introduced by C. Richard [1]; see S.H. Pine [2] too.

References

1. Richard C (1972) Second Thesis, University of Nancy/France
2. Pine SH (1972) J. Chem. Ed. 49: 664

PAS

Acronym for Photoacoustic Spectroscopy

Photoacoustic spectroscopy (PAS) is an increasingly used method for infrared spectra (see IR) of samples that are hard to prepare as transparent films, or have internal light scattering or are coated onto opaque or strongly light-scattering substrates. The PAS technique is therefore complementary to ATR (see p 26) and diffuse reflectance and has the important advantage of needing no sample preparation. The signal-to-noise ratio for PAS is rather low, so longer scanning time is required compared to transmission FT-IR (see p 116). The photoacoustic effect is simply the generation of an acoustic signal by a sample exposed to modulated light. If the solid sample in the PAS cell absorbs a particular infrared frequency, it will respond by generating an acoustic signal at the particular audio frequency that corresponds to the incident IR frequency [1]. The result of absorption by the sample of several different frequencies is a PAS signal containing several different audio frequencies [2–5]. It should be recognized that light reflected or scattered without any frequency translation cannot heat up the sample, therefore such samples of this type which are difficult to study by other methods are particularly suited to the PAS technique [1]. The application of FT-IR (see p 116) instrumentation to PAS arises from the need for a high signal-to-noise ratio and the multiplex advantage is very helpful in this regard.

References

1. Koenig JL (1984) Fourier transform infrared spectroscopy of polymers, Springer, Berlin Heidelberg New York (Advances in Polymer Science, vol 54)
2. Rosencwaig A (1973) Science 181: 697
3. Pao YH (1977) Optoacoustic spectroscopy and detection, Academic, New York
4. Rosencwaig A (1978) Adv. Electron. Electron Phys, 46: 208
5. Rosencwaig A (1980) Photoacoustics and photoacoustic spectroscopy, Wiley, New York

PCCE
Abbreviation of **P**roton **C**oupling **C**onstant **E**xtraction

PCCE is a novel computer method for obtaining all the proton hyperfine coupling constants in an electron spin resonance (see ESR) spectrum [1]. This analyzing method works well on spectra which are poorly resolved, and thus appears to complement the recently developed correlation approach [2–4] used for high-resolved ESR spectra. No prior knowledge of the radical structure is needed, and proton splittings are clearly discriminated from spin-1 splittings. The PCCE procedure is similar in concept to the theoretical treatment recently proposed by Dracka [8], and elements of PCCE have also been suggested by Newton and co-workers [5], Stone and Maki [6], and Brumby [7]. The program itself consists of two parts: SEARCH, to identify the proton hyperfine splittings, and SEPARATE, to simplify the ESR spectrum. It is readily programmed and run on a small laboratory computer.

References
1. Motten AG, Duling DR, Schreiber J (1987) J. Magn. Reson. 71: 34
2. Jackson RA (1983) J. Chem. Soc., Perkin Trans. II: 523
3. Al-Wassil AI, Eaborn C, Hudson A, Jackson RA (1983) J. Organomet. Chem. 258: 271
4. Jackson RA, Rhodes CJ (1984) J. Chem. Soc. Chem. Commun. 19: 1278
5. Newton R, Schultz KF, Elofson RM (1966) Can. J. Chem. 44: 752
6. Stone EW, Maki AH (1963) J. Chem. Phys. 38: 1999
7. Brumby S (1979) J. Magn. Reson. 34: 318
8. Dracka O (1985) J. Magn. Reson. 65: 187

PCJCP
Abbreviation of **P**hase-**C**orrected coupled **J** **C**ross-**P**olarization Spectra

In nuclear magnetic resonance (see NMR), when dealing with J cross-polarization spectra, the occurrence of phase and multiplet anomalies has been noted and the phase-corrected JCP (see JCP) sequence, PCJCP, has been introduced [1,2] to purge coupled NMR spectra of phase anomalies or artifacts.

References
1. Chingas GC, Bertrand RD, Garroway AN, Moniz WB (1979) J. Am. Chem. Soc. 101: 4058
2. Chingas GC, Garroway AN, Bertrand RD, Moniz WB (1981) J. Chem. Phys. 74: 127

PCR
Abbreviation of **P**hoto-**C**onductive **R**esonance

PCR is a new electron spin resonance (see ESP and EPR)-based technique. A theoretical analysis has been offered and results have been summarized recently [1].

Reference
1. Haneman D (1984) Prog. Surf. Sci. 15: 85

PCS
Abbreviation of **P**rotonated **C**arbon **S**uppression

In 1979 Opella and Frey [1] presented a very useful pulse sequence for removing lines from proton-decoupled solid-state nuclear magnetic resonance (see NMR) carbon-13 spectra resulting from protonated carbons.

Their method—abbreviated PCS—, based on an experiment developed by Alla and Lippmaa (2), is applicable to spectra obtained by cross-polarization (see CP) enhancement or as a free induction decay (see FID) with or without magic-angle spinning (see MAS).

This procedure, variously referred to as dipolar dephasing [3,4] or interrupted decoupling [5], involves switching-off the used decoupler for a set time t_D after the creation of carbon-13 (or other rare nuclei) transverse magnetization. During this period, the protonated carbon-13 magnetization vectors are dephased by couplings to the protons (or other abundant nuclei). This time period is normally in the order of 40 to 100 μs [1,4] for carbons in organic solids.

However, phase distortions result from the well-known fact that the delay t_D introduces a first-order phase shift across the solid-state spectrum. Owing to the high complexity of this spectrum, manual correction of the phase shift would be quite difficult and probably nothing less than arbitrary.

To overcome this problem, an improvement has been proposed by Murphy [5]. He used the original pulse sequence [1] with the addition of a 180° carbon-13 pulse introduced into the centre of the delay. As shown previously by Stoll et al. [6] and Bodenhausen et al. [7] in somewhat different applications, this pulse has the effect of refocusing the rotating frame precession of the non-protonated carbons, responsible for the phase shift.

Recently K.R. Carduner [8] has presented a method for the implementation of PCS with the total suppression of spinning sidebands (see TOSS).

References
1. Opella SJ, Frey MH (1979) J. Am. Chem. Soc. 101: 5854
2. Alla M, Lippmaa E (1976) Chem. Phys. 37: 260
3. Murphy PD, Cassady TJ, Gerstein BC (1982) Fuel 61: 1233
4. Alemany LB, Grant DM, Alger TD, Pugmire RJ (1983) J. Am. Chem. Soc. 105: 6697
5. Murphy PD (1983) J. Magn. Reson. 52: 343
6. Stoll ME, Vega AJ, Vaughan RW (1976) J. Chem. Phys. 65: 4093
7. Bodenhausen G, Stark RE, Ruben DJ, Griffin RG (1979) Chem. Phys. Lett. 67: 424
8. Carduner KR (1987) J. Magn. Reson. 72: 173

PENIS
Acronym for **P**roton-**E**nhanced **N**uclear **I**nduction **S**pectroscopy

The combination of high-power dipolar decoupling and cross-polarization (see CP) in the field of solid-state nuclear magnetic resonance (see NMR) has been introduced by Pines and co-workers [1] and sometimes named PENIS.

One can start principally an acquisition of the conventional free induction decay (see FID) of rare spins S (for example carbon-13) under high-power dipolar decoupling from the abundant spins I (here: hydrogen-1). I-S cross-polarization

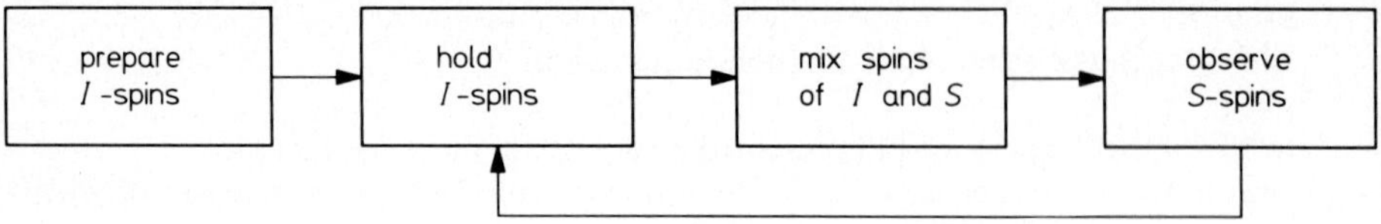

Fig. 21

involves magnetization transfer (see MT) under enhancement via the protons. But, much more practically PENIS may be rationalized by Fig. 21.

Reference

1. Pines A, Gibby MG, Waugh JS (1973) J. Chem. Phys. 59: 569

PET
Abbreviation of **P**ositron **E**mission **T**omography

When using radioactive tracers, PET became an interesting method in the field of computer-assisted imaging procedures such as X-ray computer tomography (see CT), nuclear magnetic resonance imaging (see NMRI) or electron spin resonance imaging (see ESRI) [1–3]. PET may be useful in studying the total spectrum of biochemical, physiological, and pharmacological processes in regional domains with relatively high sensitivity [4] before morphological manifestation of diseases. On the other hand, PET have to compete with the combination of NMRI and the so-called *in vivo* NMR spectroscopy.

References

1. Ell PJ, Holman BL (eds) (1982) Computed emission tomography, Oxford University Press, New York
2. Reivich M, Alavi A (eds) (1985) Positron emission tomography, Liss, New York
3. Phelps ME, Mazziotta JC, Schelbert (eds) (1985) Positron emission tomography, Raven, New York
4. Stöcklin G (1986) Nachr. Chem. Tech. Lab. 34: 1057; see also Stöcklin G, Wolff AP (eds), Radiochemistry related to life sciences, Spec. Issue Radiochim Acta 30: (1982) and 34: (1983)

PFG-NMR
Abbreviation of **P**ulsed **F**ield **G**radient-**N**uclear **M**agnetic **R**esonance

PFG-NMR is a well established method in the field of diffusion measurements [1,2], and its implementation on a modern high-resolution PFT spectrometer (see PFT) permits the simultaneous measurement of diffusion coefficients of individual components in a multicomponent mixture [3].

In the PFG-NMR method, developed by Steijskal and Tanner, two magnetic field gradients, each of the duration δ, are inserted into the standard spin-echo sequence $(90° - \tau - 180°)$. The first gradient pulse is applied between the 90 and 180° pulses and the second between the 180° pulse and the spin-echo.

Therefore, this procedure has also been called field-gradient spin-echo (see FGSE).

The diffusion coefficient D is related to the amplitude of the spin-echo by the following equation

$$\ln R = -\gamma^2 G^2 \delta^2 (\Delta - \delta/3)\, D \qquad (1)$$

where R is the ratio of the echo amplitude in the presence of the magnetic field gradient to the amplitude in the absence of the gradient. In Eq. (1), γ is the magnetogyric ratio, G is standing for the magnitude of the magnetic field gradient, and Δ means the time between the starting points of the gradient pulses. For convenience the echo amplitude is measured rather than a ratio of the amplitudes. This results only in a change of the intercept in a plot of $\ln R$ versus G^2. In the FOURIER transform version of the experiment, data collection starts at the peak of the spin-echo. FT then produces the usual high-resolution spectrum except that the amplitudes of the peaks show attenuation as described by Eq. (1) [3].

For further details in diffusion and self-diffusion measurements see the acronyms FGSE, SGSE and PGSE in this dictionary.

Recently it has been pointed out, that PFG-NMR and holographic relaxation spectroscopy (HRS) may be competetive techniques for the measurement of tracer or self-diffusion coefficients.

PFG-NMR and HRS have been compared experimentally in a study of tracer diffusion rates for the photochromic compounds azobenzene and aminoazo-benzene in acetone and dioxane [4]. It has been shown, that these different techniques are applicable to exactly the same range of diffusion rates but can provide complementary information. The *cis* and *trans* forms of azobenzene can distinguished in the NMR experiment, and the tracer diffusion coefficient for the *cis* form has been found to be nearly 5% lower than for the corresponding *trans* form. In the HRS experiment a single diffusion coefficient has been determined from the rate of decay of the laser-induced (see LASER) concentration pattern of *cis*-azobenzene. Results obtained by these different methods agree, and the magnitudes of the errors involved are similar. It has been concluded that, in general, HRS will be a much more sensitive method, but only NMR permits the diffusing species to be identified [4].

References

1. Steijskal EO, Tanner JE (1965) J. Chem. Phys. 42: 288
2. Steijskal EO (1972) Adv. Mol. Relaxation Processes 3: 27
3. James TL, McDonald GG (1973) J. Magn. Reson. 11: 58
4. Lever LS, Bradley MS, Johnson CS Jr (1986) J. Magn. Reson. 68: 335

PFSEPR
Acronym for **P**ulse **F**ield-**S**weep **E**lectron **P**aramagnetic **R**esonance

A new technique in the field of electron spin resonance (see ESR or EPR) spectroscopy which enables the extraction of the hyperfine information from inhomogeneously broadened EPR lines to be made [1]. For this purpose a low power microwave pulse method has been developed and primarily applied to interesting bioinorganic systems such as bisimidazole ferric porphyrin systems and the Cu_A signal of cytochrome c oxidase. Saturating, hole-burning microwave pulses are followed by a field sweep (see SWEEP) in order to monitor the spread of saturation away from the original hole. Satellite holes have been observed whose splittings have been assigned to the hyperfine-coupled nitrogens, protons, and deuterons. Hyperfine splittings of the iron-liganding ^{14}N atoms of imidazole and pyrrole have been found, too. The authors [1] have studied in detail the dependence of the spectral features on experimental parameters like microwave power, pulse time, field sweep, 100 kHz modulation, and magnetic field. It has been pointed out that sensitivity is an order of magnitude greater than in CW-ENDOR (see p 64) and an order of magnitude less than in continuous wave electron paramagnetic resonance (CW-EPR). The spectral resolution may be of the order of 1.0 MHz.

Reference
1. Falkowski KM, Scholes CP, Taylor H (1986) J. Magn. Reson. 68: 453

PFT
Acronym for **P**ulse **FOURIER T**ransformation used in Nuclear Magnetic Resonance (see NMR) Spectroscopy

General principles of pulsed excitation (see Fig. 22) in contrast to the older continuous wave techniques (see CW) are clearly explained in two monographs [1,2]. FOURIER transformation (see FT and FFT) of a free induction decay (see FID) leads to a normal absorption spectrum (see Fig. 23). The combination of PFT, PND (see p 220) and signal averaging techniques [3] is a most powerful technique in NMR of nuclei with low natural abundance and magnetogyric ratio like carbon-13 or nitrogen-15.

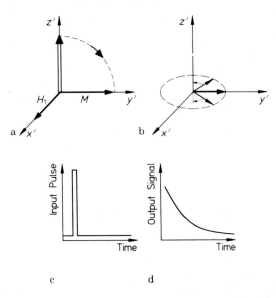

Fig. 22. a) A 90° pulse along x′ rotates magnetization M to the y′ axis; **b)** M decrease as magnetic moments dephase; **c)** input signal, the 90° pulse, corresponding to (a); **d)** FID, corresponding to (b).

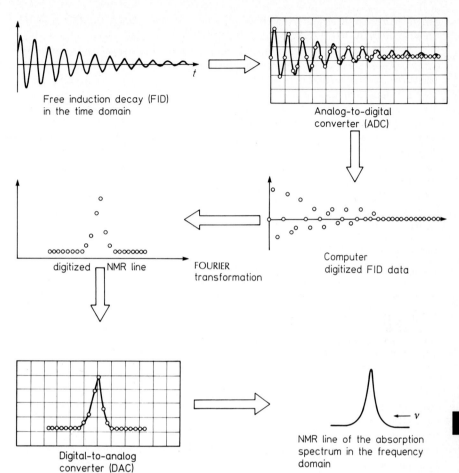

Fig. 23. Scheme of data flow in an PFT experiment in NMR spectroscopy after obtaining the FID

References

1. Farrar TC, Becker ED (1971) Pulse and Fourier transform nuclear magnetic resonance, Academic, London
2. Shaw D (1976) Fourier NMR, Elsevier, Amsterdam
3. Ernst RR (1966) Adv. Magn. Reson. 2: 1

PGSE
Acronym for **P**ulsed-**G**radient **S**pin-**E**cho

PGSE is a technique used in nuclear magnetic resonance (see NMR) for diffusion measurements. The method requires a pulsed NMR spectrometer with the possibility for creating a uniform calibrated magnetic field gradient in the region of the sample under investigation [1–3]. The magnetic field gradient is usually generated by an electric current through a set of coils centered on the sample. These coils may be of the opposed HELMHOLTZ design [3] or else of the quadrupole type [4]. PGSE measurements require a fast response by the coil current power supply and switch, combined with precise current regulation and also precise timing of the current pulses. Time and phase stability of the spin-echo (see SE) depends on physical rigidity of the gradient coils and probe assembly, too. Advances in these areas [5] have eliminated many of the early difficulties in PGSE work. The magnetic field gradient may be calibrated by direct "mapping" the magnetic field [5], by a first-principles calculation [4], by relating the duration of the spin-echo for a given sample diameter to the current of the coil [1], e.g. from measurements of the spin echo envelope during off-resonance FT-PGSE experiments [6], or ultimately by echo attenuation measurements in samples of known diffusivity. Essentially fully automatic operation of SGSE work (see p 252) has been achieved [7], and is, at present, almost attained [5,8] for routine PGSE measurements.

References

1. Carr HY, Purcell EM (1954) Phys. Rev. 94: 630
2. Stejskal EO, Tanner JE (1965) J. Chem. Phys. 42: 288
3. Tanner JE (1966) Ph.D. Thesis, University of Wisconsin
4. Zupančič I, Pirš J (1976) J. Phys. (London) E9: 79
5. Callaghan PT, Trotter CM, Jolley KW (1980) J. Magn. Reson. 37: 247
6. Hrovat MI, Wade CG (1981) J. Magn. Reson. 44: 62; (1981) J. Magn. Reson. 45: 67
7. Cantor DM, Jonas J (1977) J. Magn. Reson. 28: 157
8. Callaghan PT, Jolley KW, Trotter CM (1980) JEOL News 16A: 48

Photo-CIDNP
Acronym for **Photo**-**C**hemically **I**nduced **D**ynamic **N**uclear **P**olarization (see CIDNP)

P

Photo-CIDNP has made remarkable progress in the field of biological macromolecules by use of high-field NMR spectrometers equipped with a superconducting magnet and a laser apparatus (see LASER) as a light source. Owing to the strong intensities of CIDNP signals, one can distinguish the photoreactive signals from the complex background. Employing the cyclic photoreaction of dyes (e.g. flavin) with aromatic amino acids, several dynamic structures of proteins in solution have been deduced from photo-CIDNP studies [1]. Recently [2–6] Japanese scientists have applied photo-CIDNP to the study of photo-induced electron-transfer reactions of porphyrin compounds, as simple models for the primary charge separation process of photosynthesis. In these studies the presaturation pulse technique and subtraction method were used to obtain clear CIDNP signals [7].

References

1. Müller F, Schagen CG van, Kaptein R (1980) Methods in Enzymology 66: 385
2. Maruyama K, Furuta H: Chem. Lett. 1986: 243
3. Maruyama K, Furuta H: Chem. Lett. 1986: 473
4. Maruyama K, Furuta H, Osuka A: Chem. Lett. 1986: 475
5. Osuka A, Furuta H, Maruyama K: Chem. Lett. 1986: 479
6. Maruyama K, Furuta H: Chem. Lett. 1986: 645
7. Schäublin S, Wokaun A, Ernst RR (1977) J. Magn. Reson. 27: 273

PINES's operator

The name for a so-called pulse cluster sequence used in nuclear magnetic resonance (see NMR)

In 1982, A. Pines and co-workers [1] introduced a very useful new building block for NMR pulse sequences—a cluster of four pulses which acts as a spin-inversion operator for protons remote from carbon-13, but which leaves protons directly bound to carbon-13 spins essentially unaffected:

protons $90° (X) - \tau - 180° (X) - \tau - 90° (-X)$

carbon-13 $180°$ (1)

where $^1J_{CH}$ is equal to 1/2.

PINES's operator was originally used in an ingenious scheme for recording homonuclear-decoupled proton resonance spectra [1], and then applied to two-dimensional nuclear magnetic resonance (see 2D-NMR), including homonuclear-decoupled heteronuclear shift correlation [2,3], heteronuclear J spectra without direct CH splittings [4,5], long-range heteronuclear shift correlation without peaks arising from direct bound protons [6,7], and a procedure of measuring and assigning proton-proton couplings [7].

A successful application of PINES's operator method depends on several factors. First, it is important that the radio-frequency pulses within the cluster achieve the desired rotations over the whole sample volume and over the chosen range of rf offsets. This can be realized either by a careful pulse-length calibration and strong rf fields, or by application of composite pulses (see p 51) [8]. In either case, the pulses should have the correct relative phases. Second, remote protons must undergo negligible divergences as a result of long-range CH coupling or homonuclear proton coupling during the delay 2τ. This condition is usually not restrictive because these coupling constants are normally an order of magnitude less than the direct carbon-proton coupling. The scheme should also be able to tolerate an appreciable variation in the values of $^1J_{CH}$ within a given spectrum. Since these coupling can vary over a range of 125 to 250 Hz, this is by no means a trivial matter.

Therefore, an examination of different schemes for compensating the effects of variations in $^1J_{CH}$ data while still retaining the desired discrimination between direct and long-range couplings, became necessary [9].

The results of this examination [9] are as follows:

a) With consideration of the analogy with the composite rf pulses of the form $90° (X) \, 180° (Y) \, 90° (X)$, the first designed version of PINES's operator that is compensated for variations in $^1J_{CH}$ (1) is given by Eq. (2):

protons $90° (X) - \tau/2 - 180° (Y) - \tau/2 - 90° (-Y) - \tau -$

carbon-13 $180°$

                $180° (Y) - \tau - 90° (Y) - \tau/2 - 180° (Y) - \tau/2 - 90° (X)$

                $180°$ $180°$ (2)

The sensitivity to $^1J_{CH}$ can be calculated [10] by using the product operator formalism (see POF).

b) An alternative procedure of compensating variations in $^1J_{CH}$ originates in its analogy to certain pulse sequences normally used for solvent suppression techniques (see SST). The so-called jump and return sequence (JRP), introduced by Plateau and Guéron [11], consists of two antiphase pulses separated by a short period of pulse free precession (PFP) τ, sometimes written in the notation $1\bar{1}$. Sklenář and Starčuk [12] have suggested a $1\bar{2}1$ sequence which is more tolerant to resonance offset effects and permits a better attenuation of broad flanks of the strong solvent and its spinning side-bands (see SSB). Turner [13] and Hore [14,15] have proposed the $1\bar{3}31$ pulse sequence, the next member of a family of pulse sequences with intensities following the binomial coefficients. It has been shown [9] that—for small flip angles—the $1\bar{1}$ sequence gives an excitation pattern, M_{xy}, which varies with the offset Δv as $\sin(\pi\Delta v\tau)$, whereas the corresponding excitation by a $1\bar{2}1$ sequence is flatter, varying as $\sin^2(\pi\Delta v\tau)$, while that of $1\bar{3}31$ is flatter still, following the curve $\sin^3(\pi\Delta v\tau)$.

c) There is a direct relationship to the $1\bar{1}$ solvent suppression sequence, and compensation could be achieved by expansion to the $1\bar{2}1$ sequence, given with Eq. (3):

protons $45°(X)-\tau-180°(Y)-\tau-90°(-X)-\tau-180°(Y)-\tau-45°(X)$

carbon-13 $180°$ $180°$ (3)

The free precession stages τ now involve motion of proton vectors on 45° cone in the rotating frame. Replacement of the 180°(Y) pulses by 180°(X) pulses will convert this sequence into one that inverts the local protons but leaves the remote protons unaffected. The action of (3) may be visualized in a different way by writing (3) as two consecutive so-called TANGO [6] sequences (see TANGO) (4):

protons $45°(X)-\tau-180°(Y)-\tau-45°(-X)45°(-X)-\tau-180°(Y)-\tau-45°(X)$

carbon-13 $180°$ $180°$ (4)

Further expansion of sequence (3) may be contemplated along the lines of $1\bar{3}3\bar{1}$ sequence as given by Eq. (5):

protons $22.5°(X)-\tau-180°(X)-\tau-67.5°(-X)-\tau-180°(X)-\tau-$

carbon-13 $180°$ $180°$

$67.5°(X)-\tau-180°(X)-\tau-22.5°(-X)$

$180°$ (5)

d) For general use in 2D NMR,PINES's operator 1 performs adequately. In the case of a very large spread of $^1J_{CH}$, use of sequence (3) has been recommended [9]. In some cases refocusing of the carbon-13 magnetization is required, then a modified version of (3) can be used as given by Eq. (6):

protons $135°(X)-\tau-180°(Y)-\tau-90°(-X)-\tau-180°(Y)-\tau-45°(-X)$

carbon-13 $180°$ $180°$ $180°$ (6)

A similar form of compensation is applicable to the TANGO sequence excitation (rather than spin inversion of remote protons, leaving protons on carbon-13 unaffected. This gives an excitation scheme that is compensated for moderate variations in $^1J_{CH}$:

protons $22.5°(X)-\tau-180°(Y)-\tau-45°(-X)-\tau-180°(Y)-\tau$
$-22.5°(X)$

carbon-13 $180°$ $180°$ (7)

References

1. Garbow JR, Weitekamp DP, Pines A (1982) Chem. Phys. Lett. 93: 504
2. Bax A (1983) J. Magn. Reson. 53: 517
3. Nakashima TT, John BK, McClung RED (1984) J. Magn. Reson. 59: 124
4. Bax A (1983) J. Magn. Reson. 52: 330
5. Rutar V (1984) J. Magn. Reson. 56: 87
6. Bauer CJ, Freeman R, Wimperis SC (1984) J. Magn. Reson. 58: 526
7. Wimperis SC (1984) Chemistry Part II Thesis, Oxford University
8. Lewitt MH, Ernst RR (1983) Mol. Phys. 50: 1109
9. Wimperis SC, Freeman R (1985) J. Magn. Reson. 62: 147
10. Sørensen OW, Eich GW, Lewitt MH, Bodenhausen G, Ernst RR (1983) Progr. NMR Spectrosc. 16: 163
11. Plateau P, Guéron M (1982) J. Am. Chem. Soc. 104: 7310
12. Sklenář V, Starčuk Z (1982) J. Magn. Reson. 50: 495
13. Turner DL (1983) J. Magn. Reson. 54: 146
14. Hore PJ (1983) J. Magn. Reson. 54: 539
15. Hore PJ (1983) J. Magn. Reson. 55: 283
16. Wimperis SC, Freeman R (1984) J. Magn. Reson. 58: 348
17. Kögler H, Sørensen OW, Bodenhausen G, Ernst RR (1983) J. Magn. Reson. 55: 157
18. Tycko R, Schneider E, Pines A (1984) J. Chem. Phys. 81: 680

PM-ENDOR
Acronym for **P**olarization **M**odulated RF Fields in **E**lectron **N**uclear **D**ouble Resonance

A new method in electron spin or paramagnetic resonance (see ESR or EPR), which can be used for the determination of the different tensor characters of parameters on a time-varying polarization direction of the linear-polarized radio-frequency field [1,2]. Application of the procedure was outlined recently by A. Schweiger [3].

References

1. Schweiger A, Günthard HH (1984) J. Magn. Reson. 57: 65
2. Forrer J, Schweiger A (1986) Rev. Sci. Instrum. 57: 209
3. Schweiger A (1986) Chimia 40: 111

PMFG
Acronym for **P**ulsed **M**agnetic **F**ield **G**radient (Spin-Echo Method)

This method is used in nuclear magnetic resonance (see NMR) for flow velocity measurements and has been developed by Steijskal [1] and Packer [2,3].

The principles of flow velocity measurements by NMR have been known for more than twenty years [4–6].

Applications of this noncontact technique in industry and in particular the possibility of noninvasive blood flow measurements have been the stimulus for numerous investigations over the last two decades [4,7]. In connection with the new NMR imaging techniques (see NMRI) for diagnostical purposes in medicine, the interest in NMR flow measuring methods has increased. The aim of these efforts is the visualization of blood flow by "velocity images" or flow images".

On the other hand, in the course of work on dynamics of electrolyte solutions Holz et al. [8,9] have recently developed methods for the measurement of ionic mobilities by applying an NMR pulsed gradient experiment in the presence of a electric field and thus in the presence of an electric current. Coherent motion along a magnetic field gradient produces a shifting of the phase of the spin-echo signal. This phase shift is commonly observed via the spin-echo amplitude in the PMFG experiment. Measurements of small velocities therefore require the observation of relatively small signal intensity changes. Holz and co-workers [10] have proposed a modification of the PMFG method, in which, in an off-resonance (see p 205) experiment, a time interval instead of an amplitude change is observed. This somewhat more direct determination of the phase shift results in higher accuracy or in a reduced number of signal accumulations, as tested in ^{7}Li resonance for drift velocities of lithium ions in aqueous LiCl (and LiCl plus CsCl) [10].

References

1. Stejskal EO (1965) J. Chem. Phys. 43: 3597
2. Packer KJ (1969) Mol. Phys. 17: 355
3. Packer KJ, Rees C, Tomlinson DJ (1972) Adv. Mol. Relax. Processes 3: 119
4. Jones DW, Child TF (1976) Adv. Magn. Reson. 8: 123
5. Singer JR (1978) J. Phys. E. Sci. Instrum. 11: 281
6. Stepišnik J (1985) Prog. NMR Spectrosc. 17: 187
7. Hemminga MA (1984) Biomedical magnetic resonance, James TL, Margulis A (eds) Radiology Research and Education Foundation, San Francisco, p 157
8. Holz M, Müller C (1982) Ber. Bunsenges. Phys. Chem. 86: 141
9. Holz M, Lucas O, Müller C (1984) J. Magn. Reson. 58: 294
10. Holz M, Müller C, Wachter AM (1986) J. Magn. Reson. 69: 108

PND

Acronym for **P**roton **N**oise **D**ecoupling

In carbon-13 resonance major progress was brought about in the mid-1960s with the introduction of proton noise decoupling [1], which resulted in a sensitivity gain of at least one or more order of magnitude and permitted henceforth recording of carbon-13 in the normal absorption mode. The advent of electronic storage devices roughly falls in the same period. Signal averaging procedures [2] are based on the statistic properties of so-called white noise. While coherent signals add up linearly when a number of scans or transients are accumulated, noise only increases with the square root of the number of passages.

References

1. Ernst RR (1966) J. Chem. Phys. 45: 3845
2. Ernst RR (1966) Adv. Magn. Reson. 2: 1

POF
Abbreviation of **P**roduct **O**perator **F**ormalism used in Nuclear Magnetic Resonance Spectroscopy

The product operator formalism [1,2] greatly simplifies the description of multipulse experiments in nuclear magnetic resonance spectroscopy and provides a physical picture of what is going on during the experiments on weakly coupled spin systems. Simplicity of the formalism follows from the systematic and consistent use of the product operators to represent the state of the spin system and from the few rules that govern the motion of the product operators. Nevertheless, applications of POF to spin systems of moderate complexity are cumbersome [3]. The behavior of large spin systems has been simulated numerically [4,5] but such approaches do not provide a physical insight [6]. Recently a modification of POF has been proposed [6]. In this modified formalism the single spin operators I_{px} and I_{py} of particular spins p are replaced by raising and lowering operators I_p^+ and I_p^- in the product operator base. Use of such a mixed operator base will simplify the formula for evolution of a spin system under scalar coupling interaction. Consequently, analysis of NMR experiments on large weakly coupled spin $(I = 1/2)$ systems becomes feasible in the mixed base. The modified POF procedure has been illustrated by the treatment of INEPT (see p 143) and DEPT (see p 69) experiments performed on spin systems with various hetero- and homonuclear coupling constants [6].

Reference
1. Sørensen OW, Eich GW, Levitt MH, Bodenhausen G, Ernst RR (1983) Progr. NMR Spectrosc. 16: 163
2. Ven FJM van de, Hilbers CW (1983) J. Magn. Reson. 54: 512
3. Sørensen OW, Ernst RR (1983) J. Magn. Reson. 51: 477
4. John BK, McClung RED (1984) J. Magn. Reson. 58: 47
5. Schenker KV, Philipsborn W von (1985) J. Magn. Reson. 61: 294
6. Blechta V, Schraml J (1986) J. Magn. Reson. 69: 293

P

POMMIE
Acronym for **P**hase **O**scillations to **M**ax**im**ize **E**diting in Nuclear Magnetic Resonance (see NMR) Spectroscopy

POMMIE in its one-dimensional version was originally developed [1,2] as an alternative to the DEPT (see p 69) method [3] of *spectral editing*. There are some advantages of the POMMIE over the DEPT method [1,2], including

a) less sensitivity to radio-frequency inhomogeneity because of the more extensive phase cycling that is achievable with POMMIE,
b) better ^{13}C–H suppression, i.e. reduced "error" signals in *subspectral editing*,
c) for POMMIE, *editing* may be realized in a number of different versions, and
d) the pulse angle in DEPT is markedly sensitive to both radio-frequency homogeneity and pulse power, whereas in POMMIE the pulse phase is to a first approximation independent of these parameters.

Bulsing and Doddrell [2] have given a detailed analysis of the POMMIE experiment on the basis of an operator formalism.

Following work on two-dimensional DEPT J (CH)-resolved *NMR spectrum editing* [3], and studies of selective 2D DEPT heteronuclear shift correlation spectroscopy (see DEPT and HETCOR) by Nakashima et al. [4,5], B. Coxon has recently extended [6] the POMMIE spectrum editing method to the 2D domain, Coxon has implemented a pulse sequence for *two-dimensional POMMIE J (CH)-resolved carbon-13 NMR spectroscopy* [6] and used it to explore three methods for the automated acquisition of data for *2D NMR spectrum editing*:

In the first two methods, sets of three 2D matrices are acquired in either sequential or interleaved modes using three values, $\phi = \pi/6$, $\pi/2$ and $5\pi/6$, respectively for the phase shift of the multiple-quantum read pulse. Computation of linear combinations of these 3 data matrices yields 2D POMMIE J (CH)-resolved CH, CH_2, and CH_3 subspectra.

In the third method, these subspectra are constructed directly during acquisition by rotation of the phase shifts of the multiple-quantum read pulse and the receiver.

The methods have been tested on selected small peptides and carbohydrate derivatives including L-threonyl-L-valyl-L-leucine hydrochloride, *N*-formyl-L-methionyl-L-leucyl-L-phenylalanine, sucrose octaacetate, methyl 2,3-di-*O*-methanesulfonyl-α-D-glucopyranoside, and 6-deoxy-1,2:3,4-di-*O*-isopropylidene-6-phthalimido-α-D-galactopyranose [6].

The results of Coxon's studies [6] showed that—if appropriate equipment is available—the 2D POMMIE procedure is easier to implement for J(CH)-resolved spectral editing than the 2D DEPT [3–5] technique. However, these methods are comparable in their effectivity to suppress residual signals in the spectra.

For further details the interested reader is referred to the original paper from B. Coxon [6].

P

References

1. Bulsing JM, Brooks WM, Field J, Doddrell DM (1984) J. Magn. Reson. 56: 167
2. Bulsing JM, Doddrell DM (1985) J. Magn. Reson. 61: 197
3. Coxon B (1986) J. Magn. Reson. 66: 230; (1986) Magn. Reson. Chem. 24: 1008
4. Nakashima TT, John BK, McClung RED (1984) J. Magn. Reson. 57: 149
5. Nakashima TT, John BK, McClung RED (1984) J. Magn. Reson. 59: 124
6. Coxon B (1988) Magn. Reson. Chem. 26: 449

PPD

Abbreviation of **P**ure **P**hase **D**etection used in Two-dimensional Nuclear Magnetic Resonance Spectroscopy (see 2D-NMR)

Conventionally, the absolute value (or magnitude) mode has been employed for the display of 2D-NMR spectra because of the possibility of relatively simple automation of calculation without phase correction. But, there are two important disadvantages:

1. The long tail of a signal impairs the inherent spectral resolution.
2. The relative sign of a spin-spin coupling constant and the sign of the nuclear OVERHAUSER effect (NOE) cannot be recognized.

To avoid both these disadvantages, phase sensitive detection was already known to be superior but it has not been widely used due to the problems with phase mixing of absorption and dispersion.

Recently the NMR group of JEOL Ltd. has described a method to represent pure phase spectra in the four quadrants [1]. This was originally proposed for 2D-NOE (see p 2) by States and co-workers [2]. Two other methods recently proposed by Marion et al. [3] (the so-called time proportional phase increment) and by Nagayama [4] (the so-called time reversal/frequency inversion) will not be described here, but the reader will find the corresponding explanations under the acronyms TPPI and TRFI in this dictionary. From the standpoints of final performance—sensitivity and data handling—these three methods are almost equivalent [5].

References

1. Jeol News 22a: 14 (1986)
2. States DJ, Haberkorn RA (1982) J. Magn. Reson. 48: 286
3. Marion D, Wüthrich K (1983) Biochem. Biophys. Res. Commun. 113: 967
4. Nagayama K (1986) J. Magn. Res. 66: 240
5. Keeler J, Neuhaus D (1985) J. Magn. Reson. 63: 454

Pre-TOCSY

is the name for a new experiment for obtaining complete two-dimensional proton nuclear magnetic (see 2D-NMR) of proteins in aqueous solution.

Pre-TOCSY [1] has been proposed to run complete 2D-NMR spectra with water-suppression by presaturation (see SST) of proteins in aqueous solution. In comparison to other techniques [2,3] pre-TOCSY can also be used with NOESY (see p 200), where so far there is no alternative approach except for the temperature variation [4], if applicable.

The pre-TOCSY experiments start with the usual saturation of the water resonance by selective irradiation during the relaxation delay, which is followed by the first non-selective pulse of the 2D pulse sequence. For protein signals near the water resonance which have been saturated, this non-selective pulse does not generate transverse magnetization. Then follows the pre-TOCSY pulse sequence, which restores some of the bleached magnetization via transfer from scalar-coupled spins. Neglecting incoherent magnetization transfer (see MT) pathways during the TOCSY (see p 284) period, no magnetization is exchanged except between scalar-coupled spins, so that the water magnetization is not restored. The rest of the 2D pulse sequences is the same as usual. While in principle, different pulse schemes might be applied for the intended purpose, the most efficient MT is achieved with one of the isotropic mixing sequences published by Braunschweiler and Ernst [5] or with the MLEV-17 (see MLEV) sequence by Bax and Davies [6].

References

1. Otting G, Wüthrich K (1987) J. Magn. Reson. 75: 546
2. Otting G, Wüthrich K (1986) J. Magn. Reson. 66: 359
3. Zuiderweg ERP (1987) J. Magn. Reson. 71: 283
4. Wüthrich K, Wider G, Wagner G, Braun W (1982) J. Mol. Biolog. 155: 311
5. Braunschweiler L, Ernst RR (1983) J. Magn. Reson. 53: 521
6. Bax A, Davis DG (1985) J. Magn. Reson. 65: 355

PROGRESS
Acronym for **P**oint-Resolved **Ro**tating **G**radient Surface-Coil Spectroscopy

PROGRESS is a new proposal to achieve efficient three-dimensional spatially localized *in-vivo* phosphorus-31 nuclear magnetic resonance (see NMR) spectroscopy by using multi-dimensional spatially selective pulses [1].

Reference
1. Bottomley PA, Hardy CJ (1987) J. Magn. Reson. 74: 550

PRONMR
Abbreviation of **Pro**cessing **N**uclear **M**agnetic **R**esonance Data

The computer program PRONMR 1 and 2 has been developed for the reduction of data from one- and two-dimensional Fourier transform nuclear magnetic resonance [1]. This software system is written in FORTRAN 77 (see FORTRAN) and is largely machine-independent. Implemented PRONMR software features are various types of apodization of FID's (see FID), routines for one- and two-dimensional Fourier transformations (see FT and FFT), phase corrections and algorithms for rapid quantization of nuclear magnetic resonance spectral features, including the presentation of data coming from two-dimensional experiments. Implementation of other existing NMR software for the analysis of NMR spectra on external computers, such as LAME or DAVINS (see p 65), the various types of programs for dynamic NMR spectroscopy (see DNMR) in such a largely machine-independent software system is possible [1].

Details of the two programs are given below:

1D data processing
Data stemming from 1D spectra can be processed on a VAX computer with PRONMR 1 software [2]. Weighting [3] of the FID's is the first step. Conventional operations such as exponential or GAUSSian multiplication, as well as convolution difference techniques (see CD), are possible. Fourier transformation of data detected in the quadrature mode (see QPD) is carried out in floating point arithmetic using the Cooley-Tukey algorithm [6]. Pure absorption mode peaks are obtained by phasing. A special subroutine for zero- and first-order phase corrections can be used. First, the intensity of the most intensive signal is optimized. This signal can be identified via the magnitude spectrum. Subsequently, the zero-order phase angle is obtained and the signal of the absorption line is displayed on the screen of the terminal. Additional corrections can be performed by a non-FORTRAN standard subroutine, if necessary. In the following step, the total spectrum is corrected with the zero-order phase angle. Finally, the frequency-dependent phase correction is carried out at a remote resonance in a similar way. Phase correction can be performed by alphanumeric input of zero- and first-order phase angles for ambient spectral ranges, too. Baseline corrections of the spectrum in the frequency domain can be carried out, and individual peaks can be eliminated by subtraction of LORENTZian profiles. This may become relevant for suppression of impurities, when NMR spectra analysis is performed with so-called automatic programs such as DAVINS (see p 65). Difference and subspectral editing of spectra are also possible.

2D data processing

Apodization in the F_1 and F_2 dimensions can be carried out with different functions such as exponential, GAUSSian, pseudo-echo, sine bell and shifted sine bell [3,4]. In the current version of PRONMR 2 software, a maximum of 1024 data points in the F_1 dimension and files of 32 K or even larger in the F_2 dimension can be FOURIER transformed. The signals of the two-dimensional matrix are presented in the absolute value mode. Although a white-wash routine and display of representative 2D stacked plots on a plotter or terminal is available, the analysis of the two-dimensional data matrices is done most economically with contour plots [5].

References

1. Benn R, Klein J (1986) Magn. Reson. Chem. 24: 638
2. Benn R, Klein J: Magn. Reson. Chem. (to be published)
3. Lindon JC, Ferrige AG (1980) Prog. Nucl. Magn. Reson. Spectrosc. 14: 27
4. a) DeMarco A, Wüthrich K (1976) J. Magn. Reson. 24: 201
   b) Bax A, Freeman R, Morris GA (1981) J. Magn. Reson. 43: 333
5. Hester RK, Ackerman JL, Neff BL, Waugh JS (1976) Phys. Rev. Lett. 36: 1081
6. Cooley JW Tukey JW (1965) Math. Comput. 19: 297

PSCE
Acronym for **P**artially **S**pin **C**oupled **E**cho

PSCE in one of the various multi-pulse techniques for the so-called *spectral editing* of carbon-13 nuclear magnetic resonance (see NMR) spectra [1,2]. Snape et al. [3–7] used the PSCE method to identify and quantify the different carbon types in coal liquefraction products. The theory and advantages of this and other multi-pulse techniques used in NMR spectroscopy have been discussed in detail in the reviews written by Turner [8] and by Benn and Günther [9].

P

References

1. Le Cocq C, Lallemand J-Y: J. Chem. Soc., Chem. Commun. 1981: 150
2. Brown DW, Nakashima TT, Rabenstein DL (1981) J. Magn. Reson. 45: 302
3. Snape CE (1982) Fuel 61: 775
4. Snape CE (1982) Fuel 61: 1165
5. Snape CE (1983) Fuel 62: 621
6. Snape CE (1983) Fuel 62: 989
7. Snape CE, Marsh MK (1985) Prepr. Pap. Am. Chem. Soc., Div. Pet. Chem. 30: 247
8. Turner CJ, (1984) Prog. Nucl. Magn. Reson. Spectrosc. 16: 311
9. Benn R, Günther H (1983) Angew. Chem., Int. Ed. Engl. 22: 350

Pseudo-COSY (or ψ-2D)
The abbreviation for Two-Dimensional (2D) Spectroscopy without an Evolution Period used in Nuclear Magnetic Resonance (see NMR)

At the present time the most useful of two-dimensional experiments may well be homonuclear correlation spectroscopy (COSY), filtered through double-quantum coherence [1], and displayed in the phase-sensitive mode [2,3].

Remember, that the essentials of 2D spectroscopy are the incorporation of a variable evolution period (t_1) and the subsequent FOURIER transformation (see FT and FFT) of the free induction decays (see FID) $S(t_1) \rightarrow S(F_1)$.

In the past, several authors [4–10] have used semi-selective pulses for COSY experiments, usually exciting an entire multiplet at a time, under retainment of the essentials of 2D-NMR as mentioned above.

Recently [11] R. Freeman and co-workers have proposed replacing t_1 incrementation by F_1 incrementation, performing only *one stage* FT $S(t_2) \rightarrow S(F_2)$. In order to distinguish this from real 2D-FT spectroscopy they named this procedure Pseudo-Cosy, ψ-2D, or ψ-COSY [11]. They have demonstrated that this apparently pedestrian mode of operation can have some important practical advantages. For further details the reader is referred to Ref. [11].

References

1. Piantini U, Sørensen OW, Ernst RR (1982) J. Am. Soc. 104: 6800
2. States DJ, Haberkorn RA, Ruben DJ (1982) J. Magn. Reson. 48: 286
3. Marion D, Wüthrich K (1983) Biochem. Biophys. Res. Commun. 113: 967
4. Pei FK, Freeman R (1982) J. Magn. Reson. 48: 519
5. Bauer C, Freeman R, Frenkiel T, Keeler J, Shaka AJ (1984) J. Magn. Reson. 58: 442
6. Canet D, Brondeau J, Marchal J-P, Nery H (1979) J. Magn. Reson. 36: 35; Brondeau J, Canet D (1982) J. Magn. Reson. 47: 159
7. Oschkinat H, Freeman R (1984) J. Magn. Reson. 60: 164
8. Kessler H, Oschkinat H, Griesinger C, Bermel W (1986) J. Magn. Reson. 70: 106
9. Brüschweiler R, Madsen JC, Griesinger C, Sørensen OW, Ernst RR (1987) J. Magn. Reson. 73: 380
10. Cavanagh J, Waltho JP, Keeler J (1987) J. Magn. Reson. 74: 386
11. Davies S, Friedrich J, Freeman R (1987) J. Magn. Reson. 75: 540

PSFT

P

Acronym for **P**rogressive-**S**aturation **F**ourier **T**ransform

PSFT is one of the common techniques for the measurement of spin-lattice relaxation times (T_1) in the field of nuclear magnetic resonance [1]. If a spin system is subjected to a set of repetitive 90° pulses, a dynamic equilibrium will eventually be established in which the saturation effect of the radio-frequency pulses and that of the relaxation balance each other. Provided that magnetization is completely transverse at the end of the pulse duration time and exclusively longitudinal just at the moment the pulse is initiated, the spin dynamics involved are governed by Eq. (1). The initial conditions is such a case demand that $M_z(0) = 0$ and therefore $A = M_\infty$, which leads to Eq. (2).

$$M_\infty - M_z = Ae^{-t/T_1} \tag{1}$$

$$M_\infty - M_z = M_\infty e^{-t/T_1} \tag{2}$$

The so-called steady-state situation is usually realized after three or four pulses, which means that the first few free induction decays (see FID) should be disregarded. The typical sequence is a one-pulse sequence

$$(90° - t)_n .$$

Apart from this, the experiment requires no other or additional provisions and the spectra are collected by simply varying the pulse interval, t. The data for T_1 are normally determined in the same manner described for the inversion-recovery experiment (see IRFT). The progressive-saturation technique lends itself in particular to situations where, for S/N reasons (see SNR), a large number of transients are to be collected coherently. The procedure, however is restricted to T_1-values that are not essentially shorter than the used acquisition time, which sets a lower limit to the pulse interval, t. PSFT-experiments will put more stringent demands on the instrumentation of a NMR spectrometer [2] and are thus more sensitive to systematic errors.

References

1. Freeman R, Hill HDW (1971) J. Chem. Phys. 54: 3367
2. Freeman R, Hill HDW, Kaptein R (1972) J. Magn. Reson. 7: 82

PD

Abbreviation of **P**ulse **D**elay

PD stands for the waiting time (normally seconds) after the end of the acquisition time and before starting the next radio-frequency pulse with given pulse width (see PW) in nuclear magnetic resonance (see NMR) as obtained under a pulsed FOURIER transform (see PFT) experiment. Sometimes, instead of PD the terminus relaxation delay, abbreviated RD. has been used.

PFA

Abbreviation of **P**ulse **F**lip **A**ngle

PFA stands for the angle (measured in degrees or radians) through which the magnetization is rotated by a radio-frequency pulse (such as a $90°$ pulse or $\pi/2$ pulse) used in nuclear magnetic resonance (see NMR) under conditions of pulsed FOURIER transform (see PFT).

P

PW

Abbreviation of **P**ulse **W**idth

PW stands for the duration time (normally microseconds) of a radio-frequency pulse used in nuclear magnetic resonance (see NMR) under conditions of pulsed FOURIER transform (see PFT).

PEM and **PZT**

Abbreviation of **P**hotoelastic **M**odulator and of **P**iezoelectric **T**ransducers

Both elements are used in the field of infra-red (see IR and FT-IR) spectroscopy. For example, these elements were applied successfully for the measurement of vibrational circular dichroism (see VCD) spectra on a grating spectrometer in (1974) [1]. In this and subsequent [2] experiments, radiation emerging from the exit slit of a monochromator was first linearly polarized and then passed into a *photoelastic modulator* (PEM) to create circularly polarized radiation in the same manner originally suggested by Grosjean and Legrand [3,4]. This device consists of an isotropic optical element—ZnSe is commonly used in the infra-red region—which is periodically compressed and expanded by one or more *piezoelectric transducers* (PZT). The beam emerging from the PEM is passed through the sample and is then focused on a photo-detector. The VCD spectrum is measured as the monochromator is scanned.

The same type of experiment can be performed on an FT-IR instrument. In this case, FELGETT's and JACQUINOT's advantages (see p 152) should improve the signal-to-noise ratio (see SNR) above the best attainable with a monochromator. The radiation from the source is modulated by a MICHELSON interferometer at frequencies between $2v\bar{v}_{max}$ and $2v\bar{v}_{min}$, the so-called FOURIER frequencies (see FF). Because the radiation is modulated twice, once by the interferometer and again by the PEM, this is sometimes referred to as *double-modulation* spectrometry. The frequency of the PEM should be at least 10 times higher than the highest FF [5], and it is preferable to separate them by an even greater amount.

Nafie et al. [6] have published some excellent VCD spectra in the past and indicated that mirror velocities of 0.37 and 0.185 cm sec^{-1} were used, so that the FF for $v\bar{v}_{max} \sim 4000$ cm^{-1} were 3.0 and 1.5 kHz, respectively.

References

1. Holzwarth G, Hsu EC, Mosher HS, Faulkner TR, Moskowitz A (1974) J. Am. Chem. Soc. 96: 251
2. Nafie LA, Keiderling TA, Stephens PJ (1976) J. Am. Chem. Soc. 98: 2715
3. Grosjean M, Legrand M (1960) C.R. Acad. Sci. (Paris) 251: 2150
4. Velluz L, Grosjean M, Legrand M (1965) Optical circular dichroism, Academic, New York
5. Nafie LA, Diem M (1979) Appl. Spectrosc. 33: 130
6. Nafie LA, Lipp ED, Zimba CG (1981) Proc. Soc. Photo-Opt. Instrum. Eng. 289: 457

P

Q-CPMG
Abbreviation of **Q**uadrupole Echo-CPMG (**C**ar-**P**urcell-**M**eiboom-**G**ill)

Q-CPMG is a variant of CPMG (see p 56) used in nuclear magnetic resonance (see NMR). The Q-CPMG pulse sequence $90_x^\circ - t - (90_y^\circ - 2t)_n$ may be used to investigate *slow motions in deuterium NMR* [1]. It has been shown recently [2] that CPMG and Q-CPMG sequences give qualitatively similar responses to the effects of finite pulse power, but, the Q-CPMG sequence is only half as sensitive to those as the CPMG sequence. Therefore it should be valid to use recommendations regarding CPMG sequences in the quadrupolar case. In particular, the recommendation of acquiring echoes over at least one decade of delay times [3] seems to be a good one in order to ensure that the effects of even/odd echo-amplitude alternation have been canceled out. One effect which has not yet been discussed is the obvious weighting of the echo-amplitude toward small ω_Q because of the "rolloff" of the radio-frequency pulse response at large ω_Q. The possibility of using ω_Q-compensated pulses [4] in this regard is currently under investigation.

References
1. Bloom M, Stermin E (1987) Biochemistry 26: 2101
2. Pratum TK (1988) J. Magn. Reson. 78: 123
3. Vold RL, Vold RR, Simon HE (1973) J. Magn. Reson. 11: 283
4. Diminovich DJ, Raleigh DP, Olejniczak ET, Griffin RG (1986) J. Chem. Phys. 84: 2556

QE
Abbreviation of **Q**uadrupole-**E**cho Spectroscopy for Deuterium

QE spectroscopy has become a powerful method in nuclear magnetic resonance of deuterium. In recent years deuterium lineshape studies have become the almost important source of information about dynamic processes in molecular crystals [1,2], liquid crystals [3], synthetic polymers [4,5], and lipids [6,7]. This is due to a fruitful combination of technical advances such as the development of QE [8] techniques and efficient computational approaches [9–12] derived from a stochastic LIOUVILLE's equation. The most ambitious deuterium lineshape calculations involve matrices of several hundred dimensions and are best executed on modern supercomputers, while the simple case of jumps between two sites can readily be carried out by a minicomputer. Recently [13] an American group of scientists has focused its interest on a class of exchange problems which is intermediate in complexity between the above mentioned two extremes, with special emphasis on methods for enhancing the computational efficiency.

Spiess and co-workers have achieved a two-dimensional deuterium spectroscopy much more recently [14].

Deuterium relaxation studies on lipids with incorporated protein have shown a substantial lowering of the chain mobility of lipids by the protein [15].

The generation of special aggregates in phospholipid double layers from two different components has been detected via deuterium resonance, too [16].

References
1. Spiess HW (1978) Rotation of molecules and nuclear spin relaxation, Springer, Berlin, Heidelberg, New York, p 55 (NMR basic principles and progress, vol 75)

2. M. Mehring (1983) Principles of High Resolution NMR in Solids, Springer, Berlin, Heidelberg, New York
3. Meier P, Ohmes E, Kothe G, Blume A, Weidner J, Eibl H-J (1983) J. Phys. Chem. 87: 4904
4. Spiess HW (1983) Colloid Polym. Sci. 261: 193
5. Hentschel D, Sillescu H, Spiess HW (1984) Polymer 25: 1078
6. Griffin RG (1981) Methods Enzymol. 72: 108
7. Davis JH (1983) Biochim. Biophys. Acta 736: 177
8. Bloom M, Davis JH, Valic MI (1980) Can. J. Phys. 58: 1510
9. Alexander S, Baram A, Luz Z (1974) Mol. Phys. 27: 441
10. Schwartz LJ, Meirovitch E, Ripmeester JA, Freed JH (1983) J. Phys. Chem. 87: 4453
11. Moro G, Freed JH (1981) J. Chem. Phys. 74: 3757
12. Campbell RF, Meirovitch E, Freed JH (1979) J. Phys. Chem. 83: 525
13. Greenfield MS, Ronemus AD, Vold RL, Vold RR, Ellis PD, Raidy TE (1987) J. Magn. Reson. 72: 89
14. Spiess HW (1987) Phys. Bl. 43: 233
15. Meier P, Sachse J-H, Brophy PJ, Marsh D, Kothe G (1987) Proc. Natl. Acad. Sci. USA 84: 3704
16. Hübner W, Blume A (1987) Ber. Bunsenges. Phys. Chem. 91: 1127

QF
Acronym for **Q**uadrupole-**E**lectric **F**ield **G**radient

In the field of nuclear magnetic relaxation we find a lot of different mechanisms operating [1]. One of them is the relaxation by quadrupole-electric field gradient (QF) interaction. The QF relaxation rate is given by Eq. (1), valid in the case of the "extreme narrowing" condition ($\omega_o \cdot \tau_c \ll 1$).

$$1/T_1^{QF} = 1/T_2^{QF} = 3/10 \; \pi^2 \frac{2I+3}{I^2(2I-1)} C_{QF}^2 \left(1+\frac{\eta^2}{3}\right)\tau_c \tag{1}$$

Meaning of the symbols in Eq. (1):

$$C_{QF} = \frac{e^2 q_{zz} Q}{h} = \text{"quadrupole coupling constant", given in Hz;}$$

q_{zz} = principal component of the field gradient tensor, sometimes simply called "*the electric field gradient*";

η = asymmetry parameter for the electric field gradient;

Q = nuclear electric quadrupole moment, given in 10^{-24} cm^2;

τ_c = molecular reorientational correlation time.

The remaining symbols of Eq. (1) have their usual meaning.

Reference
1. Wehrli FW, Wirthlin T (1976) Interpretation of carbon-13 NMR spectra, Heyden, London, Chap 4

QPD
Abbreviation of **Q**uadrature **P**hase **D**etection

QPD is the almost commonly used detection method in nuclear magnetic resonance spectroscopy (see NMR) with modern spectrometers [1]. QPD has a

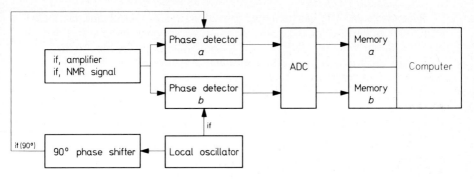

Fig. 24. Scheme of QPD. ADC stands for analog-to-digital converter and if for intermediate-frequency

lot of advantages over the older simple phase detection. Figure 24 demonstrates QPD schematically.

Reference
1. Stejskal EO, Schaefer J (1974) J. Magn. Reson. 14: 160

QSEX
Acronym for **Q**uadrature **S**elective **Ex**citation

In 1978 QSEX was proposed as an alternative to normal excitation in nuclear magnetic resonance (see NMR) spectroscopy [1].

Reference
1. Morris GA, Freeman R (1978) J. Magn. Reson. 29: 433

Q

QUAT
Abbreviation of **Qua**ternary-Only Carbon Spectrum

A number of nuclear magnetic resonance (see NMR) multiple-pulse techniques have been developed whereby the different carbon types within a complex molecule or mixture can be readily identified via spectral editing.

These different pulse sequences can be classified into spin-echo (see SE) and polarization transfer (see PT and MT) techniques. Most important in the field of SE techniques are the attached proton test (see APT), the gated spin-echo (see GASPE), the part-coupled spin-echo (see PSCE), and the quaternary-only carbon-13 spectrum, introduced by a group of scientist in 1982 and abbreviated as QUAT [1].

Although the SE and PT procedures give comparable information in regards to carbon-species identification, they differ in theory and in complexity of operations for correct quantitative determination of different carbon types. The theory and advantages of these and other multiple-pulse techniques have been discussed in reviews written by Turner [2] and by Benn and Günther [3].

Recently Netzel [4] has presented a combination of QUAT and the DEPT technique (see DEPT) for the quantitation of carbon types in fossil-fuel-derived oils. He found that the DEPT/QUAT technique is preferable to the most commonly used quantitative SE experiments because of time sharing for data acquisition and reduction.

References

1. Bendall MR, Pegg DT, Doddrell DM, Johns SR, Willing RI: J. Chem. Soc., Chem. Commun. 1982: 1138
2. Turner CJ (1984) Prog. Nucl. Magn. Reson. Spectrosc. 16: 311
3. Benn R, Günther H (1983) Angew. Chem., Int. Ed. Engl. 22: 350
4. Netzel DA (1987) Anal. Chem. 59: 1775

QWIKCORR

Artificial name for a special pulse sequence used in nuclear magnetic resonance (see NMR)

A two-dimensional carbon-13/proton correlated spectrum obtained under using the pulse sequence developed by Reynolds et al. [1] has been named as QWIKCORR by Krishnamurthy and Casid recently [2].

References

1. Reynolds WF, Hughs DW, Perpick-Dumont M, Enriquez RG (1985) J. Magn. Reson. 64: 304
2. Krishnamurthy VV, Casid JE (1987) Magn. Reson. Chem. 25: 837

Q

RAMAN Spectroscopy

is based on inelastic light scattering predicted theoretically by SMEKAL in 1923 [1,2] and detected experimentally five years later by RAMAN [3,4]. The inelastic light scattering has been called therefore the *SMEKAL-RAMAN effect* [2], the spectroscopy based on this effect was given the name RAMAN spectroscopy.

RAMAN spectroscopy is, from the experimental point of view, a spectroscopic method operating in the region of visible light (see VIS)—but not confined to it— and belongs mechanistically clear to infrared (see IR) or virbrational spectroscopy.

Selection rules met regarding the change in dipole moments (IR) or polarizabilities (RAMAN) made these spectroscopies complementary to each other.

In the field of polymer spectroscopy, RAMAN spectroscopy was applied sucessfully after the introduction of laser (see LASER) excitation, which removes some of the difficulties connected with strong elastic light scattering of polymer samples, using the intensive and highly monochromatic radiation which is typical of lasers [5–7]. However, in some early work, RAMAN spectra obtained using classical excitation sources were used to verify STAUDINGER's theory of linear macromolecules [8,9].

It has been demonstrated that when a Nd: YAG laser is applied to excite the RAMAN effect, fluorescence is eliminated. This results from the absence of electronic transitions at the Nd: YAG laser wavelength of 1.06 micrometers. A further advantage comes from the fact that at these relatively low photon energies, sample heating, and subsequent photodegradation is unlikely. FT-RAMAN spectroscopy (see FT-RS) in the near-infrared (see NIR) presents some interesting challenges in addition to the advantages discussed above. As is well-known, RAMAN lines have an intensity proportional to the 4th power of the exciting frequency. Comparison of the excitation with a conventional argon-ion laser at 488 nm to that of the Nd: YAG line at 1.06 μm shows that the anticipated decrease in sensitivity is a factor of 22.6 with the Nd: YAG laser. The so-called *noise equivalent power* (see NEP) of NIR detectors is normally several orders of magnitude higher than that of the photo-multiplier tubes used in conventional RAMAN spectroscopy. In FT-RAMAN spectroscopy (see FT-RS), there is a multiplex disadvantage, since statistical noise of the exciting radiation scattered onto the detector is transformed into noise at all frequencies of the RAMAN spectrum. But, recently a lot of problems have been solved and efficient FT-RAMAN accessories have been constructed which allow RAMAN measurements in the NIR region with FT-spectrometers to obtain spectra with reasonable signal-to-noise ratios (see SNR). The new technique is, in particular, suitable for samples which cannot be studied by conventional RAMAN spectroscopy due to fluorescence problems.

References

1. Smekal A (1923) Naturwissenschaften 11: 873
2. Kohlrausch KWF (1931) Der Smekal-Raman-Effekt, Springer, Berlin
3. Raman CV, Krishnan KS (1928) Nature 121: 501
4. Long DA (1977) Raman spectroscopy, McGraw-Hill, New York
5. Hendra D (1974) In: Hummel DO (ed) Polymer spectroscopy, Verlag Chemie, Weinheim, p 151

6. Cutler DJ, Hendra PJ, Fraser G (1980) In: Dowkins PV (ed) Developments in polymer characterisation, Applied Science Publ., vol 2, chap 3
7. Lascombe J, Huong PV (eds) (1982) Raman Spectroscopy—Linear and Nonlinear, Proc. of the 8th Internatl. Conf. on Raman Spectroscopy, Bordeaux/France, Wiley and Heyden, Chichester (London)
8. Signer R, Weiler J (1932) Helv. Chim. Acta. 15: 649
9. Mitushima S-I, Morino Y, Inoue Y (1937) Bull. Chem. Soc. (Japan) 12: 136

RAYLEIGH's criterion

has been used to define the resolution of an infrared (see IR) spectrometer.

Originally RAYLEIGH's criterion was used to define the resolution obtainable from a diffraction-limited grating spectrometer, the *instrument line shape* (see ILS) of which may be represented by a function of the form $sinc^2x$. By RAYLEIGH's criterion, two adjacent spectral lines of equal intensity, each with a $sinc^2x$ ILS, are considered to be just resolved when the center of the one line is at the same frequency as the first zero value of the ILS of the other one. Under this condition, the resultant curve has a dip of approximately 20% of the maximum intensity in the spectrum. However, if the same criterion is applied to a line having a sinc x ILS, it is found that the two lines are not resolved.

Another resolution criterion in IR spectroscopy is the so-called *full width at half-height* (see FWHH) criterion.

RCT

Acronym for **R**elayed **C**oherence **T**ransfer used in NMR

This method has been employed to establish the remote connectivity between nuclei which are themselves not directly coupled, but which are coupled to a common spin partner. The technique has been used successfully to find out long-range connectivities in typical homonuclear proton spin systems [1–5] and in spin systems involving heteronuclei, too: $X = {}^{31}P$, ${}^{13}C$, or ${}^{15}N$ [6–13]. RCT is generally considered as a two-dimensional method unless some special forms of selective excitation can be used in order to isolate a single nucleus in a spin system of interest [13,14].

References

1. Eich G, Bodenhausen G, Ernst RR (1982) J. Am. Chem. Soc. 104: 3731
2. Wagner G (1983) J. Magn. Reson. 55: 151
3. Braunschweiler L, Ernst RR (1983) J. Magn. Reson. 53: 521
4. Oschkinat H, Freeman R (1984) J. Magn. Reson. 60: 164
5. Macura S, Kumar NG, Brown LR (1984) J. Magn. Reson. 60: 99
6. Bolton PH (1982) J. Magn. Reson. 48: 336
7. Bolton PH, Bodenhausen G (1982) Chem. Phys. Lett. 89: 139
8. Bax A (1983) J. Magn. Reson. 53: 149
9. Kessler H, Bernd M, Kogler H, Zarbock J, Sørensen OW, Bodenhausen G, Ernst RR (1983) J. Am. Chem. Soc. 105: 6944
10. Kogler H, Sørensen OW, Bodenhausen G, Ernst RR (1983) J. Magn. Reson. 55: 157
11. Delsuc MA, Guittet E, Trotin N, Lallemand JY (1984) J. Magn. Reson. 56: 163
12. Neuhaus D, Wider D, Wagner G, Wüthrich K (1984) J. Magn. Reson. 57: 164

13. Field LD, Messerle BA (1985) J. Magn. Reson. 62: 453
14. Eich G (1982) Diploma Thesis, ETH Zurich

RELAY

Artificial name for special homonuclear and heteronuclear correlation
experiments in nuclear magnetic resonance (see NMR)

RELAY experiments [1,2] became very important in establishing *neighboring
relations* between protons and heteronuclei such as carbon-13, phosphorus-31,
etc. as shown in Fig. 25.

   Principal aspects of RELAY experiments may be illustrated by a scheme (see
Fig. 26) for a structural fragment of three carbons and protons represented by

$$C(1)H(a) - C(2)H(b) - C(3)H(c) .$$

Part (a) of Fig. 27 describes the pulse sequence used in the heteronuclear version of
the RELAY experiment [2] whereas part (b) of this figure represents a pulse
sequence of a homonuclear RELAY experiment [1]. Variations of these pulse
sequences were published later on [3].

   It must be pointed out that the most attractive aspect of RELAY experiments
is indeed the possibility of detecting carbons in their direct connectivity, but, the
poor sensitivity in comparison to other common methods may be a severe
limitation.

   Practical applications in the field of heteronuclear experiments have been
increasing [3,4].

   In the field of homonuclear spin systems RELAY experiments have received
more interest, too [5–7].

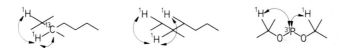

Fig. 25. Neighboring relations which can be manifested using RELAY experiments

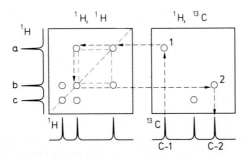

Fig. 26. RELAY experiment of the fragment $C(1)H(a) - C(2)H(b) - C(3)H(c)$

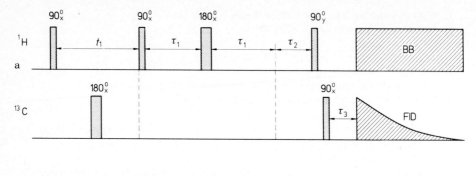

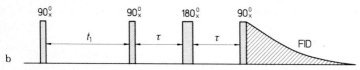

Fig. 27. Pulse sequences for heteronuclear RELAY experiments; BB = broad-band decoupling; FID = free induction decay

In a recent communication, Otter et al. [8] have reported the appearance of unexpected "cross-peaks" in homonuclear RELAY experiments. This paper prompted another group of scientists to study carefully strong coupling effects in the homonuclear RELAY experiment [9], with application to leucine spin systems of octanoyl-acyl carrier protein.

References

1. Eich G, Bodenhausen G, Ernst RR (1982) J. Am. Chem. Soc. 104: 3731
2. Bolton PH, Bodenhausen G (1982) Chem. Phys. Lett. 89: 139; Bolton PH (1982) J. Magn. Reson. 48: 336
3. Kessler H, Bernd M, Kogler H, Zarbock J, Sørensen OW, Bodenhausen G, Ernst RR (1983) J. Am. Chem. Soc. 105: 6944
4. Bigler P, Ammann W, Richarz R (1984) Org. Magn. Reson. 22: 109; Turner CJ (1984) Org. Magn. Reson. 22: 531; Musmar MJ et al. (1985) J. Heterocycl. Chem. 22: 219
5. Wesener JR, Schmitt P, Günther H (1984) Org. Magn. Reson. 22: 468; Wesener JR, Günther H (1985) J. Am. Chem. Soc. 107: 1537
6. Wagner G (1983) J. Magn. Reson. 55: 151; Bax A, Drobny G (1985) J. Magn. Reson. 61: 306; Fesik SW, Perun TJ, Thomas AM (1985) Magn. Reson. Chem. 23: 645
7. Morris GA, Richards MS (1985) Magn. Reson. Chem. 23: 676
8. Otter A, Kotovych G (1986) J. Magn. Reson. 69: 187
9. Kay LE, James P-J, Prestegard JH (1987) J. Magn. Reson. 72: 392

reverse POMMIE

Acronym for **R**everse **P**hase **O**scillations to **M**aximize **E**diting in Nuclear Magnetic Resonance (see NMR)

J.M. Bulsing and D.M. Doddrell [1] have presented a detailed study about the utility of *reverse polarization pulse trains* (or sequences) with particular interest on their potential for solvent suppression (see SST) [2–14].

They described briefly the various pulse trains for reverse polarization transfer terms of a simplified operator formalism introduced in former papers [15,16] and showed that the *reverse INEPT sequence* (see INEPT) has a built-in self-compensatory characteristic which makes it to the favoured procedure. As an example of biological interest, the metabolism of ^{13}C-labeled acetaldehyde by alcohol dehydrogenase from *Drosophila melanogaster* has been studied successfully.

The demands of *spectral editing* suggest the trial of a third type of pulse train which they have dubbed *revrse POMMIE* [1]. For comparison, the interested reader is referred to the acronym POMMIE in this dictionary.

Reverse POMMIE is indeed closely related to DEPT (see there) in its structure but it utilizes more fully high-order coherence. This allows spectral editing to become dependent on pulse phase rather than on pulse angle. One possible *reverse POMMIE pulse train* was discussed [9] and its successful application to the formaldehyde detoxification in *Escherichia coli* was demonstrated in the same year [17].

But, in general, *reverse POMMIE pulse trains* take the form given by Eq. (1) [1].

$$\pi/2 \; [C,X] - (2J)^{-1} - \pi/2 \; [H,i] \; \pi/2 \; [H,j] \; \pi \, [C] - (2J)^{-1}$$

$$- \pi/2 \; [C,y] \ldots \ldots \pi [H, \sigma] - (2J)^{-1}$$

$$- \pi/2 \; [H, \pm k], \text{ acquire } {}^1H, \; \{\text{decouple } {}^{13}C\} \tag{1}$$

Here, in Eq. (1), a general phase σ for the proton refocusing pulse is considered, but for further details, explanations, and detailed meanings in Eq. (1) the reader is referred to the original papers [1,16].

References

1. Bulsing JM, Doddrell DM (1986) J. Magn. Reson. 68: 52
2. Maudsley AA, Ernst RR (1977) Chem. Phys. Lett. 50: 368
3. Maudsley AA, Müller L, Ernst RR (1977) J. Magn. Reson. 28: 463
4. Freeman R, Mareci TH, Morris GA (1981) J. Magn. Reson. 42: 341
5. Bendall MR, Pegg DT, Doddrell DM (1981) J. Magn. Reson. 45: 8
6. Shaka AJ, Freeman R (1982) J. Magn. Reson. 50: 502
7. Bendall MR, Pegg DT, Doddrell DM, Field J (1983) J. Magn. Reson. 51: 520
8. Brooks WM, Irwing MG, Simpson SJ, Doddrell DM (1984) J. Magn. Reson. 56: 521
9. Bulsing JM, Brooks WM, Field J, Doddrell DM (1984) Chem. Phys. Lett. 104: 229
10. Müller L (1979) J. Am. Chem. Soc. 101: 4481
11. Bodenhausen G, Ruben DJ (1980) Chem. Phys. Lett. 69: 185
12. Redfield AG (1983) Chem. Phys. Lett. 96: 537
13. Bax A, Griffey RH, Hawkins BL (1983) J. Magn. Reson. 55: 301
14. Live DH, Davis DG, Agosta WC, Cowburn D (1984) J. Am. Chem. Soc. 106: 6104
15. Lynden-Bell RM, Bulsing JM, Doddrell DM (1983) J. Magn. Reson. 55: 128
16. Bulsing JM, Doddrell DM (1985) J. Magn. Reson. 61: 197

R

RJCP
Acronym for **R**efocused **J** **C**ross-**P**olarization

RJCP is a variant [1] of the original J cross-polarization strategy (see JCP) [2,3] arising from algebraic analysis of several double-resonance situations in nuclear magnetic resonance (**DR** and **NMDR**).

The sensitivity dependence on Jt of the polarization transfer efficiency in the JCP experiment is significantly reduced by the RJCP method [1,4]. In this method, the spin system under investigation is made to evolve alternately under conditions of matched radio-frequency fields and deliberate mismatch. The RJCP sequence may be viewed [4] as comprising five time periods: during the first, third, and fifth periods, (t, 2t, t seconds), evolution is under matched fields (assumed to be on-resonance), while during the second period mismatch is induced by blanking one radio-frequency channel; during the fourth periods mismatch of the opposite sign is realized by blanking the other channel. Mismatch during the two periods is in fact $\pm\omega$, which is much larger than J (normally by a factor of 20 to 400) and is made operative for a time τ such that $\omega\tau = \pi/2$. Evolution under the RJCP sequence may be computed precisely from the relevant equations of motion with and without mismatch.

The key step in the RJCP experiment has been shown [4] to involve rotations in two-dimensional operator spaces, as a result of which a part of the magnetization is "frozen", leading to an overall reduced time dependence. This fact and the higher polarization transfer efficiency of RJCP make is to an attractive method [4] in NMR.

Applications of these results [4] to spectral editing, multiple-quantum spectroscopy (see MQ-NMR), and zero-quantum spectroscopy (see ZQS) are being actively pursued in Chandrakumar's laboratory [4].

References

1. Chingas GC, Garroway AN, Bertrand RD, Moniz WB (1981) J. Chem. Phys. 74: 127
2. Chandrakumar N (1985) J. Magn. Reson. 63: 202
3. Chandrakumar N, Visalakshi GV, Ramaswamy D, Subramanian S (1986) J. Magn. Reson. 67: 307
4. Chandrakumar N (in preparation) J. Magn. Reson., private communication

R

RMS
Acronym for **R**oot-**M**ean Square

In molecular spectroscopy one uses the *RMS* procedure for *error calculations* in fitting processing of data.

For example, in nuclear magnetic resonance (see NMR) spectroscopy a complex proton spectrum can be analyzed by a quantum-mechanical treatment by using computer programs (see LAOCOON and DAVINS) for an iterative parameter and spectral line fit.

The frequency deviation between the calculated and experimentally measured line may be defined as Δf. Then values of $\Delta f \cdot \Delta f$, the *squared error* in line fitting, are accumulated. The sum of all $(\Delta f)^2$ values is divided by the number of all lines involved in the process and the square root from that is extracted, to give the so-called *RMS error*. In such iterative procedures the RMS error may be seen as a control value for the "goodness" of the stepwise controlled iteration cycles.

In LAOCN 3 [1] a special subroutine controls whether further iterations are to be made by evaluating the RMS error in fitting, determining whether it has been reduced by more than 1% from the previous value, and whether the chosen maximum number of iterations have been performed [2].

The deeply interested reader is referred to a recently published monograph written by G. Hägele and his group [3] dealing with problems and solutions arising from simulation and automated analysis of nuclear magnetic resonance one-dimensional spectra.

References

1. The third revised version of LAOCOON (see p 155).
2. See comments of de Tar DF (1968) In: Computer programs for chemistry, W.A.Benjamin, New York, vol 1, chap 3
3. Hägele G, Engelhardt M, Boenigk W (1987) Simulation and automated analysis of nuclear resonance spectra, VCH Verlagsgesellschaft, Weinheim.

ROESY
Acronym for **R**otating-Frame **OVERHAUSER** **E**nhancement **S**pectroscopy

Recently, A.A. Bothner-By and co-workers proposed a new method for the measurement of homonuclear nuclear OVERHAUSER enhancement factors under spin-locked conditions, which they named somewhat strangely the CAMELSPIN (see p 37) experiment [1]. A. Bax and D.G. Davis [2] suggested the alternative name ROESY as an acronym for this experiment because of its similarity with the closely related two-dimensional NOESY experiment [3,4] (see 2D-NOE and NOESY).

The ROESY method seems to be particularly suitable for molecules that have a motional correlation time, τ_c, near to the condition $\omega\tau_c = 1$, where ω stands for the angular LARMOR frequency. In this case the laboratory-frame NOE effect is near to zero, whereas the rotation-frame NOE—under spin-locked conditions—is always positive and will increases monotonically for increasing values of τ_c, and can be significantly large under those experimental conditions [2].

Practical aspects of the two-dimensional transverse NOE spectroscopy has been analyzed: First, as in the case for the NOESY experiment, it is very advantageous to record the ROESY spectrum in the 2D absorption mode [2]. As will be shown below, there is an additional advantage to the use of absorption-mode spectra: distinction between positive and negative peaks facilitates the identification of certain types of artifacts. A two-dimensional hypercomplex FOURIER transformation (see FT), as outlined by Müller and Ernst [5] and States and co-workers [6] is conveniently used in this type of experiment [2]. A simple description of this procedure has been presented by A. Bax [7]. A minimum of four experiments with phase cycling is required for this method. In practice, however, it is much more efficient to use a 16-step experiment, which incorporates CYCLOPS phase cycling [7,8] for suppression of quadrature artifacts (see CYCLOPS). Upon application of the ROESY experiment A. Bax and D.G. Davis [2] found two different types of spurious cross peaks, both due to coherent transfer between scalar coupled spins: The first type is due to the fact that the long spin-lock pulse also acts as a mixing pulse for antiphase magnetization. The second source of artifacts arises from a Hartmann-Hahn-type spurious transfer (see HAHA and HOHAHA).

The effects of frequency offsets in ROESY have been analyzed recently and a modified pulse sequence has been proposed [9] which eliminates offset effects and improves the signal-to-noise ratio (see SNR).

In contrast to earlier expectations, it has been found that even for molecular weights as high as 7000 the ROE peak intensities can match the NOE peak intensities [10]. ROESY is therefore an attractive alternative or complement to the commonly used NOESY experiment.

But, it has been shown that the interpretation of ROESY spectra can be severely complicated by a variety of types of relayed magnetization transfer [11].

References

1. Bothner-By AA, Stephens RL, Lee J, Warren CD, Jeanloz RW (1984) J. Am. Chem. Soc. 106: 811
2. Bax A, Davis DG (1985) J. Magn. Reson. 63: 207
3. Jeener J, Meier BH, Bachmann P, Ernst RR (1979) J. Chem. Phys. 71: 4546
4. Macura S, Ernst RR (1980) Mol. Phys. 41: 95
5. Müller L, Ernst RR (1979) Mol. Phys. 38: 963
6. States DH, Haberkorn RA, Ruben DH (1982) J. Magn. Reson. 48: 286
7. Bax A (1983) Bull. Magn. Reson. 7: 167
8. Hoult D, Richards RE (1975) Proc. R. Soc. London Ser. A 344: 311
9. Griesinger C, Ernst RR (1987) J. Magn. Reson. 75: 261
10. Redwine OD, Wüthrich K, Ernst RR unpublished measurements, see Ref [1]
11. Farmer BT II, Macura S, Brown LR (1987) J. Magn. Reson. 72: 347

RPT
Abbreviation of **R**everse **P**olarization **T**ransfer

Reverse polarization transfer methods [1–13] have been used in nuclear magnetic resonance (see NMR)—besides other applications—as solvent suppression techniques (see SST). The various pulse trains available for RPT can be described in terms of a simplified operator formalism introduced by Doddrell and co-workers [14,15]. This formalism is similar to several other recent schemes [16–18] and it has been assumed [19] that such techniques are now generally accepted for modeling the evolution of spin coherence within weakly coupled systems through a multiple-pulse sequence. Recently a detailed study has been made of the utility of RPT pulse sequences with particular reference to their potential for solvent (water) suppression [19]. It has been demonstrated that the reverse INEPT sequence (see INEPT) has a built-in self-compensatory characteristic which makes this pulse sequence the favoured procedure. Water suppression factors of up to 800,000 : 1 have been achieved with dilute samples in neat H_2O. As an example of biological interest, the metabolism of ^{13}C-labeled acetaldehyde by alcohol dehydrogenase from *Drosophila melanogaster* has been studied [19]. The interested reader will find a detailed discussion of the experimental requirements in Doddrell's paper [19].

References

1. Maudsley AA, Ernst RR (1977) Chem. Phys. Lett. 50: 368
2. Maudsley AA, Müller L, Ernst RR (1977) J. Magn. Reson. 28: 463
3. Freeman R, Mareci TH, Morris GA (1981) J. Magn. Reson. 42: 341
4. Bendall MR, Pegg DT, Doddrell DM (1981) J. Magn. Reson. 45: 8
5. Shaka AJ, Freeman R (1982) J. Magn. Reson. 50: 502
6. Bendall MR, Pegg DT, Doddrell DM, Field J (1983) J. Magn. Reson. 51: 520
7. Brooks WM, Irving MG, Simpson SJ, Doddrell DM (1984) J. Magn. Reson. 56: 521
8. Bulsing JM, Brooks WM, Field J, Doddrell DM (1984) Chem. Phys. Lett. 104: 229

9. Müller L (1979) J. Am. Chem. Soc. 101: 4481
10. Bodenhausen G, Ruben DJ (1980) Chem. Phys. Lett. 69: 185
11. Redfield AG (1983) Chem. Phys. Lett. 96: 537
12. Bax A, Griffey RH, Hawkins BL (1983) J. Magn. Reson. 55: 301
13. Live DH, Davis DG, Agosta WC, Cowburn D (1984) J. Am. Chem. Soc. 106: 6104
14. Lynden-Bell RM, Bulsing JM, Doddrell DM (1983) J. Magn. Reson. 55: 128
15. Bulsing JM, Doddrell DM (1985) J. Magn. Reson. 61: 197
16. van de Ven FJM, Hilbers CW (1983) J. Magn. Reson. 54: 512
17. Packer KJ, Wright KM (1983) Mol. Phys. 50: 797
18. Sørensen OW, Eich GW, Levitt MH, Bodenhausen G, Ernst RR (1983) Progr. NMR Spectrosc. 16: 163
19. Bulsing JM, Doddrell DM (1986) J. Magn. Reson. 68: 52

R

SC
Acronym for Scalar Coupling

Relaxation by scalar coupling [1,2] is one of the four different mechanisms dominating carbon-13 spin-lattice relaxation in nuclear magnetic resonance (see NMR). This relaxation mechanism can occur if the carbon-13 nucleus is scalar coupled (with the coupling constant J) to a second spin S (with $S \geq 1/2$) and the coupling is modulated by either chemical exchange (the so-called "SC relaxation of the first kind") or the relaxation of spin S, e. g. if $S > 1/2$, (so-called "SC relaxation of the second kind"). In this case spin splittings will disappear and single lines are observed. The SC mechanism for carbon-13 may be described by (1) and (2):

$$1/T_1^{SC} = \frac{8\pi^2 J^2 S(S+1)}{3}\left[\frac{\tau_{SC}}{1+(\omega_C-\omega_S)^2\tau_{SC}^2}\right] \tag{1}$$

$$1/T_1^{SC} = \frac{4\pi^2 J^2 S(S+1)}{3}\left[\tau_{SC}+\frac{T_{SC}}{1+(\omega_C-\omega_S)^2\tau_{SC}^2}\right] \tag{2}$$

The symbols τ_{SC} and ω_S in Eqs, (1) and (2) designate the relaxation time of the interacting nucleus S and its LARMOR frequency, respectively. Note the following conditions:

$\tau_{SC}=\tau_e$, if exchange time $\tau_e \ll T_1$ of either spin ("first kind");

$\tau_{SC}=T_1^S$ (relaxation time of spin S), if $T_1^S \ll \tau_e$, $1/2\pi J$ ("second kind").

For further details and typical examples see Reference [3] and references therein.

References
1. Lyerla JR, Grant DM (1972) In: McDowell CA (ed) Int. Rev. Science, Phys. Chem. Series Medical and Technical Publishing, Chicago, vol 4, chap 5
2. Lyerla JR, Levy GC (1974) In: Levy GC (ed) Topics in carbon-13 nuclear magnetic resonance spectroscopy, Wiley-Interscience, New York, vol 1, chap 3
3. Wehrli FW, Wirthlin T (1976) Interpretation of carbon-13 NMR spectra, Heyden, London, chap 4

S

S. COSY
Acronym for Scaled Correlation Spectroscopy

Like COSS (see p 53) S. COSY is a recently proposed pulse scheme for realizing chemical shift (δ) scaling along the f_1 dimension of a two-dimensional correlated nuclear magnetic resonance (see 2D-NMR and COSY) spectrum [1,2]. Both the pulse sequences of COSS and S. COSY may be characterized as relatively simple modifications of the so-called *constant time experiment* [3,4] and have been given for resolution enhancement in COSY.

The combined use of S. COSY and COSS has been proposed much more recently [5] after a detailed analysis of the phase characteristics of both these chemical-shift scaling sequences. In obtained pure-phase shift-scaled NMR

COSY spectra, one finds a net coherence transfer (see CT) occurring between the involved spins.

The product operator formalism (see POF) [6,7] can be used as theoretical background for these techniques.

References

1. Ravikumar M, Sheth A, Hosur RV (1986) J. Magn. Reson. 69: 418
2. Hosur RV, Ravikumar M, Sheth A (1986) paper presented (p 572) at the 23rd Congress Ampère on Magnetic Resonance and Related Phenomena, Rome.
3. Bax A, Freeman R (1981) J. Magn. Reson. 44: 542
4. Rance M, Wagner G, Sørensen OW, Wüthrich K, Ernst RR (1984) J. Magn. Reson. 59: 250
5. Hosur RV, Sheth A, Majumdar A (1988) J. Magn. Reson. 76: 218
6. Sørensen OW, Eich GW, Levitt MH, Bodenhausen G, Ernst RR (1983) Prog. NMR Spectrosc. 16: 163
7. Naskashima TT, McClung RED (1986) J. Magn. Reson. 70: 187

SCS
Acronym for **S**ubstituent **C**hemical **S**hift

SCS-values measured on many monosubstituted benzenes have been well documentated for different kinds of solvents [1]. Additive use of these data leads to rough calculation or to a first estimate of carbon-13 NMR spectra of polysubstituted benzenes. In the meantime, somewhat different SCS-data have become available for naphthalenes, too [2]. Uncritical use of benzene data in the field of heteroaromatics would be dangerous. Stereochemical effects in "crowded" polysubstituted benzenes limit the additive use of the typical SCS-values from monosubstituted benzenes.

References

1. Ewing DF (1979) Org. Magn. Reson. 12: 499
2. Hansen PE (1979) Org. Magn. Reson. 12: 109

S

SD
Abbreviation of **S**elective **D**ecoupling

SD, sometimes called *selective double resonance*, is a commonly used method in nuclear magnetic double resonance (NMDR). In proton magnetic resonance SD has been applied for *assignment* purposes in the case of complex and overlapping multiplets and used for determination of *relative signs* of scalar coupling constants J (see CC) [1]. In contrast to the "normal" spin-decoupling one has to use a weaker amplitude of the second radio-frequency field, B_2, that is of the order of magnitude of J ($\gamma B_2 \simeq J$). The interested reader will find a survey of double resonance phenomena, their experimental procedures and ranges of application in the literature [2,3].

In carbon-13 NMR spectroscopy the terminus technicus SD is sometimes used in contrast to broad-band decoupling (see BB) if one observes the carbon-13 spectrum when irradiating with the single-frequency of a certain proton resonance for assignment purposes.

References

1. Günther H (1980) NMR spectroscopy. An introduction, Wiley, Chichester, chap 9; first published in 1973 by Georg Thieme Verlag, Stuttgart
2. Philipsborn W von (1971) Angew. Chem. 83: 470; (1971) Angew. Chem. Int. Ed. Engl. 10: 472
3. Reynolds GF (1977) J. Chem. Educ. 54: 390

SD

Abbreviation of Spectral Diffusion

The saturation dynamics of an inhomogeneous spin system in nuclear magnetic resonance (see NMR) is affected to a large extent by the effectiveness of SD mechanisms which tend to spread the spin excitations over the entire line profile.

SD may originate in several processes: the cross-relaxation (CR) between spins belonging to different packets [1–4], the time fluctuations of the local field caused by time variations of the field sources [5–11], the spatial migration of spin centers from one site to another with a different environment [12], and the velocity-changing collisions [13]. All these different mechanisms result in a random modulation of the resonance frequencies of the generic spin or pseudo-spin center, so that an excitation initially localized in a given position of the line spreads either progressively or abruptly over the entire line profile.

The saturation kinetics of an inhomogeneous spin system in the presence of SD mechanisms has been theoretically investigated by calculating the transient response of the system to a resonant step-modulated excitation [14].

References

1. Bloembergen N, Shapiro S, Pershan PS, Artmann JO (1959) Phys. Rev. 114: 445
2. Buishvili LL, Zviadadze MD, Kutsishvili GR (1969) Sov. Phys. JEPT 29: 159
3. Clough S, Scott CA (1968) J. Phys. C1: 919
4. a) Korb JP, Muruani J (1980) J. Magn. Reson. 37: 331
   b) Korb JP, Muruani J (1980) J. Magn. Reson. 41: 247
   c) Korb JP, Muruani J (1981) Phys. Rev. B23: 971, 5700
5. Herzog B, Hahn EL (1956) Phys. Rev. 103: 148
6. Mims WB, Nassau K, McGee JD (1961) Phys. Rev. 123: 2059
7. Klauder JR, Anderson PW (1962) Phys. Rev. 125: 912
8. Mims WB (1968) Phys. Rev. 168: 370
9. Hu P, Hartmann SR (1974) Phys. Rev. B9: 1
10. Pomotsev VV (1975) Sov. Phys. Solid State 17: 1146
11. De Voe RG, Wokaun A, Rand SC, Brewer RG (1983) Phys. Rev. B23: 3125
12. Wolf EL (1966) Phys. Rev. 142: 555
13. Kofman AG, Burshtein AI (1979) Sov. Phys. JETP 49: 1019
14. Boscaino R, Gelardi FM, Mantegna RN (1986) J. Magn. Reson. 70: 251

SDDS

Acronym for Spin Decoupling Difference Spectroscopy in Nuclear Magnetic Resonance (see NMR)

Two-dimensional NMR experiments (see 2D-NMR) such as homonuclear correlation spectroscopy (see COSY) easily resolve the problem of the coupling pathway determination, but only a few 2D experiments report on *which* coupling constant is common to two multiplets. In this regard, spin decoupling difference

S

spectroscopy (SDDS) [1] is a powerful, yet infrequently used, one-dimensional technique. The principles of the method and some refinements have been adequately described [1], and elegant examples of its use exist in the literature [2]. The experiment is, however, complicated by a number of factors which limit its use. Recently, M.A. Bernstein [3] presented a paper, in order to delineate ways by which the experiment may be optimally performed, particularly when the irradiated signals resonate in very crowded spectral regions. It was pointed out, that BLOCH-SIEGERT shifts [4] are a major source of artifacts in SDDS. Bernstein suggested ways by which these complications may be recognized and minimized. In very crowded spectral regions, the possibility of "spin-tickling" with either the on- or off-resonance experiment exists. This results in the appearance of additional lines in the coupled proton's spectrum [5]. Spin-tickling effects should be minimized [3] by taking the same precautions as those for BLOCH-SIEGERT shifts (see BSS on p 34).

References

1. Sanders JKM, Mersh JD (1982) Progr. Nucl. Magn. Reson. 15: 353 and further references cited therein
2. See, for example, Hall LD, Sanders JKM (1980) J. Am. Chem. Soc. 102: 5703
3. Bernstein MA (1985) Magn. Reson. Chem. 23: 992
4. Martin ML, Delpeuch J-L, Martin GL (1980) Practical NMR spectroscopy, Heyden, London, p 202
5. See Ref [4] p 217

SDR
Abbreviation of **Spin-Dependent Resonance**

SDR is a new technique arising from the field of electron spin resonance (see ESP and EPR). An attempt for the theoretical understanding has been proposed and results of SDR have been summarized [1].

Reference

1. Haneman D (1984) Prog. Surf. Sci. 15: 85

S

SE
Abbreviation of **Spin-Echo** used in Nuclear Magnetic Resonance

Early in the history of nuclear magnetic resonance (see NMR) it was shown [1,2] that it is possible to produce a so-called spin-echo, a refocusing of the nuclear magnetization which appears at a time τ after the second radio-frequency pulse in a two-pulse sequence $90° - \tau - 180°$. The angles refer to the rotations of the nuclear magnetization vector from its original direction along the magnetic field. This two-pulse sequence—or variations of it—have been used to measure spin-spin relaxation times (T_2). Spin-echos are important in the field of diffusion and self-diffusion measurements, too. For further details see the explanation of the acronyms FGSE, PGSE, and SGSE.

References
1. Hahn EL (1950) Phys. Rev. 80: 580
2. Carr HY, Purcell EM (1954) Phys. Rev. 94: 630

SECOSS
Acronym for **S**pin **E**cho **Co**rrelation with **S**hift **S**caling used in NMR

SECOSS is one of a series of new pulse sequences for enhancing resolution by frequency scaling along the ω_1 (or f_1) axis in a normal spin-echo correlation experiment (see SECSY) [1]. Arbitrary scaling factors can be chosen dependent upon the extent of overlap of peaks in the conventional SECSY spectrum.

In this connection, resolution R has been defined [1] as

$$R = \frac{1}{L}(\delta - L) \tag{1}$$

and the resolution enhancement parameter, E, as

$$E = \frac{1}{L}\left[\frac{\alpha\delta - \beta L}{\alpha} - (\delta - L)\right] \quad \text{for uncoupled} \tag{2}$$

and for coupled spins as

$$E = \frac{1}{L}\left[\frac{\alpha\delta - \beta L - \gamma J}{\alpha} - (\delta - L - J)\right] \tag{3}$$

In Eqs. (1) to (3) symbols have the following meaning:

$\delta = |\omega_k - \omega_l|$, the separation between chemical shifts (see CS) of the spins k and l,

L = half of the sum of the linewidths at half heights, and

α, β, and γ are the scaling factors for chemical shifts (see CS), linewidths, and coupling constants J (see CC).

S

Reference

1. Hosur RV, Ravikumar M, Sheth A (1985) J. Magn. Reson. 65: 375

SECSY
Acronym for **S**pin **E**cho **C**orrelated **S**pectroscopy used in NMR

One of the earliest proposals [1] for modifying Jeener's [2] original two pulse sequence was the SECSY-experiment with the following sequence (1):

$$90_{\Phi_1} - t_1/2 - 90_{\Phi_2} - t_1/2 - \text{acquire} \tag{1}$$

using the phase cycling scheme (2):

$$\Phi_1 = (03212130)$$
$$\Phi_2 = (0_4 1_4 2_4 3_4) \tag{2}$$

the modulation frequencies during t_1 are the differences between the chemical shifts of coupled nuclei. Although this can lead to a reduction in size of the data matrix needed if all couplings are between spins with similar shifts, the principal reason for the popularity of SECSY was that it could be carried out using commercial software intended for J-spectroscopy. Since it is rarely possible to take advantage of any reduction in size of data matrix with SECSY, and it has a lower sensitivity than Jeener's sequence [2], it is now rarely used [3].

References

1. Nagayama K, Wüthrich K, Ernst RR (1979) Biochem. Biophys. Res. Commun. 90: 305
2. Aue WP, Bartholdi E, Ernst RR (1976) J. Chem. Phys. 64: 2229
3. Morris GA (1986) Magn. Reson. Chem. 24: 371

SEFT
Acronym for Spin-Echo FOURIER Transform

In spin systems without coupling constants, e.g. proton-decoupled carbon-13 systems in nuclear magnetic resonance (see NMR), it is possible to enhance the sensitivity by recycling the magnetization. One can exploit the steady-state magnetization that arises if a string of $\pi/2$-pulse is used [1–4], or alternatively observe series of echos excited by a Carr-Purcell (see CP) pulse sequence [5] with the so-called SEFT method [6,7]. SEFT is an alternative approach to the driven equilibrium FOURIER transform (see DEFT).

References

1. Bradford R, Clay C, Strick E (1951) Phys. Rev. 84: 157
2. Kaufmann J, Schwenk A (1976) Phys. Lett. 24A: 115
3. Schwenk A (1968) Z. Phys. 213: 482
4. Schwenk A (1970) Phys. Lett. 31A: 513
5. Carr HY, Purcell EM (1954) Phys. Rev. 94: 630
6. Waugh JS (1970) J. Mol. Spectrosc. 35: 298
7. Allerhand A, Cochran DW (1970) J. Am. Chem. Soc. 92: 4482

SEMINA

S

Abbreviation of SEMUT (see p 248) editing of INADEQUATE (see p 141) nuclear magnetic resonance (see NMR) spectra

SEMUT editing of INADEQUATE carbon-13 NMR spectra became a powerful tool in deducing [1,2,6] carbon-carbon connectivities from the spectra. The general technique has been dubbed SEMINA and different pulse sequences has been proposed [3,4]. It has been demonstrated [3] that a SEMINA pulse sequence with the same sensitivity as the conventional INADEQUATE experiment is recommended for separation of $^{13}CH_n$–$^{13}CH_m$ satellite spectra into two subspectra according to n + m being even or odd. Combined with the information obtained from a common SEMUT GL (see p 248) or DEPT GL edited carbon-13 spectrum a SEMINA experiment is very useful in construction of the carbon skeleton of molecular frameworks. A pulse sequence yielding simplifications beyond the n + m even/odd level (see above) has been also proposed [3,4]. The proposed SEMINA experiments have clear advantages over the reported [5] combination of the SEFT (see p 247) technique with the INADEQUATE experiment both with respect to information content and experimental performance.

References

1. Richarz R, Ammann W, Wirthlin T (1981) J. Magn. Reson. 45: 270
2. Freeman R, Frenkiel T, Rubin MB (1982) J. Am. Chem. Soc. 104: 5545
3. Sørensen OW, Sørensen UB, Jakobsen HJ (1984) J. Magn. Reson. 59: 332
4. Sørensen UB, Jakobsen HJ, Sørensen OW (1985) J. Magn. Reson. 61: 382
5. Benn R (1983) J. Magn. Reson. 55: 460
6. Buddrus J, Bauer H (1987) Angew. Chem. 99: 642; (1987) Angew. Chem. Int. Ed. Engl. 26: 625

SEMUT
Acronym for **S**ubspectral **E**diting using a **Mu**ltiple **Q**uantum **T**rap

This pulse sequence [1] yields edited proton-decoupled subspectra with an acceptable (and identical) accuracy provided the range of $^1J_{CH}$ values for the sample is small, i.e. within 10% of the $^1J_{CH}$ used in timing the delays. In the case of samples with a large spread in $^1J_{CH}$ values $(120 < {}^1J_{CH} < 220\ Hz)$ the *GL* procedure must be incorporated: *DEPT GL* and *SEMUT GL* [2] to suppress J cross-talk between subspectra.

References

1. Bildsøe H, Dønstrup S, Jakobsen HJ, Sørensen OW (1983) J. Magn. Reson. 53: 154
2. Sørensen OW, Dønstrup S, Bildsøe H, Jakobsen HJ (1983) J. Magn. Reson. 55: 347

SEMUT-GL
Acronym for **SEMUT-G**rand **L**uxe [1,2]

S

The basic idea of "editing" carbon-13 nuclear magnetic resonance (see NMR) spectra by separating signals belonging to CH_n groups with n = 0, 1, 2, and 3 has been refined, notably to allow for variations of the magnitude of the J-coupling. Such a refinement is the *SEMUT-GL* sequence given by O.W. Sørensen et al. [1]. This experiment starts with carbon-13 magnetization and allows the observation of *all* carbons including the quarternary ones. SEMUT-GL has been documented, for example, for the editing of proton-decoupled carbon-13 spectra of a mixture of brucine and 2-bromothiazole, which has a spread of J-couplings from approx. 125 to 192 Hz. In this case one has to note the clean separation of the signals associated with different CH_n fragments of the components and the absence of any cross-talk.

The interested reader is referred to the original literature [1] for a discussion of the relative merits of SEMUT-GL and other various sequences used in this field.

References

1. Sørensen OW, Dønstrup S, Bildsøe H, Jakobsen HJ (1983) J. Magn. Reson. 55: 347
2. SEMUT itself stands for **S**ubspectral **E**diting using a **Mu**ltiple **Q**uantum **T**rap (see SEMUT).

SELEX
Acronym for **Sel**ective Spin-**Ex**change

A new method used in high-resolution solid-state nuclear magnetic resonance (see NMR). Several selective two-dimensional spin-exchange pulse sequences has been described [1]. Two-dimensional homonuclear spin-exchange experiments are being widely used in both solid-state and solution NMR [2]. They are capable of providing information about the structure, dynamics, or spectroscopic properties of molecules. Abundant proton spin-exchange measurements have proven useful in structural studies of proteins and DNA in solution [3]. Proton magnetic resonance (see PMR) benefits from high sensitivity, relatively narrow lines, chemical-shift dispersion among sites, and the ability to obtain assignments from related two-dimensional experiments. Applications of both abundant [4] and dilute (^{13}C, ^{15}N) [5–12] two-dimensional spin-exchange measurements to solid-state nuclear magnetic resonance are at an early stage in development. Uniform ^{15}N labeling of proteins and DNA is an inexpensive and easy way of obtaining spectroscopically important samples [13] and ^{15}N spin-exchange measurements have been performed on both proteins [8] and DNA. These and related natural-abundance carbon-13 nuclear magnetic resonance experiments on peptides [10] demonstrate considerable promise for dilute-spin-exchange measurements in structure elucidation studies. To utilize uniformly labeled or natural-abundance samples rather than selectively labeled biopolymers methods for spectral simplification and assignment aids are necessary. The same is true in the case of synthetic polymers or in mixtures with complex individual spectra. SELEX has been tested recently [1] on solid (^{15}N$_{1,3}$) cytosine. The results presented indicate that it should be possible to develop even more complex and helpful pulse sequences by combining elements of homo- and heteronuclear dipolar spectroscopy in solid-state nuclear magnetic resonance.

References

1. Morden KM, Opella SJ (1986) J. Magn. Reson. 70: 476
2. Jeener J, Meier BH, Bachmann P, Ernst RR (1979) J. Chem. Phys. 71: 4546
3. Wemmer D, Reid B (1985) Annu. Rev. Phys. Chem. 36: 105
4. Caravatti P, Neuenschwander P, Ernst RR (1985) Macromolecules 18: 119
5. Caravatti P, Deli JA, Bodenhausen G, Ernst RR (1982) J. Am. Chem. Soc. 104: 5506
6. Szeverenyi NM, Sullivan MJ, Maciel GE (1982) J. Magn. Reson. 47: 462
7. Szeverenyi NM, Bax A, Maciel GE (1983) J. Am. Chem. Soc. 105: 2579
8. Cross TA, Frey HM, Opella SJ (1982) J. Am. Chem. 104: 7471
9. Bronniman CE, Szeverenyi NM, Maciel GE (1983) J. Chem. Phys. 79: 3694
10. Frey MH, Opella SJ (1984) J. Am. Chem. Soc. 106: 4942
11. De Jong AE, Kentgens APM, Veeman WS (1984) Chem. Phys. Lett. 109: 337
12. Harbison GS, Raleigh DP, Herzfeld J, Griffin RG (1985) J. Magn. Reson. 64: 284
13. Cross TA, Diverdi JA, Opella SJ (1982) J. Am. Chem. Soc. 104: 1759

S

SERS
Acronym for **S**urface-**E**nhanced **RAMAN S**cattering (or Spectroscopy)

The discovery of the intensive RAMAN spectrum (see RAMAN Spectroscopy) of pyridine adsorbed on a roughened silver electrode [1] has opened a new branch in vibrational spectroscopy: so-called surface-enhanced RAMAN spectroscopy or scattering (SERS). For reviews see the literature [2–4]. The enhancement of RAMAN scattering by several orders of magnitude (typically 10^3 to 10^6) makes it possible to selectively probe the vibrational pattern of adsorbates on the surface of metals such as Ag, Au, Pt, Pd, and Ni. This phenomenon is currently under intensive investigation, [5,6]. It is expected that this technique will be established as powerful tool in surface sciences for the study of metal/molecule adsorption processes [7]. Electrodes, sols, and island films may be the different morphologies in this field. In view of catalyst poisoning by sulfur compounds, understanding the adsorption process of mercaptans on the catalytic surface is of practical import- ance [8]. 1-Propanethiol has therefore been studied in silver sol as a prototype for catalyst poisoning [5]. On the other hand, if the excitation wavelength coincides with an electronic transition of the adsorbate, molecular resonance RAMAN (RR) and SERS can combine to give surface-enhanced resonance RAMAN scattering (SERRS) so that the limit of detection is further reduced [9–15].

For a long time most of the experimental work in the field was done with small organic and inorganic molecules and ions. But, in the last few years this technique has been successfully extended to macromolecules of biological interest [6].

Smulevich and Feis [16] have recently published the SERS spectra of adriamycin, 11-Desoxycarminomycin, their model chromophores and their com- plexes with DNA.

References

1. Fleischmann M, Hendra PJ, McQuillan AJ (1974) Chem. Phys. Lett. 26: 163
2. Chang RK, Furtak TE (eds) (1982) Surface-enhanced RAMAN scattering, Plenum, New York
3. Birke R, Lombardi JR, Sanchez LA (1982) In: Electrochemical and spectrochemical studies of biological redox components, Chapter 4, American Chemical Society, Washington DC (Advances in Chemistry Series 4)
4. Chang RK, Laube BL (1984) CRC Crit. Rev. Solid State Mat. Sci. 12: 1
5. Joo TH, Kim K, Kim MS (1986) J. Phys. Chem. 90: 5816
6. Hildebrandt P, Stockburger M (1986) J. Phys. Chem. 90: 6017, and more literature references cited therein.
7. Joo TH, Kim K, Kim MS (1985) Chem. Phys. Lett. 117: 518
8. Oudar J (1980) Catal. Rev. Sci. Eng. 22: 171
9. Jeanmaire DL, van Duyne RP (1977) J. Electroanal. Chem. Interface Sci. 84: 1
10. Lippitsch ME (1981) Chem. Phys. Lett. 79: 224
11. Watanabe T, Pettinger P (1982) Chem. Phys. Lett. 89: 501
12. Baranov AV, Bobovich YS (1982) Opt. Spectrosc. (Engl. Transl.) 52: 231
13. Kneipp K, Hinzmann G, Fassler D (1983) Chem. Phys. Lett. 99: 503
14. Siiman O, Lepp A, Kerker M (1983) J. Phys. Chem. 87: 5319
15. Hildebrandt P, Stockburger M (1984) J. Phys. Chem. 88: 5935
16. Smulevich G, Feis A (1986) J. Phys. Chem. 90: 6388

S

SESET
Acronym for **S**emi-**S**elective **E**xci**t**ation

SESET has been used in nuclear magnetic resonance (see NMR) to get informa- tion on both direct and remote connectivities in homonuclear coupling networks [1–3]. Recently [4] an extension of this experimental method has been introduced under the acronym SESET-RELAY (see p 251).

References

1. Millot C, Brondeau J, Canet D (1984) J. Magn. Reson. 58: 143
2. Brondeau J, Canet D (1982) J. Magn. Reson. 47: 159
3. Brondeau J, Canet D (1982) J. Magn. Reson. 47: 419
4. Santro J, Rico M, Bermejo FJ (1986) J. Magn. Reson. 67: 1

SESET-RELAY
Acronym for Semiselective Excitation in an RELAY (see p 235) Experiment

In nuclear magnetic resonance two-dimensional (see 2D) methods became more and more routine procedures. On the other hand special methods or combinations of these has been published. SESET-RELAY is such a typical case and has been described [1] as an extension of the recently introduced [2–4] SESET (see p 250) experiment. The proposed method simultaneously provides information on both direct and remote connectivity. Being a one-dimensional method, it allows an important reduction in measuring time in comparison to standard two-dimensional experiments. Feasibility of the experimental procedure has been demonstrated for a sample of 1,2-propanediol and an acid extract from rat brain [1].

References

1. Santro J, Rico M, Bermejo FJ (1986) J. Magn. Reson. 67: 1
2. Millot C, Brondeau J, Canet D (1984) J. Magn. Reson. 58: 143
3. Brondeau J, Canet D (1982) J. Magn. Reson. 47: 159
4. Brondeau J, Canet D (1982) J. Magn. Reson. 47: 419

SESFOD
Acronym for Selective Excitation with Single-Frequency Off-resonance Decoupling

SESFOD was proposed for various purposes in nuclear magnetic resonance (see NMR) in 1979 [1].

References

1. Martin GE, Matson JA, Turley JC, Weinheimer AJ (1979) J. Am. Chem. Soc. 101: 1888

SFORD
Acronym for Single Frequency Off Resonance Decoupling

Method used in carbon-13 nuclear magnetic resonance to distinguish between quarternary carbons, CH, CH_2, and CH_3 groups via the multiplicities of the signals [1]. With retention of the NOE the spacings of the multiplets correspond to the so-called reduced coupling constants depending on decoupler frequency and power. This kind of decoupling technique is less time consuming than the gated decoupling experiment (see GD) and has the advantage of an almost non-overlap of the multiplets.

Reference

1. Ernst RR (1966) J. Chem. Phys. 45: 3845

SGSE
Acronym for Steady-Gradient Spin-Echo

SGSE is a technique used in nuclear magnetic resonance (see NMR) for diffusion measurements. The parameter to be measured in spin-echo diffusion experiments is the amplitude A of the echo. Its magnitude depends [1] on the spin-spin relaxation time T_2 and on the diffusion coefficient D. For an SGSE experiment employing a $90° - \tau - 180°$ radiofrequency pulse sequence in the presence of a steady field gradient of magnitude G_0 [2],

$$A(2\tau) = A(0)\exp - [2\tau/T_2 + D(2\gamma^2 G_0^2 (2\tau)^3)/3] \qquad (1)$$

where γ is the magnetogyric ratio of the nucleus at resonance. Because the transverse magnetization decay may not be strictly exponential and T_2 may not be precisely known, it is much more usual to avoid dealing with the first term in the exponential by conducting SGSE experiments at fixed τ, varying G_0. If two or more molecular species are simultaneously at resonance, with distinct T_2 and D values, the single exponential in Eq. (1) must be replaced by a weighted sum. The weights representing the relative numbers of nuclei per species [3,4]. Since the spin-echo time duration is inversely proportional [1,2] to G_0—an effect independent of diffusivity—, there are practical limits to the magnitude of G_0: the echo cannot be detected as it becomes shorter than the inverse audio band width of the NMR receiver. Therefore, the lowest value of D measurable by an SGSE experiment are on the order of $D \approx 5 \times 10^{-8} \, cm^2 \, sec^{-1}$. Another two orders of magnitude may be gained by arranging for the magnetic field gradient to have a small magnitude G_0 during the radiofrequency pulses and the echo, and a much larger strength $G + G_0$ for duration δ after the 90° pulse and again after the 180° pulse [5,6].

References

1. Hahn EL (1950) Phys. Rev. 80: 580
2. Carr HY, Purcell EM (1954) Phys. Rev. 94: 630
3. McCall DW, Douglas DC, Anderson EW (1963) J. Polym. Sci. A 1: 1709
4. McCall DW, Huggins CM (1965) Appl. Phys. Letters 7: 153
5. Stejskal EO, Tanner JE (1965) J. Chem. Phys. 42: 288
6. Tanner JE (1966) Ph.D. Thesis, University of Wisconsin

SHARP
Acronym for Sensitive, Homogeneous and Resolved Peaks

A. Pines and co-workers [1] have described experiments which provide sharp chemical shift spectra, with or without scalar coupling constants, for heteronuclear spin systems in inhomogeneous fields. The typical features of such experiments in nuclear magnetic resonance (see NMR) led then to give this method the acronym SHARP.

This adds to an array of conceptual developments, starting with the work of Maudsley and Ernst [2] which has enlarged the applicability of heteronuclear magnetic resonance as a probe of complex molecules and mixtures.

Sensitivity has been improved by using coherence transfers with protons [2–28], allowing experiments for which the signal intensity is independent of the gyromagnetic ratio γ_S of the "dilute" heteronucleus [4,5,8,11–18].

Selectivity is improved as well in these methods since the heteronuclear coupling operates as a label to distinguish the magnetizations of different heteronuclear systems from one another and from other spin groupings within a molecule.

Resolution, however, was the major factor limiting the application of these heteronuclear techniques to topical and *in vivo* spectroscopy and in general to samples where spatial magnetic field inhomogeneity prevents precise shift measurements.

The basic concepts for these two-dimensional [29–31] experiments with SHARP will briefly described here: Each experiment begins with longitudinal proton magnetization and prepares a coherence involving the carbon (heteronucleus) transition operator S_+. Each incorporates a heteronuclear coherence transfer echo [4,12,13] in which linear combinations of the gyromagnetic ratios γ_I and γ_S of the protons and carbons (heteronuclei) will determine the rates of de- and rephasing in the inhomogeneous static fields. In analogy to earlier techniques designed for measuring chemical shift differences in inhomogeneous fields [13,32,33], the echo occurs at the end of the evolution period t_1, which consists of two or more subintervals whose lengths are incremented proportionally. In each case the evolution period is terminated by transfer of the echoed coherence to transverse proton magnetization for detection. FOURIER transformation (see FT) with respect to t_1 gives the homogeneous spectrum.

Thus the sequences presented by Pines and his co-workers [1] combine the sensitivity advantages of recent heteronuclear double-quantum experiments [14,15] with the resolution advantage of heteronuclear coherence-transfer echoes [4,13] and add all the options of heteronuclear and homonuclear [21,22] decoupling. In complex systems it will be desirable to collapse all multiplet structures to improve the resolution of the chemical shift information. This can be realized by incorporating bilinear rotation decoupling (see BIRD) [21,22].

References

1. Gochin M, Weitekamp DP, Pines A (1985) J. Magn. Reson. 63: 431
2. Maudsley AA, Ernst RR (1977) Chem. Phys. 50: 368
3. Maudsley AA, Müller L, Ernst RR (1977) J. Magn. Reson. 28: 463
4. Maudsley AA, Wokaun A, Ernst RR (1978) Chem. Phys. Lett. 55: 9
5. Müller L (1979) J. Am. Chem. Soc. 101: 4481
6. Morris GA, Freeman R (1979) J. Am. Chem. Soc. 101: 760
7. Bodenhausen G, Ruben DJ (1980) Chem. Phys. Lett. 69: 185
8. Burum DP, Ernst RR (1980) J. Magn. Reson. 39: 163
9. Minoretti A, Aue WP, Rheinhold M, Ernst RR (1980) J. Magn. Reson. 40: 175
10. Bendall MR, Pegg DT, Doddrell DM (1981) J. Magn. Reson. 45: 8
11. Pegg DT, Doddrell DM, Bendall MP (1982) J. Chem. Phys. 77: 2745
12. Yen YS, Weitekamp DP (1982) J. Magn. Reson. 47: 476
13. Weitekamp DP, Garbow JR, Pines A (1982) J. Chem. Phys. 77: 2870
14. Bax A, Griffey RH, Hawkins BL (1983) J. Am. Chem. Soc. 105: 7188
15. Bax A, Griffey RH, Hawkins BL (1983) J. Magn. Reson. 55: 301

16. Pegg DT, Doddrell DM, Bendall MR (1983) J. Magn. Reson. 51: 264
17. Bendall MR, Pegg DT, Doddrell DM (1983) J. Magn. Reson. 52: 81
18. Live DH, Davis DG, Agosta WC, Cowburn D (1984) J. Am. Chem. Soc. 106: 6104
19. Freeman R, Mareci TH, Morris GA (1981) J. Magn. Reson. 42: 341
20. Pegg DT, Doddrell DM, Bendall MR (1982) J. Chem. Phys. 77: 2745
21. Garbow JR, Weitekamp DP, Pines A (1982) Chem. Phys. Lett. 93: 504
22. Bax A (1983) J. Magn. Reson. 53: 517
23. Rutar V (1979) J. Am. Chem. Soc. 101: 4481
24. Rutar V (1983) J. Am. Chem. Soc. 105: 4059
25. Bax A (1983) J. Magn. Reson, 52: 330
26. Rutar V (1984) J. Magn. Reson. 58: 132
27. Wilde JA, Bolton PH (1984) J. Magn. Reson. 59: 343
28. Rutar V (1984) In: Müller KA, Kind R, Roos J (eds) Proceedings of the XXII. Congress
    Ampère on Magnetic Resonance and Related Phenomena, Zürich, p 572
29. Aue WP, Bartholdi E, Ernst RR (1976) J. Chem. Phys. 64: 2229
30. Bax A (1982) Two-dimensional NMR, Delft University Press, Delft
31. Ernst RR, Bodenhausen G, Wokaun A (1987) Principles of NMR in one and two dimensions,
    Clarendon, Oxford
32. Weitekamp DP, Garbow JR, Murdoch JB, Pines A (1981) J. Am. Chem. Soc. 103: 3578
33. Garbow JR, Weitekamp DP, Pines A (1983) J. Chem. Phys. 79: 5301

SHIM

An artificial name for homogeneity optimization procedures used in nuclear magnetic resonance (see NMR) for all kinds of magnets

SHIM or *shimming* have been originally used only to describe the mechanical procedure which brings the two pole caps of a magnet into an optimal position one to the other resulting in a good homogeneity of the magnetic field.

Today, in NMR we call all the different kinds of homogeneity optimization methods as shim or shimming procedures. In order to get high-resolution on a spectrometer equipped with an electromagnet it is necessary to have homogeneity correction coils. These coils are placed on the magnet or on the probe assembly (*"magnet shimming"* or *"probe shimming"*) of the spectrometer system. The homogeneity correction coils compensate the field gradients arising from those of the three coordinates (x,y,z), and those of higher order ($y^2, y^3, y^4, x^2 - y^2, x^2 - z^2$, etc.), too. The parameters for that are controlled by DC voltages in the "shim coils" and have to be carefully tuned by precision potentiometers.

In the case of superconducting magnets, produced by using solenoids operating at temperatures of liquid helium, we have normally two shimming procedures: The so-called *"low temperature* or *superconducting shims"* at the first build-up of the magnetic field, which have to be set using the correction coils located on the solenoid, and the so-called "room temperature shims" placed on the probe, like those of an electromagnet system.

On some modern NMR spectrometers "probe shimming" procedures have been automated, and therefore have been called *"autoshim"*.

Concurrent in vivo imaging and spectroscopy or chemical-shift imaging [1,2] would be benefit greatly from improved field-shimming procedures. Most conventional *autoshim software* [3] emulates the way a human operator would proceed, in that a search algorithm is applied to change shim settings iteratively, which can be a time-consuming method. Following the basic concept of Romeo and Hoult [4] a new approach to automatic shimming has recently been proposed

[5]. It is a fast, algebraic technique based on field maps computed by phase measurements within known CARR-PURCELL echo sequences (see CP-T). The *correction* computation itself is executed by using an algorithm which compensates field inhomogeneities within the sample by a linear combination of all used shim control functions. The proposed procedure is non-iterative, thus avoiding certain instability problems and enhancing speed performance. Within this new approach two algorithms have been discussed: (1). a CHEBYCHEV's approach which minimizes the maximum field deviation and (2). a least-squares procedure [5].

References

1. Maudsley AA, Hild SK, Perman WH, Simon SE (1983) J. Magn. Reson. 51: 147
2. Haselgrove JC, Subramanian VH, Leigh JS, Gyulai L, Chance B (1983) Science 220: 1170
3. See Conover WW (1984) In: Levy G (ed) Topics in carbon-13 NMR spectroscopy, vol 4, chap 2, Wiley, New York
4. Romeo F, Hoult DI (1984) Magn. Reson. Med. 1: 44
5. Prammer MG, Haselgrove JC, Shinnar M, Leigh JS (1988) J. Magn. Reson. 77: 40

SHO
Abbreviation of **S**imple **H**armonic **O**scillator

SHO's are used in modelling methods in the field of non-linear spectral analysis such as the digital FOURIER transformation method (see DFT) for nuclear magnetic resonance (see NMR) spectra.

SHRIMP
Acronym for **S**calar **H**eteronuclear **R**ecoupled **I**nteractions by **M**ultiple **P**ulse

The evolution of nuclear spin systems under mutual coupling, whether involving dipolar or scalar interactions, is of crucial importance in coherence-transfer experiments in nuclear magnetic resonance (see NMR). Recently, attention has been renewed toward mixing schemes that directly lead to net coherence-transfer, which for heteronuclear spin systems depend on either the so-called HARTMANN/HAHN strategy (see HAHA), or SHRIMP [1–5]. Evolution under an isotropic mixing HAMILTONian (see IM) during spin-locking has also been employed in the homonuclear situation, leading to the well-known TOCSY (see p 284) experiment [6] where coherence-transfer by a multiple relay is achieved thanks to collective modes of evolution of the spins during mixing [7]. In general, collective modes are characterized by the conservation of one or both components of the total transverse spin angular momentum of the coupled spin system and lead to common frequencies of evolution for the entire spin system [8,9]. In the cross-polarization (see CP) experiment, only the spin-locked component of the total transverse spin momentum is conserved, while in the SHRIMP or IM strategy, both transverse components are conserved. Operator equations have been published in the literature to describe the evolution of AX [5,6,8] and AX_2 [8,9] spin systems. Recently [7] the behavior of spin-1/2 AX_3 and A_2X_2 systems evolving under CP and IM conditions has been analyzed in

S

detail by computing the time development of the density operator by commutator algebra.

References

1. Hartmann SR, Hahn EL (1962) Phys. Rev. 128: 2042
2. Müller L, Ernst RR (1979) Mol. Phys. 38: 963
3. Chingas GC, Garroway AN, Bertrand RD, Moniz WB (1981) J. Chem. Phys. 74: 127
4. Weitekamp DP, Garbow JR, Pines A (1982) J. Chem. Phys. 77: 2870
5. Caravatti P, Braunschweiler L, Ernst RR (1983) Chem. Phys. Lett. 100: 305
6. Braunschweiler L, Ernst RR (1983) J. Magn. Reson. 53: 521
7. Chandrakumar N et al. (1986) J. Magn. Reson. 67: 307
8. Chandrakumar N et al. (1985) J. Magn. Reson. 63: 202
9. Chandrakumar N, Subramanian S (1985) J. Magn. Reson. 62: 346

SIMPLEX

This is the name for an Algorithm used in Numerical Analysis of Electron Paramagnetic Resonance (see EPR) Spectra

After recording an EPR spectrum, a common aim is to calculate a theoretical spectrum in its best agreement with the experimental one. A lot of computer programs have been written for this purpose [1–6], but the algorithms which have been used so far require the evaluation of derivatives of the error function with respect to the parameter adjustment. These derivatives may be estimated numerically, but in the experience of other scientists [7], these estimates become progressively more unreliable as the number of adjustable spectral parameters increases.

One way of avoiding this problem is to analyze, instead of the EPR spectrum itself, a so-called significance plot calculated from the spectrum [8,9]. But, this strategy also has its own disadvantages.

These considerations have led a group of scientists [7] to investigate the possibilities of the SIMPLEX algorithm [10–12], which does not require derivatives. They have found the algorithm to function well, even with as many as 17 parameters for adjustment in the numerical analysis of EPR spectra.

References

1. Marquardt DW, Bennett RG, Burrell EJ (1961) J. Mol. Spectrosc. 7: 269
2. Bauder A, Myers RJ (1968) J. Mol. Spectrosc. 27: 110
3. Fischer L (1971) J. Mol. Spectrosc. 40: 414
4. Heinzer J (1971) QCPE 11: 197
5. Heinzer J (1974) J. Magn. Reson. 13: 124
6. Barzaghi M, Simonetta M (1983) J. Magn. Reson. 51: 175
7. Beckwith ALJ, Brumby S (1987) J. Magn. Reson. 73: 252
8. Brumby S (1980) J. Magn. Reson. 39: 1
9. Brumby S (1980) J. Magn. Reson. 40: 157
10. Nelder JA, Mead R (1965) Comput. J. 7: 308
11. Daniels RW (1978) An introduction to numerical methods and optimization techniques, North-Holland, New York, p 183
12. Caceci MS, Cacheris WP, Byte, May 1984: 340 Letters concerning this article appeared in Byte, Sep. 1984: 22, and Byte, Oct. 1984: 16

SIR-ΔNOE
Acronym for Selective Inversion-Recovery Difference Spectroscopy (Δ) of the Nuclear OVERHAUSER Effect (or Enhancement)

It has been found that SIR-ΔNOE with truncated relaxation delays is a superior method for minimizing secondary cross-relaxation effects (= spin diffusion) in studies of protein-bound conformations of small molecules [1].

The selective inversion-recovery NOE experiment is the classic experiment, given by I. Solomon [2] and has generally been designated as a transient NOE. To avoid confusion some authors [1] have used the acronym SIR-ΔNOE.

A brief consideration of the NOE experiment using the preparation/evolution/ detection (PED as abbreviation) period scheme provides a rationale for the authors [1] choice:

$$\Delta NOE = \{ PD - P1\,(t_1) - \tau - {}^{ns}\theta \, AQ \, FID\,(t_2) \}_n \, \text{minus CONTROL}$$

$$P \qquad E \qquad D$$

In this scheme, A corresponds to the spin subject to selective perturbation or saturation. For cross-relaxation, SOLOMON's equation [2] with the sign convention of Kalk and Berendsen [3] is valid. Also note that the decoupler pulse is designated as $P1$ and thus B_1 refers to its field. $\gamma B_1 t_1 = {}^s\theta$, the flip angle for the selective pulse. All other symbols or abbreviations in the scheme have the well-known meanings.

For an alternative approach to SIR-ΔNOE the interested reader is referred to the acronym ETB-NOE in this dictionary.

References
1. Anderson NH, Nguyen KT, Eaton HL (1985) J. Magn. Reson. 63: 365
2. Solomon I (1955) Phys. Rev. 99: 559
3. Kalk A, Berendsen HJC (1976) J. Magn. Reson. 24: 343

SKEWSY
Acronym for Skewed Exchange Spectroscopy

SKEWSY is a new experiment that is capable of measuring exchange and cross-relaxation rates [1] in nuclear magnetic resonance (see NMR) spectroscopy. SKEWSY requires no lineshape analysis, and all spectral information is theoretically obtainable from the cross-peaks, thus avoiding the (normally) crowded diagonal region of the two-dimensional spectrum. Like NOESY, SKEWSY utilizes a randomized mixing period to eliminate zero-quantum coherence (see ZQC) and is therefore well suited to proton magnetic resonance spectroscopy and useful in determinations of NOE's. In this experiment J-coupling effects are absent and a back-transformation method can be used to identify a relay in a three-spin system [1].

Reference
1. Bremer J, Mendz GL, Moore WJ (1984) J. Am. Chem. Soc. 106: 4691

SLAP
Acronym for **S**ign **La**beled **P**olarization Transfer

Geminal carbon-carbon coupling constants (^{2}Jc, c) are usually small in magnitude, and their signs had to be estimated by analogy considerations, as clear experimental values were not available.

Such data have become accessible from natural abundance samples only recently, owing to the development of the SLAP experiment [1] and its refinements [2–4].

Figure 28(a) shows the original pulse sequence of the SLAP experiment [1], whereas Fig. 28(b) demonstrates the sensitivity-enhanced modification of the SLAP sequence (4).

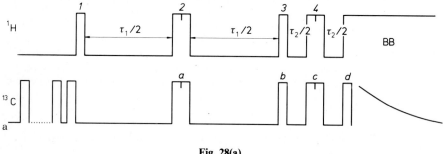

Fig. 28(a)

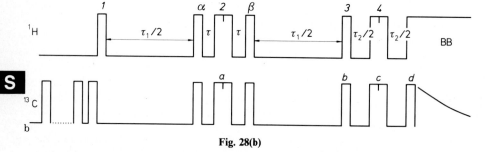

Fig. 28(b)

Recently [5] one-and two-bond C,C couplings and the signs of ten of the two-bond coupling constants have been determined from the INADEQUATE (see p 141) and SLAP spectra of four acetylenes and one allene. The data have been discussed theoretically [5].

References
1. Sørensen OW, Ernst RR (1983) J. Magn. Reson. 54: 122; for corrections to the SLAP experiment see ref. [3].
2. Lambert J, Wilhelm K, Klessinger M (1985) J. Magn. Reson. 63: 189
3. Sørensen OW, Ernst RR (1985) J. Magn. Reson. 63: 219
4. Lambert J, Klessinger M (1987) J. Magn. Reson. 73: 323
5. Lambert J, Klessinger M (1987) Magn. Reson. Chem. 25: 456

SMART
Acronym for **S**imulation of **Ma**gnetic **R**esonance Pulse Experiments in the Time Domain

SMART is a general purpose pulse experiment simulation program which generates time-domain signals [1]. It must be pointed out that a number of valuable pulse experiment simulation programs reported before SMART put emphasis either on numerical density matrix calculations in the time-domain [2,3], or in the frequency-domain [4], or on symbolic formula manipulations [5,6]. SMART uses numerical density matrix calculations in the time-domain. It features a CARTESian product-operator interpreter to handle arbitrary HAMILTONians (see p 122) without limitations on the spin quantum number in different situations. A high-level pulse experiment simulation language has been tailored to a wide variety of one- or two-dimensional homo- and heteronuclear magnetic resonance experiments using radio-frequency pulses, delays, and phase-cycling schemes. Illustrative simulations of two-dimensional nuclear magnetic resonance (see 2D-NMR) experiments such as E.COSY (see p 84) and time reversal of scalar J couplings with a strongly coupled three-spin system and electron spin-echo modulation experiments (see ESEEM) have been presented [1].

References
1. Studer W (1988) J. Magn. Reson. 77: 424
2. Piveteau D, Delsuc MA, Lallemand J-J (1986) J. Magn. Reson. 70: 290
3. Kay LE, Scarsdale JN, Hare DR, Prestegard HJ (1986) J. Magn. Reson. 68: 515
4. Widmer H, Wüthrich K (1986) J. Magn. Reson. 70: 270
5. Bernassau JM, Boissiere F, Thomas JF (1986) Comput. Chem. 10: 253
6. Nakashima TT, McClung RED (1986) J. Magn. Reson. 70: 187

SMILE
Acronym for **SWIFT** (see SWIFT) **M**ethod for **I**n Vivo **L**ocalized **E**xcitation in Nuclear Magnetic Resonance (see NMR)

SMILE has been proposed [1] as a method for localization based on selective suppression under using the SWIFT [2] (see SWIFT) excitation procedure.
   In preliminary experiments, it has been shown that SMILE

a) leads to a good reliable localization while minimizing T_2 or T_2^* losses, thus maintaining maximal signal-to-noise ratio (see SNR) from the volume or region of interest (VOI and ROI),
b) achieves localization in two chopped acquisitions enabling one to optimize magnetic field homogeneity in the ROI, and
c) shows immunity to gradient eddy-current effects.

In the opinion of the authors [1] this last point should enhance the transfer of this sequence to different NMR spectrometers and imagers.

References
1. Chew WM, Chang L-H, Flaming DP, James TL (1987) J. Magn. Reson.75: 523
2. Hsu AT, Hunter WW Jr, Schmalbrock P, Marshall AG (1987) J. Magn. Reson. 72: 75

SNAPTAG
Acronym for **S**mall **N**utation **A**ngle **P**ulse **T**wins with **A**lternated **G**radients

This new approach [1] allows the design of improved selective 180° radio-frequency pulses for magnetization inversion and phase reversal. The proposed technique [1] may be useful for practical nuclear magnetic resonance imaging (see NMRI), particularly in multiple echo sequences, for volume-localized NMR (see NMR) spectroscopy, and for flow measurements, where accurate 180° pulses are crucial. In comparison to other attempts [2–4] to get correct selective 180° radio-frequency pulses the new approach seems to be very fruitful in both areas of magnetization inversion and phase reversal.

References
1. Yan H, Gore JC (1987) J. Magn. Reson. 71: 116
2. Lurie DJ (1985) Magn. Reson. Imaging 3: 235
3. Silver MS, Joseph RI, Hoult DI (1984) J. Magn. Reson. 59: 347
4. Nishimura DG (1985) Med. Phys. 12: 413.

SNR
Abbreviation of **S**ignal-to-**N**oise **R**atio
A suitable—but not the only possible—measurement for sensitivity is the signal-to-noise ratio (S/N), which is defined by Eq. (1) [1].

$$S/N = \frac{\text{peak signal voltage}}{\text{rms noise voltage}} \qquad (1)$$

In Eq. (1) rms stands for root-mean-square. Noise frequencies with periods longer than the spectral trace do not appear as "noise"; they merely cause a variable baseline shift which is unimportant in most nuclear magnetic resonance measurements. But, it is necessary to subtract this "nonobservable" noise in the definition of S/N [1]. This causes the signal-to-noise ratio to depend on the length of the trace. The "observable" noise increases with the length of the trace. For white noise, it is always a minor correction which can be neglected. In the case of strong low-frequency noise, it may be convenient to use Eq. (2) as the definition.

$$S/N = \sqrt{2}\,\frac{\text{peak signal voltage}}{\text{rms value of the difference of two}} \qquad (2)$$
$$\text{noise voltages one line width apart}$$

The S/N criterion is appropriate if the problem is to determine the presence or absence of a resonance line and if no information about the line shape must be retrieved. This is typical for the analysis of complicated spectra with much fine structure, like high-resolution NMR and ESR.

The basic definition of S/N assumes a signal function with a single extremum such as a LORENTZian or GAUSSian signal. It must be changed in the case of a line shape with two extrema, like the dispersion mode signal or the derivative of a LORENTZian or GAUSSian signal. In such a case one has to use Eq. (3).

$$S/N = \frac{(\text{maximum of signal voltage}) - (\text{minimum of signal voltage})}{\text{rms value of the difference of the noise voltages at } t_1 \text{ and } t_2} \qquad (3)$$

The mean square value of the difference can be expressed by the mean square value of the noise voltage and by its correlation function $R_n(t)$:

$$\overline{(n(t_1)-n(t_2))^2} = \overline{2n(t)^2} - 2R_n(t_1-t_2) \tag{4}$$

In the case of white random noise, Eq.(5) is valid (2).

$$S/N = \frac{(\text{maximum of signal voltage})-(\text{minimum of signal voltage})}{\sqrt{2}\cdot(\text{rms value of noise voltage})} \tag{5}$$

From the earliest pulsed FOURIER transform (see PFT) NMR experiments performed by Ernst and Anderson in 1966 [3], the conditions required to optimize the SNR were examined. In samples with long spin lattice relaxation times (T_1) of the order of hours it becomes particularly necessary to optimize the experimental parameters so that an adequate SNR may be accumulated at a reasonable rate.

With the increasing numbers of high-resolution solid-state studies via dilute spin-1/2 systems where there are no efficient sources of relaxation, such as paramagnetic impurities, or nuclei to cross-polarize to, systems with long T_1's being more frequently encountered. Calculations so far have not been specifically concerned with a situation where the total measuremental time may be only a few times T_1 and are not strictly valid in this limit as they assume constant magnetization throughout the total experiment [4,5].

Recently [6], a reminder has been given of the existing ideas for SNR optimization together with a new calculation procedure of SNR using a discrete summation.

A new technique that improves the effective sensitivity of NMR by means of post-detection signal processing has been presented [7]. This technique will be of value in obtaining localized spectra, allowing a smaller voxel volume for more accurate localization. The procedure takes advantage of the a priori knowledge of relaxation times and resonant frequencies of the spectral lines to be measured and uses vector-space methods to determine the corresponding concentrations. With use of this technique the extent of SNR improvement is limited only by the spin-spin relaxation time (T_2). In particular, under the conditions of a typical NMR experiment, this method results in an eightfold increase in SNR over that obtainable by conventional algorithms using FOURIER transformation.

S

References

1. Ernst RR (1965) Rev. Sci. Instr. 36: 1689
2. Ernst RR (1966) Sensitivity enhancement in magnetic resonance. In: Waugh JS (ed) Advances in magnetic resonance, vol 2, Academic, New York
3. Ernst RR, Anderson WA (1966) Rev. Sci. Inst. 37: 93
4. Waugh JS (1970) J. Mol. Spec. 35: 298
5. Christensen KA, Grant DM, Shulman EM, Walling C (1974) J. Phys. Chem. 78: 1971
6. Dupree R, Smith ME (1987) J. Magn. Reson. 75: 153
7. Noorbehesht B, Lee H, Enzmann DR (1987) J. Magn. Reson. 73: 423

Soft Pulses

Name for Special Pulses used in Nuclear Magnetic Resonance (see NMR)

Two-dimensional NMR spectroscopy today is routinely used (see 2D-NMR) for assignment of resonances in NMR spectra and for structure elucidation [1,2].

Homonuclear correlation spectroscopy (see COSY) and nuclear OVERHAUSER enhancement spectroscopy (see NOESY), for example, are the two essential procedures in the structural analysis of biopolymers in solution [3].

Recently [4] it has been proposed to employ frequency-selective "soft" radio-frequency pulses in order to circumvent the limitations of the methods cited above. When excitation and detection are focused to a small area of the 2D-NMR spectrum, high-resolution can be achieved by concentrating all available data points in this region.

It is noteworthy, that other selective experiments derived from 2D-NMR spectroscopy have been published by Kessler and co-workers [5]. These proposed experiments, however, result in one-dimensional spectra corresponding to the so-called cross-sections through 2D-NMR spectra. The new kind of experiment [4], on the other hand, will deliver true 2D-NMR spectra of restricted frequency ranges.

References

1. Ernst RR, Bodenhausen G, Wokaun A (1987) Principles of nuclear magnetic resonance in one and two dimensions, Clarendon, Oxford
2. Chandrakumar N, Subramanian S (1986) Modern techniques in high-resolution FT-NMR, Springer, Berlin Heidelberg New York
3. Wüthrich K (1986) NMR of proteins and nucleic acids, Wiley-Interscience, New York
4. Brüschweiler R, Madsen JC, Griesinger C, Sørensen OW, Ernst RR (1987) J. Magn. Reson. 73: 380
5. Kessler H, Oschkinat H, Griesinger C, Bermel W (1986) J. Magn. Reson. 70: 106

SP

Abbreviation of Selective Pulses used in Nuclear Magnetic Resonance (see NMR)

The selective excitation of specific resonances with typically 90° ($\pi/2$) or 180° (π) pulses has been considered by several authors in the past [1–3]. In nuclear magnetic resonance imaging (see NMRI), attention has been focused on the use of amplitude-modulated pulses rather than "tailored excitation" [4] or complicated pulse sequences [5–7]. However, while reasonable advantages were realized using 90° ($\pi/2$) pulses, no sharply selective 180° (π) pulse waveform was published before 1984.

In 1984 experimental results for 90° and 180° pulses, obtained using wave-forms derived from solutions of the BLOCH-RICCATI equation [8] were presented [9]. A full theoretical analysis of these results has been given since 1984.

References

1. Lauterbur PC, Kramer DM, House WV, Chen C-N (1975) J. Am. Chem. Soc. 97: 6866
2. Mansfield P, Maudsley AA, Baines T (1976) J. Phys. E9: 271
3. Levitt MH (1983) J. Magn. Reson. 50: 95
4. Tomlinson BL, Hill HDW (1973) J. Chem. Phys. 59: 1775
5. Levitt MH, Freeman R (1981) J. Magn. Reson. 43: 502
6. Levitt MH (1983) J. Magn. Reson. 48: 234
7. Levitt MH (1983) J. Magn. Reson. 50: 95

8. Silver MS, Joseph RI, Hoult DI (1983) Proceedings of the 2nd Annual Meeting of The Society of Magnetic Resonance in Medicine, San Francisco
9. Silver MS, Joseph RI, Hoult DI (1984) J. Magn. Reson. 59: 347

SPHINX
Acronym for Simulation of Pulse Sequences in High-Resolution Nuclear Magnetic Resonance Experiments

SPHINX is an computer-assisted approach [1] to visualize the effects of different pulse sequences in modern techniques of nuclear magnetic resonance (see NMR) which are especially used in structure elucidation of macromolecules such as proteins and nucleic acids [2,3].

References
1. Widmer H, Wüthrich K (1986) J. Magn. Reson. 70: 270
2. Ernst RR, Bodenhausen G, Wokaun A (1987) Principles of nuclear magnetic resonance in one and two dimensions, Clarendon, Oxford
3. Wüthrich K (1986) NMR of proteins and nucleic acids, Wiley, New York

SPI
Abbreviation of Selective Populations Inversion

SPI is a one-dimensional experiment in nuclear magnetic resonance (see NMR) [1] and is used like SPT (see p 267) [2] in the field of heteronuclear "δ,δ-spectroscopy".

Let us consider an AX spin system of an abundant nucleus (such as a proton) and rare nucleus (such as cabon-13, nitrogen-15, silicon-29, and metal nuclei, etc.). Via selective inversion of BOLTZMANN's distribution for two energy niveaus which are connected by one line of the A nucleus one can produce a situation that for almost all transitions of the nucleus X the spin-polarization of the A nucleus is relevant (see Fig. 29). Intensity enhancement for the X nucleus corresponds to the quotient of the different magnetogyric ratios, γ_A/γ_X.

In the case of the "absorption"-line one observes a fully realized enhancement. The "emission"-line enhancement of the intensity is equal to $1 - (\gamma_A/\gamma_X)$. Normally, because $\gamma_A/\gamma_X > 1$, we will therefore find a better signal-to-noise ratio (see SNR), too. The (a) part of Fig. 29 symbolizes the energy niveau scheme of an AX spin system of an abundant nucleus (A) and a rare nucleus (X) under normal conditions. The (b) section of Fig. 29 shows the perturbed situation after inversion to the population of states (1) and (3), experimentally realized by a selective 180° pulse applied on line A1. Part (c) of Fig. 29 shows a vector diagram for the A lines after the selective 180° pulse on A1. An analogous diagram is visualized for the corresponding X lines of the AX spin system in section (d) of the figure. The comparison of (e) and (f) of Fig. 29 demonstrates the enhancement of the SNR of the X signal (f) in part (e).

S

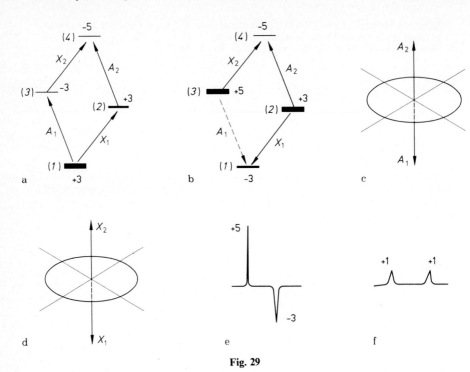

Fig. 29

References

1. Pachler KGR, Wessels PL (1973) J. Magn. Reson. 12: 337
2. Sørensen S, Hansen RS, Jakobsen HJ (1974) J. Magn. Reson. 14: 243

S

Spin-Tickling

This is the name for a special phenomenon arising from nuclear magnetic double resonance (NMDR)

If weak B_2 fields—of the order of magnitude defined by the relation $B_2 \approx \Delta$, where Δ stands for the half-width of a spectral line—are used in NMDR one can observe "Spin-Tickling" effects.

This experiments is helpful in the assignment of protons in proton magnetic resonance spectra of varying complexity.

In fact, on irradiation, the corresponding transition in the energy level diagram, if B_2 has the frequency of a particular resonance signal. This leads to a splitting of each line in the spectrum that has an energy level in common with the perturbed transition.

This has been well demonstrated by Hoffmann and Forsén [1] for the AB spin system of the two protons in 2-bromo-5-chlorothiophene as shown in Fig. 30.

A plausible explanation for this effect states thus as a result of the perturbation, the eigenvalues E_1 and E_3 of the spin system mix with one another and two

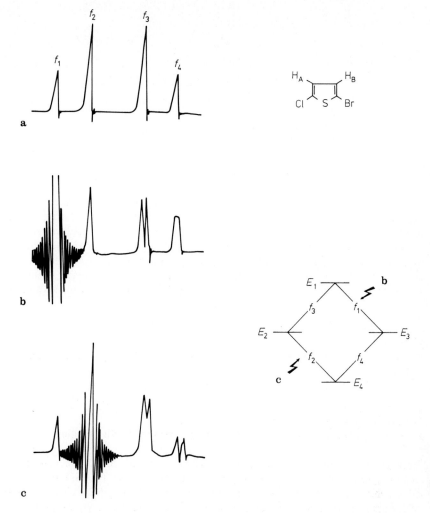

Fig. 30. Spin-tickling experiment in the AB system of 2-bromo-5-chlorothiophene (after Hoffman and Forsén [1]).

transitions then become possible. The new transition, then, practically corresponds to the previously forbidden double-quantum transition $E_4 \rightarrow E_1$. It must be pointed out that the pathway in which the different transitions are connected in the energy level diagram manifests itself in this experiment.

One has to differentiate so-called *progressively* connected transitions in which the three eigenvalues change their energies continuously (e.g. f_2 and f_3) and so-called *regressively* connected ones in which the central eigenvalue is larger or smaller than the energies of initial and final states (f_2, f_4 or f_1, f_3).

The initial and final states of a *progressively* linked pair differ in their total spin by two units ($\Delta m_T = 2$). In the case of a *regressively* connected pair, Δm_T is equal to zero.

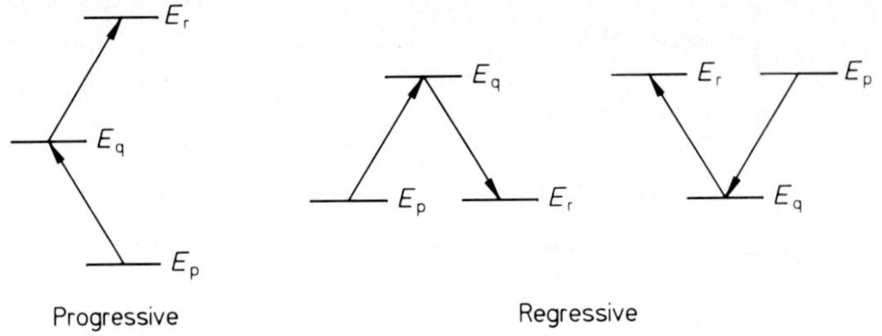

Progressive

Regressive

Fig. 31

Figure 31 explains these connections schematically: It has been shown experimentally that *regressively* linked transitions are sharply split (or resolved), while for the *progressively* connected transitions, splitting is less well defined.

The spin-tickling experiment provides a possibility of *testing* an energy level diagram in NMR spectra. Besides the above mentioned assignment of helping in the analysis of ^{1}H-NMR spectra, this NMR experiment is particularly useful for the determination of the relative signs of spin-spin coupling in systems with more than two protons.

Reference

1. Hoffman RA, Forsén S (1966) In: Emsley JW, Feeney J, Sutcliffe LH (eds) Progress in nuclear magnetic resonance spectroscopy, vol 1, Pergamon, Oxford

SPLASH Imaging
Acronym for **S**pectroscopic **F**lash (fast **l**ow **a**ngle **sh**ot) NMR **Imaging**

Recently [1] a readout-time-encoding method, analogous to the echo-time-encoding technique [2], has been applied to one-dimensional FLASH imaging for the creation of one-dimensional spectroscopic flash imaging (SPLASH imaging). SPLASH imaging can be extended to other modalities for the interesting localized spectroscopy. Using a surface coil and a single read gradient perpendicular to the coil, the readout-time-encoding SPLASH technique will result in a 2D representation with depth-resolution and high-resolution chemical shift information. This procedure can be further extended to two-dimensional SPLASH imaging of a selected plane. In addition, this technique and a 3D SPLASH version provide the possibility of applying water-suppression pulses for proton SPLASH imaging. Further work dealing with another variant of 2D SPLASH imaging has been reported [3].

References

1. Haase A, Matthaei D (1987) J. Magn. Reson. 71: 550
2. Park HW, Cho ZH (1986) Magn. Reson. Med. 3: 448
3. Ströbel B, Ratzel D (1986) Book of Abstracts, 5th Annual Meeting of the Society of Magnetic Resonance in Medicine, Montreal, p 664

SPT
Abbreviation of **S**elective **P**opulation **T**ransfer

SPT is a special kind of nuclear magnetic double resonance (NMDR) in a pulse experiment with very small decoupler power (γB_2 corresponds to the linewidth of a multiplet signal) [1–4]. The pulse sequence can be written schematically with

$$^1H \text{ dec.:}\quad DO - P2_X(\text{sel.}) - DO - DO \tag{1}$$

$$^1H \text{ or } X \text{ obs.:}\quad D1 - PW - FID \tag{2}$$

where symbols have the following meaning:

(1) = situation in the decoupler channel,
(2) = situation in the observe channel,
DO = decoupler gated off,
P2 = a selective low-power pulse (constant or alternate phase) to saturate or invert one transition,
D1 = a relaxation delay,
PW = arbitrary pulse length (or width),
FID = free induction decay.

Under distinct conditions this experiment will cause a total inversion of a line in the spectrum, then one speaks about selective population inversion (see SPI).

On the other hand one can combine SPT with a difference spectroscopy method: the SPT experiment follows a normal acquisition [5,6]. The resulting FID's (see FID) will be subtracted from each other. After transformation we will see only the variations in intensity by SPT, which sometimes have been called Pseudo-INDOR [7].

References
1. Pachler KGR, Wessels PL (1973) J. Magn. Reson. 12: 337
2. Sørensen S, Hansen RS, Jakobsen HJ (1974) J. Magn. Reson. 14: 243
3. Dekker TG, Pachler KGR, Wessels RL (1976) Org. Magn. Reson. 8: 530
4. Linde SA, Jakobsen HJ, Kimber BJ (1975) J. Am. Chem. Soc. 97: 3219
5. Chalmers AA, Pachler KGR, Wessels PL (1974) J. Magn. Reson. 15: 415
6. Bundgaard T, Jakobsen HJ (1975) J. Magn. Reson. 18: 209
7. Feeney J, Partington PJ: J. Chem. Soc., Chem. Commun. 1973: 611

SR
Acronym for **S**pin-**R**otation Interaction

SR is one of the four different mechanisms dominating carbon-13 spin-lattice relaxation in nuclear magnetic resonance (see NMR). In addition to dipolar relaxation (see DD), where H_{loc} has its origin in the magnetic dipole of neighbouring nuclei with $I \neq 0$, the molecule itself may have a magnetic moment. This originates from a modulation of the magnitude and direction of the angular momentum vector associated with the rotation of the molecule [1]. The resulting molecular magnetism is independent of the symmetry of the electronic charge distribution. The induced magnetic moment increases as the molecular motion becomes faster. Relaxation via the SR mechanism may be more efficient for small, rapidly tumbling molecules and at higher temperatures, especially in the vapor phase. For a carbon-13 in a cylindrically symmetric environment, the SR relaxation rate $1/T_1^{SR}$ is given by Eq. (1) [2].

$$1/T_1^{SR} = (2\,kT/3\,\hbar^2)\,I_m\,(C_{\parallel}^2 + 2\,C_{\perp}^2)\,\tau_{SR} \tag{1}$$

In Eq. (1), I_m designates the moment of inertia of the molecule relative to axis m. $C_{\parallel}$ and $C_{\perp}$ are the spin-rotation constants parallel and perpendicular to the symmetry axis, respectively, and τ_{SR} is the angular momentum correlation time. All other symbols of Eq. (1) have their usual meaning. The angular momentum correlation time is a measure of the persistence of a given angular momentum. SR is best identified as a dominating relaxation mechanism via the temperature dependence of relaxation rates. Whereas dipolar relaxation rates (see DD) decrease with increasing temperature, the opposite behaviour is observed in the case of SR. If the condition $\tau_{SR} \ll \tau_c$ is fulfilled, the two correlation times can be converted into one another by HUBBARD's relation [3], given with Eq. (2).

$$\tau_c \tau_{SR} = I_m/6\,kT \tag{2}$$

References

1. Carrington A, McLachlan AD (1967) Introduction to magnetic resonance, chap 11, Harper and Row, New York
2. Deverell C (1970) Mol. Phys. 18: 319
3. Hubbard P S (1963) Phys. Rev. 131: 1155

SRS and RAS
Abbreviation of **S**pecular **R**eflection **S**pectrometry and of **R**eflection-**A**bsorption Spectrometry

Both methods are used in the field of infrared (see IR and FT-IR) spectroscopy.

In the past, the term *specular reflectance* was applied to two different types of IR measurements. One applies to the measurement of the reflectance spectrum of a flat, clean surface of an absorbing material, either inorganic or organic. The other one involves a double-pass through a thin surface film on a reflective-typically metallic-surface.

Today, it is now becoming common practice to call this type of measurement *reflection-absorption* spectrometry (RAS), whereas the former method is still known as *specular reflection* spectrometry (SRS).

For further details of both the methods the interested reader is referred to the literature [1].

Reference

1. Griffiths PR, de Haseth JA (1986) In: Elving PJ, Winefordner JD (eds) Fourier transform infrared spectrometry, J. Wiley, New York

SRFT
Acronym for **S**aturation-**R**ecovery **F**ourier **T**ransform

SRFT is one of the common procedures for the measurement of spin-lattice relaxation times (T_1) in the field of nuclear magnetic resonance [1,2].

S

The initial complete elimination of magnetization, not only along the z-axis, but also in the x,y-plane, is peculiar to this method. This is accomplished, for example, by a 90° pulse followed by a typical field-gradient (see FG) pulse along the z-aixs [2], which has a dispersing effect on M_{xy}. After a suitable time $t \lesssim T_1$, magnetization has partially recovered and can be detected in an almost analogous manner from a FOURIER transform of the free induction decay (see FID) following a 90° pulse.

The pulse sequence required can thus be written as

$$(90° - HSP - t - 90°)_n$$

in which the acronym HSP stands for a so-called homogeneity spoiling pulse.

References

1. Markley JL, Horsley WJ, Klein MP (1971) J. Chem. Phys. 55: 3604
2. McDonald GC, Leigh JS (1973) J. Magn. Reson. 9: 358

SSB

Abbreviation of **S**pinning **S**ide **B**and in Nuclear Magnetic Resonance (see NMR) Spectroscopy

In all kinds of high-resolution NMR experiments the sample tube (made from glass) is rapidly spun around its long axis by means of a small air turbine (the so-called "spinner"). This has the effect of improving the field homogeneity because, as a result of the macroscopic rotation of the sample tube, the individual nuclei are exposed to a time-averaged value of the external magnetic field H_o, the magnitude of which varies within certain limits over the total sample volume. As a result of rapid spinning the sample tube SSB's appear on both sides of the principal signal (or center band) and at equal distance from it. The difference in frequency between the central signal and the SSB's corresponds to the rotational frequency of the glass-made sample tube.

In the case of high-resolution NMR experiments on solids incorporated into "rotors" running with high-speed around the so-called magic angle (see MAR, MAS, and MASS) we are confronted with the same problem of SSB's. According to the chemical shift anisotropy (see CSA) here the observed SSB's and their "structures" have implicit information. On the other hand, it is often helpful to use total-sideband-suppression techniques (see TOSS) in solid-state NMR in qualitative and quantitative aspects.

S

SS-INADEQUATE
Acronym for **S**uper-**S**implified **I**ncredible **N**atural **A**bundance **D**ouble **Qua**ntum Transfer Experiment

A one-dimensional selective version of the original INADEQUATE experiment (see INADEQUATE) has been proposed recently by D. Canet and co-workers [1] and abbreviated with SS-INADEQUATE.

The simple pulse sequence (see Eq. (1)) allows selective polarization transfer from a nucleus A (on resonance) towards any nucleus X, J coupled to A, provided that τ is properly varied around J/2 with co-addition of the relevant free induction decays (see FID).

$$(\pi/2)_x - \tau - (\pi/2)_{-x}(\pi/2)_\phi(\text{acq.})_\psi \qquad (1)$$

With an appropriate phase cycling of ϕ and ψ, parent signals in a natural abundance carbon-13 spectrum are eliminated. This consequently leads to an almost un-ambiguous determination of all carbon-carbon connectivities (CCC) of a molecule. In addition, a universal vectorial representation aimed at multi-pulse sequence analysis has been proposed and applied to the SS-INADEQUATE experiment [1].

Reference
1. Canet D, Brondeau J, Boubel JC, Retournard A (1987) Magn. Reson. Chem. 25: 798

SS-NMR
This acronym or abbreviation is sometimes used for **S**olid **S**tate-**N**uclear **M**agnetic **R**esonance

NMR of solids requires special techniques which are explained under their individual acronyms or abbreviations in this dictionary. Here, we will discuss only some new aspects of SS-NMR.

Applications of three new solid-state NMR techniques have been described recently [1]. Two of them are located in the two-dimensional frequency space, the third in the three-dimensional ordinary space [1].

The first technique allows detection and analysis of ultraslow motions in solids via carbon-13 magic angle spinning (see MAS) NMR [2]. Using this method, slow helix rotations over 200° in poly (oxymethylene) have been investigated [3].

Nutation angle NMR (see NNMR) can be used to detect and analyze changes in the local symmetry around quadrupolar nuclei [4]. For example, the hydration of NaA-zeolite can be studied in detail by NNMR [5].

NMR imaging (see NMRI) of rigid components in special materials is often hampered by the very large solid-state NMR linewidth. To overcome this problem one can use line-narrowing techniques such as multiple-pulse decoupling or MAS.

Standard NMRI techniques can be employed when instead of static or pulsed field gradients a field gradient is applied that rotates synchronously with the magic angle spinner [6].

Until recently no suitable *chemical shift* (see CS) *standard* or reference compound had been devised for SS-NMR spectroscopy. Axelson has reviewed the work done in this area [7] so far.

A suitable *standard for SS-NMR* must meet several important criteria:

- it should exhibit a resonance that is quantitative, narrow, and distinct;
- it should be grindable, non-volatile, and chemically inert, with a high solubility in some solvents so it can be easily removed from the sample under SS-NMR investigation;

– in addition, it should display a chemical shift (see CS) that is independent of the magnetic field strength [8]; and
– have favorable relaxation properties and cross-polarization (see CP) dynamics for use in BLOCH's decay and CP analyses.

Very recently [9] tetrakis (trimethylsilyl)silane (TKS) has been proposed as a suitable reference compound that meets all the above mentioned criteria.

References

1. Veeman WS (1986) paper presented at the 8th European Experimental NMR Conference, June 3–6, 1986 in Spa/Belgium
2. de Jong AF, Kentgens APM, Veeman WS (1984) Chem. Phys. Lett. 109: 337
3. Kentgens APM, de Jong AF, de Boer E, Veeman WS (1985) Macromolecules 18: 1045; Kentgens APM, de Boer E, Veeman WS (1987) J. Chem. Phys. 87: 6859
4. Kentgens APW, Lemmens JJM, Geurts FFM, Veeman WS (1987) J. Magn. Reson. 71: 62
5. Tijink GAH, Janssen R, Veeman WS (1987) J. Am. Chem. Soc. 109: 7301
6. Cory DG, Rischweim AM, Veeman WS (1988) J. Magn. Reson. 80: 259
7. Axelson DE (1985) Solid state nuclear magnetic resonance of fossil fuels: An experimental approach, Multiscience, Montreal, Canada
8. van der Hart DL (1986) J. Chem. Phys. 84: 1196
9. Muntean JV, Stock LM, Botto RE (1988) J. Magn. Reson. 76: 540

SSNOEDS
Acronym for Steady-State Nuclear OVERHAUSER Effect Difference Spectroscopy

SSNOEDS [1] is an acronym which is sometimes used instead of NOE-DIFF (see p 199) in nuclear magnetic resonance.

Reference

1. Sanders JKM, Mersh JD (1982) Prog. Nucl. Magn. Reson. 15: 353, and reference cited therein

SST
Abbreviation of Solvent-Suppression Techniques

The difficulties associated with intensive signals in FOURIER transform NMR spectra (see FT and NMR) are widely recognized and a plethora of so-called *solvent-suppression techniques* has been suggested. All these different techniques will be explained in this dictionary under their originally termed acronyms or abbreviations. But, one approach that has gained some favor in the last few years involves selective excitation while using trains of "*hard*" pulses interspersed with delays [1–9] will be discussed here in detail, because recently [10] a paper dealing with the computer-optimized *solvent suppression* has come to our knowledge. The fundamental idea is to return the solvent magnetization most accurately to the $+z$ axis of the coordination system involved while exciting spins at different chemical shifts into the xy plane. In this way the strong solvent signal does not reach the receiver and dynamic range problems are avoided. These pulse sequences have been designed by geometric insight, consideration of FT of the pulse train, a rotation operator treatment, computer simulation, trial and error, and most combinations thereof [6]. Two new *"hard"* pulse sequences for *solvent suppression* have been discovered by an iterative variation of pulse lengths to achieve a desired pattern of excitation [10]. Lineshape distortions and baseline roll [3] that result from phase corrections might be avoided by maximum entropy (see ME or MEM) reconstruction of the spectrum [11–14]. This possibility is currently under investigation. Computer optimization [15] would seem to be a valuable method, perhaps the only efficient method, for devising new pulse sequences under circumstances where their complexity defies more conventional approaches [10].

Much more recently [16] a new approach has been described for *water suppression* in one- and two-dimensional NMR (see also WST in 2D-NMR), generating absorption-mode spectra that are free of baseline distortions. This procedure involves the use of a 1-1 *"hard"* pulse as a *reading pulse*, followed by a 1-1 *refocusing pulse* which is phase cycled to get greatest possible water suppression. Examples of this method for water suppression have been demonstrated for a 10 mg sample of a decapeptide in 90% H_2O by recording two-dimensional NOE (see 2D-NOE), 2D-HOHAHA, and heteronuclear proton/nitrogen-15 shift correlation spectra [16].

A simple modification for the elimination of phase distortions as a characteristic of a "binomial" SST has been published [17] recently.

References

1. Plateau P, Guéron M (1982) J. Am. Chem. Soc. 104: 7310
2. Sklenár V, Starčuk Z (1982) J. Magn. Reson. 50: 495
3. Plateau P, Dumas C, Guéron M (1983) J. Magn. Reson. 54: 46
4. Turner DL (1983) J. Magn. Reson. 54: 146
5. Clore GM, Kimber BJ, Gronenborn AM (1983) J. Magn. Reson. 54: 170
6. Hore PJ (1983) J. Magn. Reson. 55: 283
7. Bleich H, Wilde J (1984) J. Magn. Reson. 56: 154
8. Starčuk Z, Sklenár V (1985) J. Magn. 61: 567
9. Starčuk Z, Sklenár V (1986) J. Magn. Reson. 66: 391
10. Hall MP, Hore PJ (1986) J. Magn. Reson. 70: 350
11. Sibisi S (1983) Nature (London) 301: 134
12. Sibisi S, Skilling J, Berenton RG, Laue ED, Staunton J (1984) Nature (London) 311: 446
13. Laue ED, Skilling J, Staunton J, Sibisi S, Berenton RG (1985) J. Magn. Reson. 62: 437
14. Hore PJ (1985) J. Magn. Reson. 62: 561
15. Shaka AJ (1985) Chem. Phys. Lett. 120: 201
16. Sklenár V, Bax A (1987) J. Magn. Reson. 74: 469
17. Galloway GJ, Haseler LJ, Marshman MF, Williams DH, Doddrell DM (1987) J. Magn. Reson. 74: 184

S

ST
Abbreviation of **S**aturation **T**ransfer

ST is one kind of magnetization transfer (see MT) experiments used in nuclear magnetic resonance (see NMR). The investigation of chemical exchange processes by ST experiments was first performed by Hoffman and Forsén [1,2]. Rapid development of the method has made it a powerful tool for the study of kinetic phenomena [3–9]. ST experiments today may be performed most efficiently by modern pulse techniques [10]. Different experiments with monoexponential as well with biexponential decay functions have been demonstrated [10]. Such experiments are very useful for the estimation of rate constants of chemical reactions in the time range of 0.1 to 1 second, where classical methods as well as relaxation kinetics are unfavorable. The application of spin saturation experiments on the hydration of pyridine-4-carbaldehyde (as a model substance for the study of elementary reaction steps of pyridoxal phosphate) has been demonstrated in detail [10].

References

1. Forsén S, Hoffman RA (1963) J. Chem. Phys. 39: 2892
2. Forsén S, Hoffman RA (1964) J. Chem. Phys. 40: 1189
3. Haslinger E, Wolschann P (1978) Monatsh. Chem. 109: 1263
4. Noggle JH, Schirmer R (1971) The nuclear Overhauser effect, Academic, New York
5. Mann BE (1977) J. Magn. Reson. 25: 91
6. Perrin CL, Johnston EE, Lollo CP, Kobrin PA (1981) J. Am. Chem. Soc. 103: 4691
7. Led JJ, Gesmar H (1982) J. Magn. Reson. 49: 444
8. Hvidt A, Gesmar H, Led JJ (1983) Acta Chim. Scand. B37: 227
9. Feigel M, Kessler H, Leibfritz D, Walter A (1979) J. Am. Chem. Soc. 101: 1943
10. Kalchhauser H, Wolschann P (1986) Monatsh. Chem. 117: 841

STNMR
Abbreviation of **S**tochastic **N**uclear **M**agnetic **R**esonance

STNMR spectroscopy [1–4] is a conceptually more difficult method than other methods in nuclear magnetic resonance (see NMR). In STNMR, a random excitation is applied, which supplies the excitation energy continuously and leads to a significantly reduced input and response power. As for pulsed excitation, the excitation spectrum is characterized by a broad frequency band, but its frequency components do not exhibit a well-defined phase relationship. For the whole length of the STNMR experiment excitation and response are continuous in time. Multi-dimensional NMR spectra which contain spectroscopic informations similar to those of two-dimensional FOURIER transform (see FT and FFT) spectra [3–9] are derived from the experimental data by multi-dimensional cross-correlation of excitation and response records in combination with FT (see MUDISM).

So far STNMR still has to establish itself in competition with all other commonly used NMR methods [10]. But, apart from its experimental simplicity, STNMR permits computation of high-resolution two-dimensional spectra at high magnetic fields, where the resolution of conventional 2D-FT spectra is limited by the needed acquisition time and data storage capacities [11,12].

For example, D. Ziessow and co-workers [18] have pointed out that MUDISM (see p 183) promises to obtain 2D-spectra with high digital resolution which, with commonly used 2D-techniques, would require the processing of 10 or more gigawords (!) data arrays.

STNMR does not utilize strong deterministic variations of excitation sequences for the generation of various experiments which is true for conventional 2D-NMR (see p 7) methods. Therefore, noise excitation is non-selective, and a good part of information is obtained via a single experiment. In particular, information dealing with spin interactions such as spin-spin couplings (see CC) [4–9,13–15], zero- and double-quantum coherences [13,16], chemical exchange

[6–9,13–15,17], and cross-relaxation [13] can be retrieved from the non-linear response of STNMR.

In STNMR separation of information is achieved not by different experiments, but by data processing after the experiment itself.

For further details, the interested reader is referred to a recently published review about scope and limitations of STNMR [10].

References

1. Ernst RR (1970) J. Magn. Reson. 3: 10
2. Kaiser R (1970) J. Magn. Reson. 3: 28
3. Blümich B (1985) Bull. Magn. Reson. 7: 5
4. Blümich B (1985) Proc. Second Conf. Mod. Methods Radiospectr., Reinhardsbrunn/DDR, p 113
5. Bartholdi E (1975) Dissertation, ETH Zürich
6. Bartholdi E, Wokaun A, Ernst RR (1976) Chem. Phys. 18: 57
7. Blümich B (1981) Dissertation, Technical University Berlin
8. Blümich B, Ziessow D (1980) Ber. Bunsenges. Phys. Chem. 84: 1090
9. Blümich B, Ziessow D (1983) J. Chem. Phys. 78: 1059
10. Blümich B (1987) Prog. NMR Spectrosc. 19: 331
11. Blümich B (1983) Stochastic NMR on the XL-200, Report of the Phys. Dep. of the University of New Brunswick, Fredericton/Canada
12. Blümich B, Kaiser R (1984) J. Magn. Reson. 58: 149
13. Blümich B, Kaiser R (1983) J. Magn. Reson. 54: 486
14. Blümich B, Ziessow D (1983) J. Magn. Reson. 52: 42
15. Blümich B, Ziessow D (1981) Nachr. Chem. Tech. Lab. 29: 291
16. Blümich B, Ziessow D (1983) Mol. Phys. 48: 955
17. Blümich B, Twesten I (1985) J. Magn. Reson. 61: 349
18. Blömer U, Genge I, Khuen A, Ziessow D: poster presented at the 8th European Experimental NMR Conference, June 3–6, Spa/Belgium 1986.

STTR-CIDNP

Acronym for Stochastic Time-Resolved Chemically Induced Dynamic Nuclear Polarization

Recently B. Blümich [1] has suggested a stochastic time-resolved CIDNP spectroscopy because a CIDNP experiment can be considered as a special case in the analysis of two-input non-linear systems. He proposed a method to obtain time-resolved CIDNP nuclear magnetic resonance (see NMR) spectra at relatively low excitation powers and high time-resolution: With binary modulation of light and radio-frequency input signals, a two-dimensional time-domain function is computed on-line from analog two-dimensional cross-correlation of input and response signals. This function is transformed into a stack of one-dimensional CIDNP spectra, which follow the time evolution of the stimulated chemical reaction.

The proposed scheme by Blümich [1] is the stochastic analogue of the JEENER type double pulse scheme [2–5]. It has been worked out in order to provide a lower power alternative to fast time-resolved flash photolysis experiments [6–8].

References

1. Blümich B (1984) Mol. Phys. 51: 1283
2. Schäublin S, Höhner A, Ernst RR (1974) J. Magn. Reson. 13: 196

3. a) Kühne RO, Schaffhauser T, Wokaun A, Ernst RR (1979) J. Magn. Reson. 35: 39;
   b) Schäublin S, Wokaun A, Ernst RR (1976) J. Chem. Phys. 14: 255
4. Schäublin S, Wokaun A, Ernst RR (1976) J. Chem. Phys. 14: 285
5. Schäublin S, Wokaun A, Ernst RR (1977) J. Magn. Reson. 27: 273
6. Miller RJ, Closs GL (1981) Rev. Scient. Instrum. 52: 1876
7. Closs GL, Miller RJ (1981) J. Am. Chem. Soc. 103: 3586
8. Closs GL, Miller RJ (1979) J. Am. Chem. Soc. 101: 1639

SUCZESS

Acronym for **Suc**cessive **Z**ero-Quantum, Single-Quantum Coherences for Spin Correlation

SUCZESS is a modification [1] of Müller's method [2] for mapping networks of spin-spin couplings by zero-quantum coherences in nuclear magnetic resonance (see NMR), which, besides retaining part of the advantages of Müller's procedure, requires fewer data.

The modification is based on the well-known principle of delayed acquisition [3,4]. SUCZESS spectra with narrow lines and a clean definition of the diagonal peaks have the additional advantage of spanning a reduced frequency range in the F_1 dimension. Furthermore, the use of a τ value of the order of $0.15/^3 J$ [5] results in even excitation of AX, AX_2, and AX_3 spin systems, largely reducing the correlation via long-range couplings, thus simplifying the habitus of the final NMR spectrum. Alternatively, all correlations can be obtained, provided that the coupling constants are resolved [6,7], by incrementing the τ value and t_1 in a concerted manner [6].

All these advantages will make SUCZESS an interesting alternative to other correlation methods, particularly when resolution or the amount of data give rise to problems.

References

1. Santoro J, Bermejo FJ, Rico M (1985) J. Magn. Reson. 64: 151
2. Müller L (1984) J. Magn. Reson. 59: 326
3. Nagayama K, Wüthrich K, Ernst RR (1979) Biochem. Biophys. Res. Commun. 90: 305
4. Nagayama K, Kumar A, Wüthrich K, Ernst RR (1980) J. Magn. Reson. 40: 321
5. Burum DP, Ernst RR (1980) J. Magn. Reson. 39: 163
6. Braunschweiler L, Bodenhausen G, Ernst RR (1983) Mol. Phys. 48: 535
7. Sørensen OW, Eich GW, Levitt MH, Bodenhausen G, Ernst RR (1983) Progr. NMR Spectrosc. 16: 163

S

SUFIR

Acronym for **Su**per**f**ast **I**nversion **R**ecovery

For the last decade, efforts have been made to shorten the time it takes to measure nuclear magnetic resonance (see NMR) longitudinal relaxation times or spin-lattice relaxation times (T_1). Recently Canet et al. [1] have proposed a so-called one-shot sequence given with Eq. (1).

$$[\pi/2(S_1) - \tau - \pi - \tau - \pi/2(S_2) - \tau]_n \tag{1}$$

This sequence seems to be a rapid method to yield T_1 values with fairly good accuracy, provided that τ lies in an interval from $0.5\,T_1$ to $3\,T_1$ and any transverse magnetization has disappeared prior to each radio-frequency pulse applied. The inverting pulse is a composite pulse (see p 51), $(\pi/2)_x (4\pi/3)_y (\pi/2)_x$. The signals S_1 and S_2 of Eq. (1) are acquired and added in separate blocks of the computer memory. T_1 values are obtained by using Eq. (2).

$$T_1 = -\tau/\ln(1 - S_2/S_1) \tag{2}$$

In addition, a phase-cycling scheme has been devised that eliminates spurious effects due either to residual transverse magnetization or to misadjusted $\pi/2$ pulses [1].

Reference

1. Canet D, Brondeau J, Elbayed K (1988) J. Magn. Reson. 77: 483

SUPER COSY
Acronym for an advanced pulse scheme in two-dimensional **Co**rrelation **S**pectroscopy (see COSY)

It has been shown [1,2] that the normal COSY experiment [3] can be very much improved by addition of fixed time delays in the evalution and detection periods. These delays remove antiphase character from the multiplets of the cross peaks and introduce antiphase character into the multiplets of the diagonal peaks, achieving many improvements in the COSY nuclear magnetic resonance spectrum (see NMR). The scheme, named SUPER COSY [1], drastically enhances the sensitivity of detection of the cross peaks under conditions of limited digital resolution, reduces the dynamic range of diagonal to cross peaks, and additionally allows J tuning of the COSY spectra for the detection of particular coupling constants [1].

Recently [7] SUPER COSY has been analyzed in detail. An important limitation is loss of signal intensity during the delay periods due to spin-spin relaxation (T_2). Therefore, it is often helpful to bring the anti-parallel components [1,2,4–6] only partially into a parallel character by using shorter delays. The analytical approach to SUPER COSY [7] deals with the optimization for maximum cross-peak intensity in a rational given experimental time. It has been found that the optimum is governed by a function of spin-spin coupling, spin-spin relaxation, and digital resolution.

References

1. Hosur RV, Chary K VR, Kumar A, Govil G (1985) J. Magn. Reson. 62: 123
2. Kumar A, Hosur RV, Chandrasekhar K (1984) J. Magn. Reson. 60: 143
3. Aue WP, Bartholdi E, Ernst RR (1976) J. Chem. Phys. 64: 2229
4. Mayor S, Hosur RV (1985) Magn. Reson. Chem. 23: 470
5. Kumar A, Hosur RV, Chandrasekhar K, Murali N (1985) J. Magn. Reson. 63: 107
6. Kumar A, Hosur RV, Chandrasekhar K, Murali N (1985) In: Govil G, Khetrapal CL, Saran A (eds) Magnetic resonance in biology and medicine, Tata McGraw-Hill, New Delhi
7. Chandrasekhar K, Kumar A (1987) J. Magn. Reson. 73: 417

SUPER RCOSY
Acronym for an advanced pulse scheme in **R**elayed **C**oherence Transfer **S**pectroscopy (RCOSY)

S

An advanced pulse scheme has been suggested [1] which removes the antiphase intensity character from the multiplets of cross peaks and introduces antiphase intensity character into the multiplets of diagonal peaks in the relayed coherence transfer spectroscopy (RCOSY) [2–4]. The scheme has been called SUPER RCOSY [1] and suggested with the aim of enhancing the intensity of cross peaks in situations of limited digital resolution.

References

1. Kumar A, Hosur RV, Chandrasekhar K, Murali N (1985) J. Magn. Reson. 63: 107
2. Eich GW, Bodenhausen G, Ernst RR (1982) J. Am. Chem. Soc. 104: 3731
3. Bolton PH, Bodenhausen G (1982) Chem. Phys. Lett. 89: 139
4. Wagner G (1983) J. Magn. Reson. 55: 151

SUPER SECSY 1 and 2
Acronym for superior pulse schemes for Spin-Echo Correlated Spectroscopy (see SECSY)

Two versions (1 and 2) of SUPER SECSY have been suggested [1] under using the product operator formalism of Sørensen et al. [2] for the calculation of the density operators in the case of two weakly coupled spins (each spin 1/2). These procedures will remove the antiphase intensity character from the multiplets of cross peaks and introduce antiphase intensity character into the multiplets of diagonal peaks in spin-echo correlated 2D spectroscopy (see SECSY) [3]. These are suggested with the aim of enhancing the intensity of cross peaks in situations of limited digital resolution.

References

1. Kumar A, Hosur RV, Chandrasekhar K, Murali N (1985) J. Magn. Reson. 63: 107
2. Sørensen OW, Eich GW, Levitt MH, Bodenhausen G, Ernst RR (1983) Prog. Nucl. Magn. Reson. Spectrosc. 16: 163
3. Nagayama K, Kumar A, Wüthrich K, Ernst RR (1980) J. Magn. Reson. 40: 321

SVD
Acronym for Singular Value Decomposition

S

SVD is an algorithm for a linear least-square fit used in different methods of non-linear spectal analysis applied to nuclear magnetic resonance (see NMR).

In principle, a spectrum is the FOURIER transformation of a time record (see FT) but, in practice, difficulties arise when the time record is contaminated with measurement noise or has data missing. In such cases the frequency domain spectrum is no longer a deterministic transformation of the time domain record. Statistic estimation procedures must be employed, and some of these are explained under their special acronyms in this dictionary.

SW
Abbreviation of Sweep Width or Spectral Width

Both abbreviations are used in nuclear magnetic resonance (see NMR) spectroscopy to define the spectral region of a spectrum. In continuous wave (see CW) operations the terminus sweep width have been used because of obtaining spectra under field or frequency sweeping (see SWEEP). When using pulsed FOURIER transform (see PFT) NMR the spectral region of interest will be abbreviated with SW standing for spectral width.

SWEEP

Field sweeping stands for systematically varying the magnetic field strength, at constant applied radio-frequency field, to bring NMR transitions of different energies successively into resonance, thereby making available an NMR spectrum consisting of signal intensity versus magnetic field strength.

Frequency sweeping stands for systematically varying the frequency of the applied radio-frequency field—or of a modulation side-band—to bring NMR transitions of different energies successively into resonance at constant magnetic field strength. The resulting NMR spectrum consists of signal intensity versus applied radio-frequency.

SWIFT
Acronym for **S**tored **W**aveform **I**nverse **F**OURIER **T**ransform

The tailord-excitation experiment in which excitation and detection are temporally separated has been denoted as SWIFT [1–3].

In particular, quadratically phase-modulated SWIFT excitation has been successfully applied for broadband selective detection or ejection (analogous to saturation in nuclear magnetic resonance (see NMR spectroscopy) in FOURIER transform ion cyclotron resonance mass spectrometry.

Recently [4] it has been demonstrated, that phase-modulated SWIFT excitation can provide water suppression (see also SST) in depth-resolved proton chemical shift spectra for the *in-vivo* study of metabolites in a spatially resolved region with a clinical nuclear magnetic resonance imaging (see NMRI) device.

S

References
1. Chen L, Wang T-CL, Ricca TL, Marshall AG (1987) Anal. Chem. 59: 449
2. Marshall AG, Wang T-CL, Ricca TL (1985) J. Am. Chem. Soc. 107: 7893
3. Wang T-CL, Ricca TL, Marshall AG (1986) Anal. Chem. 58: 2935
4. Hsu AT, Hunter WW, Schmalbrock P, Marshall AG (1987) J. Magn. Reson. 72: 75

SYMMETRY
is not an acronym or abbreviation, but a concept in molecular spectroscopy arising from mathematics

Symmetry, a major source of simplification in many physical problems, is common in nuclear magnetic resonance (see NMR) applications [1–5].

NMR pulse symmetry has been analyzed recently [6]. It has been pointed out that derived symmetry rules fall into two principal categories: (a) comparisons between the magnetization responses of isochromats experiencing different applied fields related by symmetry, and (b) analyses of the rotation experienced by a single isochromat in a field of high symmetry. In the field of category (a), for example, it has been shown (i) that symmetric responses require only that the magnetic field B_1 to be real, not that it be symmetric, and (ii) that the flip angle response to a fixed-phase pulse applied in the presence of a time-varying gradient field is unaffected by time reflection. A very simple group-theoretical formalism has been developed and extensions of the procedure to the symmetry analysis of composite pulses and multiple selective pulses has been indicated [6].

Ernst and co-workers have recently [7] discussed the possibility of a computer-assisted analysis of nuclear spin systems based on local symmetry in two-dimensional NMR spectra. This paper [7] may be considered as an extension of the proposed simple and time-efficient cluster analysis [8], which is suitable for the localization of cross-peaks. The chemical shift centers, as well as coupling constants, can be determined based on symmetry properties of the individual cross-peaks [7]. After finishing this work, the authors [7] became aware of related approaches by Bodenhausen et al. [9] and Hoch et al. [10]. Recently [11] Williamson and Neuhaus have discussed the symmetry concept in nuclear OVERHAUSER enhanced spectra (NOE) in detail.

References

1. Levitt MH, Ernst RR (1982) J. Magn. Reson. 48: 234
2. Counsell C, Levitt MH, Ernst RR (1985) J. Magn. Reson. 63: 133
3. Murdoch JB (1985) In: Proceedings 4th Meeting Soc. Magn. Reson. Med., p 1033
4. Shaka AJ, Keeler J, Smith MB, Freeman R (1985) J. Magn. Reson. 61: 175
5. Hore PJ (1983) J. Magn. Reson. 55: 283
6. Ngo JT, Morris PG (1987) J. Magn. Reson. 74: 122
7. Meier BU, Mádi ZL, Ernst RR (1987) J. Magn. Reson. 74: 565.
8. Mádi Z, Meier BU, Ernst RR (1987) J. Magn. Reson. 72: 584
9. Bodenhausen G, Pfändler P, Novic M, Oschkinat H, Wimperis S, Jaccard G, Eggenberger U, Limat D (1987) In: Proceedings, 28th Experimental NMR Conference, Asilomar, CA, April 5–9
10. Hoch JC, Poulson FM, Hengyi S, Kjaer M, Dudvigsen S (1987) In: Proceedings, 28th Experimental NMR Conference, Asilomar, CA, April 5–9
11. Williamson MP, Neuhaus D (1987) J. Magn. Reson. 72: 369

S

TA
Abbreviation of **T**ime-**A**veraging in Nuclear Magnetic Resonance (see NMR)

Optimum utilization of the available performance time for an NMR experiment can be an important factor in the attempt to maximize its sensitivity. But, two reasons may limit the achievable signal-to-noise ratio (see SNR):

1. Low-frequency noise: The power spectral density of the noise in common spectrometers often increases strongly at very low frequencies. If a long-term measurement is applied to enhance the sensitivity, the influence of this low-frequency noise increases with decreasing sweep rate (see SWEEP) and may limit the achievable SNR.
2. Relaxation times: The height and the line shape of NMR signals depend on the sweep rate. This is true for continuous wave spectra (see CW). It is, in general, possible to get stronger signals at higher sweep rates. This situation is caused by saturation effects due to relaxation which become important at low sweep rates.

Time-averaging (TA) is in practice the summation of different traces in a digital storage. All important and more technical aspects of TA have been analyzed in detail by R.R. Ernst [1].

Recently [2] a simple method for time-averaging NMR free induction decays (see FID) in non-stable magnetic fields has been reported by K. Roth. This procedure is based on a root-mean-square (see RMS) addition of each individual scan and its HILBERT transformation (see HT) and allows an easy determination of line width information and the improvement of the signal-to-noise ratio (see SNR) by data accumulation.

In a standard NMR experiment the FID for one type of nuclei in a homogeneous field is given by Eq. (1) where f_0 is the resonance frequency, T_2 the spin-spin relaxation time, and p_0 the relative phase.

$$S(t) = S_0 \exp(-t/T_2)\cos(2\pi f_0 t + p_0) \ . \tag{1}$$

Co-addition of two independent FID's only leads to a SNR improvement of the frequency and the phase are identical in both transients. K. Roth has pointed out that field stabilization, at least for simple NMR spectra, is in principle not necessary. By using a HT and suitable data manipulation, TA of individual NMR experiments with varying resonance frequency and/or phase is possible. The HT is usually defined in the frequency-domain [3]. In extension of this definition, HT of the NMR time-domain "signal" given in Eq. (1) results in Eq. (2).

$$\hat{S}(t) = \vec{H}[S(f)] = -S_0 \exp(-t/T_2)\sin(2\pi f_0 t + p_0) \ . \tag{2}$$

Numeric HT of digitized data can be calculated via successive application of two forward FOURIER transformations [4] and can be performed by using any commercially available NMR software. Summation of the squares of Eqs. (1) and (2) and calculation of the square root leads to Eq. (3).

$$[S^2(t) + \hat{S}^2(t)]^{1/2} = S_0 \exp(-t/T_2). \tag{3}$$

References

1. Ernst RR (1966) Sensitivity enhancement in magnetic resonance. In: Waugh JS (ed) Advances in magnetic resonance, Academic, New York, vol 2, p 1

2. Roth K (1987) J. Magn. Reson. 73: 343
3. Titchmarsh EE (1948) Introduction to the theory of Fourier integrals, University Press, Oxford; Helstrom CW (1960) Statistical theory of signal detection, Pergamon, Oxford; Marshall AG (ed) (1982) Fourier, Hadamard, and Hilbert transforms in chemistry, Plenum, New York
4. Bartholdi E, Ernst RR (1973) J. Magn. Reson. 11: 9; Roth K, Kimber BJ (1985) Magn. Reson. Chem. 23: 832

TANDEM-SEFT
Acronym for TANDEM-Spin Echo FOURIER Transform

This experiment has been suggested by Günther et al. for signal selection in the case of CH_mD_n fragments [1] in nuclear magnetic resonance (see NMR) spectroscopy and can be seen as a special application or modification of the originally given spin-echo experiment by E. Hahn in 1950 [2]. The präfix TANDEM stands for both decoupling in the proton and deuteron frequency channels as shown in Fig. 32.

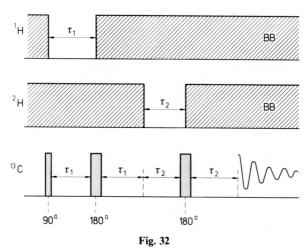

Fig. 32

In Fig. 32 τ_1 stands for the modulation with $^1J_{C,H}$, τ_2 means modulation with $^1J_{C,D}$, and BB is the abbreviation of broadband proton decoupling.

TANDEM-SEFT [1] was analyzed carefully by Wesener in 1985 [3].

References
1. Wesener JR, Schmitt P, Günther H (1984) J. Am Chem. Soc. 106: 10
2. Hahn EL (1950) Phys. Rev. 80: 580
3. Wesener JR (1985) Dissertation, University of Siegen/Germany

TANGO
Acronym for Testing for Adjacent Nuclei with a Gyration Operator

The development of new two-dimensional nuclear magnetic resonance (see 2D-NMR) pulse sequences has been significantly augmented by the introduction of

T

two new clusters of pulses which may be regarded as a unit [1]. The first such operator was the BIRD pulse sequence [2] (see BIRD) given by Eq. (1):

$$
\begin{array}{ll}
{}^{1}\text{H} & 90^{\circ}_{+x} - \tau - 180^{\circ}_{+x} - \tau - 90^{\circ}_{+x} \\
{}^{13}\text{C} & 180^{\circ}
\end{array} \tag{1}
$$

More recently, Wimperis and Freeman [3] have suggested yet another such operator which they have christened TANGO. TANGO in its originally form is shown here by Eq. (2) in which it acts as a 360° pulse for distant protons and as a 90° pulse for adjacent protons.

$$
\begin{array}{ll}
{}^{1}\text{H} & 135^{\circ}_{+x} - \tau - 180^{\circ}_{+x} - \tau - 45^{\circ}_{+x} \\
{}^{13}\text{C} & 180^{\circ}
\end{array} \tag{2}
$$

At lot of modifications of the TANGO sequence have been published by Freeman and co-workers [4,5]. Among these modifications there is a sequence (3) which acts as a 90° pulse for distant protons, leaving protons directly bound to carbon-13 inverted:

$$
\begin{array}{ll}
{}^{1}\text{H} & 45^{\circ}_{+x} - \tau - 180^{\circ}_{+y} - \tau - 45^{\circ}_{-x} \\
{}^{13}\text{C} & 180^{\circ}
\end{array} \tag{3}
$$

R. Freeman and co-workers [5] have incorporated Eq. (3) into a pulse sequence for long-range heteronuclear chemical-shift correlation as an initial excitation "pulse" while another group have used Eq. (3) as the 90° pulse in the magnetization transfer step of a similar long-range heteronuclear chemical-shift correlation experiment [6].

References

1. Zektzer AS, John BK, Castle RN, Martin GE (1987) J. Magn. Reson. 72: 556
2. Garbow JR, Weitekamp DP, Pines A (1982) Chem. Phys. Lett. 93: 504
3. Wimperis SC, Freeman R (1984) J. Magn. Reson. 58: 348
4. Wimperis SC, Freeman R (1985) J. Magn. Reson. 62: 147
5. Bauer C, Freeman R, Wimperis SC (1984) J. Magn. Reson. 58: 526
6. Zektzer AS, Quast MJ, Linz GS, Martin GE, McKenny JD, Johnston MD Jr, Castle RN (1986) Magn. Reson. Chem. 24: 1083

T

TDPAC or TDPAD
Acronym for **T**ime-**D**ifferential **P**erturbed **A**ngular **C**orrelation or **D**istribution

The study of nuclear quadrupole interactions (see NQCC and NQR), i.e. of the strength, symmetry, and orientation of the electric field gradient (see EFG) tensor, yields almost valuable information on charge density distribution on an atomic scale. Apart from the current conventional methods of nuclear magnetic resonance (see NMR) and MÖSSBAUER's spectroscopy there is yet another technique: the so-called "time-resolved perturbed γ-γ-angular correlation spectroscopy" or TDPAC[1] which has been announced as a new analytical method with

---

[1] Please note, that in integral PAC spectroscopy the coincidence count-rate is measured as a function of the angle between detector without any time resolution. But, in TDPAC or TDPAD spectroscopy the events are resolved in the time elapsed between the arrival of the first particle in one and the arrival of the second in the other detector.

mutual applications in chemistry, materials science, and biophysical chemistry [1]. The application of TDPAC spectroscopy to chemically oriented questions has been reviewed in several papers [2–7]. Recently, much more modern TDPAC studies have become possible which open a wider area of applications in chemistry [1]. In context with NQI advantages and disadvantages of this technique compared to NMR and ME have been discussed carefully [1]. Biological applications have been reviewed extensively in a recent article written by Bauer [8].

References

1. Lerf A, Butz T (1987) Angew. Chem. 99: 113; (1987) Angew. Chem. Int. Ed. Engl. 26: 110
2. Rinneberg HH (1979) At. Energy Rev. 17: 477
3. Forker M, Vianden RJ (1983) Magn. Reson. Rev. 7: 275
4. Vargas JI (1972) MTP Int. Rev. Sci. Inorg. Chem. Ser. One 8: 45
5. Adloff JP (1978) Radiochim. Acta 25: 57
6. Shirley DA, Haas H (1972) Ann. Rev. Phys. Chem. 23: 385
7. Haas H, Shirley DA (1973) J. Chem. Phys. 58: 3339
8. Bauer R (1985) Quart. Rev. Biophys. 18: 1

TMR
Abbreviation of **T**opical **M**agnetic **R**esonance

To date, only six of the many proposed surface coil localization procedures have been successfully applied for phosphorous-31 nuclear magnetic resonance (see NMR) studies *in vivo*.

Among these methods so-called topical magnetic resonance (TMR) [1,2] became an interesting procedure.

However, TMR has to compete with other interesting methods such as depth-resolved surface-coil spectroscopy [3–5] (see DRESS), volume selective excitation [6–9] (see VSE), rotating-frame zeugmatography [10–16], depth pulses [17–22], and image-selected *in vivo* spectroscopy [23,24] (see ISIS). Recently [25] we have learned somewhat more about scope and limitations of these different methods.

References

1. Hanley PE, Gordon RE (1981) J. Magn. Reson. 45: 520
2. Gordon RE, Hanley PE, Show D (1982) Prog. NMR Spectrosc. 14: 1
3. Bottomley PA, Foster TB, Darrow RD (1984) J. Magn. Reson. 59: 338
4. Bottomley PA (1985) Science 229: 769
5. Bottomley PA, Smith LS, Leue WM, Charly C (1985) J. Magn. Reson. 64: 347
6. Aue WP, Müller S, Cross TA, Seelig J (1984) J. Magn. Reson. 56: 350
7. Aue WP, Müller S, Seelig J (1985) J. Magn. Reson. 61: 392
8. Cross TA, Müller S, Aue WP (1985) J. Magn. Reson. 62: 87
9. Müller S, Aue WP, Seelig J (1985) J. Magn. Reson. 65: 530
10. Cox SJ, Styles P (1980) J. Magn. Reson. 40: 209
11. Haase A, Malloy C, Radda GK (1983) J. Magn. Reson. 55: 164
12. Garwood M, Schleich T, Matson GB, Acosta G (1984) J. Magn. Reson. 60: 268
13. Styles P, Scott CA, Radda GK (1985) Magn. Reson. Med. 2: 402
14. Hoult DI (1979) J. Magn. Reson. 33: 183
15. Garwood M, Schleich T, Ross BD, Watson GB, Winters WD (1985) J. Magn. Reson. 65: 239; Garwood M, Schleich T, Bendall MR, Pegg DT (1985) J. Magn. Reson. 65: 510
16. Metz KR, Briggs RW (1985) J. Magn. Reson. 64: 172

17. Bendall MR, Gordon RE (1985) J. Magn. Reson. 53: 365
18. Bendall MR (1983) Chem. Phys. Lett. 99: 310
19. Bendall MR, Aue WP (1983) J. Magn. Reson. 54: 149
20. Bendall MR (1984) In: James TL, Margulis AR (eds) Biomedical Magnetic Resonance, Radiology Research and Education Foundation, San Francisco, p 99
21. Bendall MR (1984) J. Magn. Reson. 59: 406
22. Bendall MR, Pegg DT (1985) Magn. Reson. Med. 2: 91
23. Ordidge RJ, Connelly A, Lohman JAB (1986) J. Magn. Reson. 66: 283
24. Lohman JAB, Ordidge RJ, Connelly A, Counsell C (1986) Soc. Magn. Reson. Med. Abstracts 3: 1091
25. Garwood M, Schleich T, Bendall MR (1987) J. Magn. Reson. 73: 191

TMS

Abbreviation of **Tetramethyl-silane**, the most important internal standard used in nuclear magnetic resonance (see CS).

TOCSY
Acronym for **T**otal **C**orrelated **S**pectroscopy

The limiting case of multiple coherence transfers in NMR is provided by the TOCSY-experiment [1], in which a given proton shows "cross-peaks" with all the spins which are part of the same coupling network, regardless of whether it has direct couplings to them. The mechanism of mixing between coherences in this experiment differs from that in COSY (see p 53) and RELAY (see p 235). Continuous strong proton irradiation is used, which effectively scales all the chemical shift differences to near zero. This causes the spins to become very strongly coupled so that if irradiation continues for a time comparable to J^{-1} all coherences from a given NMR spin system will mix. TOCSY may be viewed therefore as the homonuclear equivalent of the heteronuclear "J-cross-polarization" experiment [2]. Although TOCSY could, in principle, be useful in sorting out moderately crowded spectra, in practice it suffers from low sensitivity [3].

References
1. Braunschweiler L, Ernst RR (1983) J. Magn. Reson. 53: 521
2. Chingas GC, Garroway AN, Bertrand RD, Moniz WB (1981) J. Chem. Phys. 74: 127
3. Morris GA (1986) Magn. Reson. Chem. 24: 371

TOE
Acronym for **T**runcated Driven Nuclear **OVERHAUSER** Enhancement Method

For a full characterization of the longitudinal relaxation (T_1) matrix, it is necessary to combine numerous experiments with different perturbations [1] in nuclear magnetic resonance (see NMR). One of the most commonly used experiments is TOE [2,3] where the recovery is monitored as a function of the length of the selective radio-frequency pulse with τ equal to zero.

References

1. Vold RL, Vold RR (1978) Prog. NMR Spectrosc. 12: 79
2. Dubs A, Wagner G, Wüthrich K (1977) Biochem. Biophys. Acta 577: 177
3. Wagner G. Wüthrich K (1979) J. Magn. Reson. 33: 675

TORO
Abbreviation of the Combination of a **TOCSY** (**To**tal Correlation Spectroscopy) and **ROESY** (**Ro**tating Frame Nuclear **OVERHAUSER** Effect Spectroscopy) Experiment in Nuclear Magnetic Resonance (see NMR)

TORO can be used in NMR studies of molecules with medium-sized molecular weights [1,2]. It can be realized by variation of the mixed pulse sequence, for example, 30 ms TOCSY (see p 284) under MLEV-17 (see p 178) followed by 170 ms of a series of pulses with small flip angles. The short TOCSY transfer yields in-phase correlations (from the COSY type) which can be used directly for the ROE transfer [1,2]. The combination of TOCSY with NOESY (see p 200) can be used quite similarly in the investigation of molecules with high molecular weights [3].

References

1. Kessler H, Gemmecker G, Haase B, unpublished results, see Ref. [2]
2. Kessler H, Gehrke M, Griesinger C (1988) Angew. Chem. 100: 507; (1988) Angew. Chem. Int. Ed. Engl. 27: 490
3. Kessler H, Gemmecker G, Steuernagel S (1988) Angew. Chem. 100: 600; (1988) Angew. Chem. Int. Ed. Engl. 27: 564

TOSS
Acronym for **T**otal **S**uppression of **S**idebands

This method [1] is used in solid-state nuclear magnetic resonance (see NMR) for suppression of spinning sidebands arising from rapid magic angle spinning (see MAS) of the sample.

The TOSS experiment has been analyzed by considering the magnetization vectors of individual spin packets in the rotating frame [2,3]. This analysis has shown that TOSS aligns the magnetization paths corresponding to different crystallites. This alignment causes the centerbands of each crystallite to be in-phase, while the sidebands have phase which lead to cancellation when averaged over a powder (vide infra). Further analytical work has shown that dramatic losses in intensity may occur when TOSS is used on very large tensors [3] and a hybrid experiment that combines scaling with sideband suppression has been proposed to circumvent the sensitivity problem [3,4].

But, at high fields this requires high spinning speeds which lead to difficulties in the application of TOSS since the spacing between two of the pulses in the sequence used becomes extremely small.

Recently [5] a new variant of the original TOSS experiment has been designed to overcome this problem. The authors have derived several sequences that use six π-pulses to eliminate sidebands. They also presented a new sequence that should be helpful for samples with short spin-spin relaxation times (T_2). Then they have demonstrated that TOSS may be combined with dipolar suppression techniques

[6,7] and yielding a spectrum consisting only of centerbands from non-protonated nuclei. But, in this field the authors have to compete with the prior published paper from K.R. Caduner [8] dealing with the combination of protonated carbon suppression (see PCS) and TOSS (see below).

D.P. Raleigh and his co-workers [5] have presented the background and theory required to understand TOSS in its complete meaning.

Recently a method for the implementation of protonated carbon suppression (see PCS) with the TOSS procedure has been proposed [8]. It is now possible to use PCS with TOSS in a manner that overcomes the problem of the first-order phase shift. This is accomplished by combining the interrupted decoupling period with the final TOSS delay.

One may therefore achieve both sideband-free solid-state carbon-13 NMR spectra as well as an almost accurate baseline, two necessary requirements for a precise quantitative analysis of changes in relative concentrations of carbon-containing functionalities [9].

But, it must be pointed out, that since the TOSS sequence requires a long time period between the creation of the transverse magnetization and acquisition (about 800 μs up to 1 ms), one has to be careful to determine precisely any differences in the spin-spin relaxation time (T_2) among the different carbons if quantitative information is desired [10].

References

1. Dixon WT, Schaefer J, Sefcik MD, Stejskal EO, McKay RA (1982) J. Magn. Reson. 49: 341
2. Olejniczak ET, Vega S, Griffin RG (1984) J. Chem. Phys. 81: 4804
3. Raleigh DP, Olejniczak ET, Vega S, Griffin RG (1984) J. Am. Chem. Soc. 106: 8302
4. Raleigh DP, Olejniczak ET, Vega S, Griffin RG (1984) Proceedings of the XXII Congress Ampère, Zürich, p 592
5. Raleigh DP, Olejniczak ET, Vega S, Griffin RG (1987) J. Magn. Reson. 72: 238
6. Opella SJ, Frey MH (1979) J. Am. Chem. Soc. 101: 5894
7. Munowitz MG, Griffin RG, Bodenhausen G, Huang TH (1981) J. Am. Chem. Soc. 103: 2529
8. Caduner KR (1987) J. Magn. Reson. 72: 173
9. Wilson MA, Pugmire RJ (1986) Trends Anal. Chem. 3: 144
10. Murphy PD, Cassady TJ, Gerstein BC (1982) Fuel 61: 1233

T

TPD
Acronym for **T**ransverse **P**hotothermal **D**eflection

TPD or mirage effect spectroscopy [1–5] is increasingly employed for ultrasensitive absorption measurements of solids and surfaces. This technique provides higher sensitivity than photoacoustic spectroscopy in an experimental configuration which is normally no more complicated. TPD is commonly performed with a laser (see LASER) source to take advantage of its high spectral brightness, monochromatic output, spatial coherence, and some combinations of these properties. Recent studies [6] demonstrate that HADAMARD transform (see HT and HT-IR) transverse photothermal deflection (TPD) provides data of similar quality to that obtained in a single-element photothermal deflection experiment. This combined technique may be considered as a practical solution to the problem of photochemical or thermal sample damage [6]. HT-TPD requires that photothermal signals be additive along a line and independent of the position along that line. These fundamental assumptions have been incorporated into the photothermal deflection theory [1,2]. The assumptions were explicitly accepted for tomographic measurements of Fournier and co-workers [7] and for preliminary HT measurements by Fotiou and Morris [8].

References

1. Jackson WB, Amer NM, Boccara AC, Fournier D (1981) Appl. Opt. 20: 1333
2. Murphy JC, Aamodt LC (1980) J. Appl. Phys. 51: 4580
3. Skumanich A, Fournier D, Boccara AC, Amer NM (1986) J. Appl. Phys. 59: 787
4. Varlashkin PG, Low MJD (1986) Appl. Spectrosc. 40: 393
5. Morris MD, Peck K (1986) Anal. Chem. 58: A811
6. Fotiou FK, Morris MD (1987) Anal. Chem. 59: 185
7. Fournier D, Lepoutre F, Boccara AC (1983) J. Phys. Colloq. C6: 479
8. Peck K, Fotiou FK, Morris MD (1985) Anal. Chem. 57: 1359

TPPI
Acronym for **T**ime-**P**roportional **P**hase **I**ncrement

An approach for solving certain problems arising from two-dimensional quadrature and absorption mode in 2D-NMR (see p 7). TPPI was proposed by Bodenhausen et al. [1] and is based on an idea first introduced by Pines and his group [2]. In this procedure, the radio-frequency phases of all preparation pulses are incremented by $90°$ for consecutive t_1 values. This causes the apparent modulation frequency to be increased (or decreased) by $(2\Delta t_1)^{-1}$, and makes it appear that all modulation frequencies are positive (or negative), and therefore allows a real (cosine) FOURIER transformation with respect to t_1. Although, at first sight, the TPPI and the so-called hypercomplex FOURIER transformation (HFT) methods seem totally different, a closer analysis by Bax [3] has shown that these procedures are almost identical.

References

1. Bodenhausen G, Kogler H, Ernst RR (1984) J. Magn. Reson. 58: 370
2. Drobny G, Pines A, Sinton S, Weitekamp D, Wemmer D (1979) Faraday Div. Chem. Soc. Symp. 13: 49
3. Bax A (1985) Bull. Magn. Reson. 7: 167

T

TQF-COSY
Acronym for **T**riple-**Q**uantum **F**iltered-**Co**rrelation **S**pectroscopy in Nuclear Magnetic Resonance (see NMR)

TQF has been used in connection with a typical COSY experiment (see COSY). TQF has been applied recently [1] to give a dramatic spectral simplification in the fingerprint region of protein spectra.

In general, a p-quantum filter can be utilized to remove peaks arising from spin systems with less than p spins [2,3].

Approaches to the analysis of the fine structure of cross-peaks in multiquantum filtered COSY spectra (see MQF-COSY) have been proposed recently [4].

References

1. Boyd J, Dobson CM, Redfield C (1985) FEBS Lett. 186: 35
2. Piantini U, Sørensen OW, Ernst RR (1982) J. Am. Chem. Soc. 104: 6800
3. Shaka AJ, Freeman R (1983) J. Magn. Reson. 51: 169
4. Boyd J, Redfield C (1986) J. Magn. Reson. 68: 67

TR-ESR
Abbreviation of **T**riple **R**esonance (in) **E**lectron **S**pin **R**esonance

In the case of a triple resonance or double ENDOR (see ENDOR) experiment in electron spin or paramagnetic resonance (see ESR or EPR) we have the simultaneous effect of two radio-frequency fields on the solid under investigation [1–3]. One of these fields with constant frequency will saturate a common ENDOR transition partially, while the frequency of the second field will be "swept" over the total ENDOR spectrum. The signal amplitude of the observed ENDOR line will be changed in this procedure characteristically [4]. Overlapped ENDOR spectra may be divided in their individual spectra by the new triple resonance technique [5]. Positive and negative signals in a triple resonance ESR spectrum will correspond directly to the relative signs of the hyperfine coupling constants of the molecule under investigation [6]. Sometimes the new method will permit a higher orientation selection in polycrystalline samples, too.

References

1. Cook RJ, Whiffen DH (1964) Proc. Phys. Soc. [London] 84: 845
2. Möbius K, Biehl K (1979) In: Dorio MM, Freed JH (eds) Multiple electron resonance spectroscopy, Plenum, New York
3. Forrer J, Schweiger A, Günthard HH (1977) J. Phys. E. 10: 470
4. Schweiger A (1986) Chimia 40: 111
5. Schweiger A (1982) Struct. Bonding [Berlin] 51: 1
6. Schweiger A, Rudin M, Günthard HH (1983) Chem. Phys. Lett. 95: 285

TRAF
Acronym for **T**ransform of **R**everse **A**dded **F**ree Induction Decays (see FID)

TRAF[1] and its modification [1,2] are new and improved apodization functions (see Apodization) used for resolution enhancement in nuclear magnetic resonance (see NMR) spectroscopy.

In a variety of different spectroscopies, many apodization functions have been used to multiply a time-domain signal, so that after a FOURIER transformation (see FT), resolution will be enhanced, in the sense that linewidths will be reduced in the frequency-domain. But, until recently, these applied functions increased either the resolution or the signal-to-noise ratio (see SNR), and always one at the expense of the other. Recently, TRAF [1] has been introduced and may be used to increase either resolution, SNR, or both simultaneously. The mathematical explanations of the function TRAF [1] and its improved version [2] have been given in detail.

---

[1] Please note, that this acronym stands also for the first four letters of the author's name.

References

1. Traficante DD, Ziessow D (1986) J. Magn. Reson. 66: 182
2. Traficante DD, Nemeth GA (1987) J. Magn. Reson. 71: 237

TRS
Abbreviation of **T**ime-**R**esolved **S**pectroscopy

TRS is used in infrared (see IR and FT-IR) spectroscopy for the investigation of time-dependent repetitive processes such as *fatigue* measurements in polymers. In that case, it is possible to do stroboscopic measurements in such a fashion as to get the complete spectrum of the sample in a time domain of microseconds [1,2]. With a suitable offset and carefully realized time sorting of the data, the complete interferogram can be obtained [3]. The mechanics of the applied technique have been documentated by Fateley and Koenig [4].

Much of the early work in time-resolved infrared spectroscopy with FT-IR instruments (see FT-IR) suffered from a time-drift of the studied sample during the accumulation process [5].

Since polymers can undergo 1000 s of cycles in *fatigue* if the applied strain level is not too high, it appears that a *fatigue* experiment on polymers can be most ideally performed utilizing TRS [6]. First studies in this field dealing with isotactic poly(propylene) which was *fatigued* at a rate of 10 Hz corresponding to elongations of one to five percent. The obtained spectra showed no shifts in frequency, but reversible changes in intensity. The results have been interpreted in terms of the presence of a so-called *smectic-like structure* as obtained by cyclic stress [4,7]. The response of some ion-containing ethylene-methacrylic acid copolymers has also been studied by TRS [8]. Changes in the dichroic ratio of the absorption band at 2673 cm^{-1} were observed as a function of strain in times as short as 200 microseconds.

References

1. Murphy RE, Cook FH, Sakai H (1975) J. Opt. Amer. 65: 600
2. Sakai H, Murphy RE (1978) Appl. Optics 17: 1342
3. Honigs DE et al. (1982) Time resolved infrared interferometry I. In: Durig JR (ed) Vibrational spectra and structure, Elsevier, New York, vol 11, p 219
4. Fateley WG, Koenig JL (1982) J. Polym. Sci., Polym. Lett. Ed. 20: 445
5. Garrison AA et al. (1980) Appl Spectrosc. 34: 399
6. Koenig JL (1984) Fourier transform infrared spectroscopy of polymers, Springer, Berlin Heidelberg New York, pp 139–140 (Advances in Polymer Science, vol 54)
7. Fateley WG, Koenig JL (1982) Time resolved spectroscopy of stretched polypropylene films. In: Fewa K (ed) Recent advances in analytical spectroscopy, Pergamon, Oxford, p 291
8. Burchell DJ, Lasch JE, Farris RJ, Hsu SL (1982) Polymer 23: 965

TSCTES
Acronym for **T**otal **S**pin **C**oherence **T**ransfer **E**cho **S**pectroscopy

TSCTES is a special method used in nuclear magnetic resonance (see NMR) as a multiple-quantum approach on dipole-coupled nuclei in anisotropic phases [1–3].

The inherent information content of multiple-quantum spectra (see MQ-NMR) is essentially important for structural investigations. However, an additional merit of dipole-coupled systems in liquid crystals has been pointed out by R.R. Ernst [4], in that they provide an ideal testing basis for multi-quantum methodology itself.

The resolution and the character of the dipolar HAMILTONian make it possible to design a great variety of new sophisticated NMR-techniques. For liquid crystals a lot of different methods like TSCTES [1–3], selective p-quantum techniques [5–7], and others [8,9] have been developed so far.

The interested reader is referred to a comprehensive review of this techniques given by Weitekamp [3].

References

1. Weitekamp DP, Garbow JR, Murdoch JB, Pines A (1981) J. Am. Chem. Soc. 103: 3578
2. Garbow JR, Weitekamp DP, Pines A (1983) J. Chem. Phys. 79: 5301
3. Weitekamp DP (1983) Adv. Magn. Reson. 11: 111
4. Ernst RR, Bodenhausen G, Wokaun A (1987) Principles of nuclear magnetic resonance in one and two dimensions, Clarendon, Oxford
5. Warren WS, Weitekamp DP, Pines A (1980) J. Chem. Phys. 73: 2084
6. Warren WS, Sinton S, Weitekamp DP, Pines A (1979) Phys. Rev. Lett. 43: 1791
7. Warren WS, Pines A (1981) J. Chem. Phys. 74: 2808
8. Drobny G, Pines A, Sinton S, Weitekamp DP, Wemmer D (1979) Faraday Div. Chem. Soc. Symp. 13: 49
9. Bodenhausen G, Vold RL, Vold RR (1980) J. Magn. Reson. 37: 93

T

UHRNMR
Abbreviation of Ultra-**H**igh-**R**esolution **N**uclear **M**agnetic **R**esonance

This terminus technicus was introduced by A. Allerhand and co-workers [1].

Later, UHRNMR was documentated in two subsequent papers [2–3]. In one of these papers [2] the nonprotonated carbon of toluene was used to show the feasibility of UHRNMR. Allerhand et al. achieved an instrumental broadening of as little as 6.6 mHz [2]. As pointed out, a nonprotonated carbon does not place a severe requirement on proton decoupling efficiency, Allerhand and co-workers [3] tested therefore the performance of the known WALTZ-16 (see p 300) proton decoupling sequence, developed by Freeman and co-workers [9,10]. They found, what WALTZ-16 is a suitable method for UHRNMR [3] and pointed out, that experimental tests by Shaka et al. [9] done under conditions of 0.25 Hz line broadening are not necessarily indicative for the powerfullness of the WALTZ-16 technique.

Later on [5], however, a caveat concerning UHRNMR was published which discussed in detail the effects of magnetic field induced alignment. These authors [5] demonstrated that, depending on the system studied, these effects can be larger than 0.01 Hz at moderate fields (approx. 7×10^4 Gauss), and should therefore be considered when some people claim accuracies of a few mHz [6–8].

Two-bond intrinsic chlorine and silicon isotope effects have been studied by using UHRNMR of protons at 500 MHz [11]. It was shown that $^{37}Cl/^{35}Cl$ isotope effects on the proton chemical shifts in chlorinated methanes are about $+0.2$ ppb (corresponding to 0.1 Hz at 500 MHz) and that the $^{29}Si/^{28}Si$ isotope effect on the ^1H shift in tetramethylsilane (see TMS) is about $+0.06 \pm 0.01$ ppb.

These observations will promote further investigations of other heavy-atom isotope effects on protons by UHRNMR. Such data are of great value in our understanding of vibrational effects on chemical shifts (see CS). These experimental values may provide comparisons with quantum-mechanical calculations [12].

Anet and Kopelevich [11] made some comments on the procedures required to obtain UHRNMR spectra, especially for protons [13]. Careful sample preparation, efficient temperature control of the probe, and magnetic field shimming (see SHIM) seem to be essential. Furthermore, they have pointed out that the so-called radiation damping effects [14–16] (which are not important for carbon-13 NMR) can lead to line broadening, even for fairly dilute proton samples, because of the high-Q and large filling factors of the probe coils of modern high-field spectrometers. Secondary effects of radiation damping have been mentioned, too.

Recently [17–18] UHRNMR has been used for the analysis of minor components (by definition: a minor component is one whose molar concentration is 0.1% or less of that of a major one) in mixtures without prior separation from major components by chromatography or other techniques. For example, UHRNMR has been applied to detect 0.01% diacetone alcohol in "pure" acetone of natural isotopic composition, with the use of the methyl resonances in the proton-decoupled carbon-13 spectrum. In addition, it was possible to use UHRNMR for the direct measurement of the rate of the aldol condensation of acetone [18].

References

1. Allerhand A, Addleman RE, Osman D (1985) J. Am. Chem. Soc. 107: 5809
2. Allerhand A, Dohrenwend M (1985) J. Am. Chem. Soc. 107: 6684
3. Allerhand A, Addleman RE, Osman D, Dohrenwend M (1985) J. Magn. Reson. 65: 361
4. Maple SR, Allerhand A (1986) J. Magn. Reson. 66: 168
5. Bastian EW, Bulthuis J, MacLean C (1986) Magn. Reson. Chem. 24: 723
6. a) Nies, H Bauer H, Roth K, Rewicki D (1980) J. Magn. Reson. 39: 521;
   b) Nies H, Rewicki D (1982) J. Magn. Reson. 46: 138
7. Schaefer T, Sebastian R, Penner GH (1985) Can. J. Chem. 63: 2597
8. Ellison SLR, Fellows MS, Robinson MJT, Widgery MJ: J. Chem. Soc., Chem. Commun.
   1984: 1069
9. Shaka AJ, Keeler J, Frenkiel T, Freeman R (1983) J. Magn. Reson. 52: 335
10. Shaka AJ, Keeler J, Freeman R (1983) J. Magn. Reson. 53: 313
11. Anet FAL, Kopelevich M (1987) J. Am. Soc. 109: 5870
12. For recent reviews, the reader is referred to:
    a) Hansen PE (1983) Annu. Rep. NMR Spectrosc. 15: 105;
    b) Jameson CL (1986) In: Specialist Reports on NMR Spectroscopy, The Royal Society of
       Chemistry, London vol. 15, pp 1–27;
    c) C. E. Hawkes (1986) In: Specialist Reports on NMR Spectroscopy, The Royal Society of
       Chemistry, London, vol 15, pp 28–80
    Please, note that a positive isotope effect corresponds to a greater shielding for the heavier
    isotope.
13. Anet FAL (1986) J. Am. Chem. Soc. 108: 1354
14. Abragam A (1961) The principles of nuclear magnetism, Oxford Univ. Press, London,
    pp 73/74
15. Bloembergen N, Pound N (1954) Phys. Rev. 95: 8
16. Bruce CR, Norberg RE, Pake GE (1956) Phys. Rev. 104: 419
17. Maple SR, Allerhand A (1987) J. Magn. Reson. 72: 203
18. Maple SR, Allerhand A (1987) J. Am. Chem. Soc. 109: 6609

UPT

Acronym for Universal Polarization Transfer

UPT is a pulse sequence suggested by Bendall and Pegg [1] for polarization transfer between nuclei with ambient spin quantum number in nuclear magnetic resonance (see NMR). As shown in Fig. 33 the technique uses two variable pulse angles θ and ϕ, which depend on spin quantum numbers and the number of nuclei. In Fig. 33 BB stands for broad-band decoupling and τ is equal to $1/2J$.

U

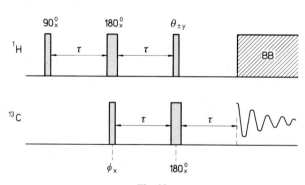

Fig. 33

In the case of a polarization transfer from deuterium to carbon-13 one obtains the following intensity functions [2]:

CD : $I = I_0(2/3)\sin(2\theta)$,

CD_2 : $I = I_0(4/9)\sin[(4\theta) + \sin(2\theta)]$,

CD_3 : $I = I_0(2/9)\sin[(6\theta) + 2\sin(4\theta) + 2\sin(2\theta)]$.

The angle ϕ is equal to $90°$ like the characteristic value in the DEPT (see p. 69) experiment.

References

1. Bendall MR, Pegg DT (1983) J. Magn. Reson. 52: 164
2. Bendall MR, Pegg DT, Tyburn GM, Brevard C (1983) J. Magn. Reson. 55: 322

UPT-SEFT
Acronym for the combination of **U**niversal **P**olarization **T**ransfer with **S**pin-**E**cho **FOURIER** **T**ransform

UPT-SEFT has been developed [1] for the total assignment of carbon-13 signals with different numbers of protons and deuterons in nuclear magnetic resonance (see NMR) spectroscopy.

UPT-SEFT may be seen as a twofold extension of a polarization transfer experiment originally reported by Rinaldi and Baldwin [2,3]. In the UPT-SEFT experiment $^2H/^{13}C$-UPT (see UPT) is combined with $^1J(^{13}C, ^1H)$ modulation of the transverse CH_nD_m magnetization (see SEFT) before recording the time domain spectrum. As shown in Fig. 34 the pulse sequence involves a $^{13}C/^1H/^2H$ triple resonance experiment with $\tau = 1/2J_{C,D}$, a time interval Δ, and broad-band (BB) decoupling. By variation of θ and Δ one obtains different subspectra with only certain signal species [4].

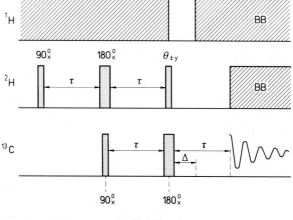

Fig. 34

References

1. Bendall MR, Pegg DT, Wesener JR, Günther H (1984) J. Magn. Reson. 59: 223
2. Rinaldi PL, Baldwin NJ (1982) J. Am. Soc. 104: 5791
3. Rinaldi PL, Baldwin NJ (1983) J. Am. Soc. 105: 7523
4. Wesener JR (1985) Dissertation, University of Siegen/Germany

UV-VIS
Abbreviation of Ultra-Violet and Visible Light Spectroscopy

Organic molecules with π-electron systems interact with the electromagnetic field of UV or VIS to absorb the resonance energy corresponding to the energy differences between ground and excited states.

The UV and VIS absorption spectra [1,2] of a variety of π-electron chromophores have been extensively studied and utilized for acquisition of chemical information.

Conjugated or isolated double or triple bonds, hetero-double bonds such as C=O or C=N–, and aromatics are well-known chromophores in organic chemistry.

In this section we want to discuss advantages of UV-VIS spectroscopy arising from *derivative spectroscopy.*

UV-VIS spectra generally arise from electronic transitions within molecules. These transitions should give rise to sharp spectral bands, but they are broadened by the contribution of vibrational and rotational energy levels to the possible transitions.

Thus, UV-VIS spectra normally consist of broad featureless absorption bands with apparently poor information. Therefore, UV-VIS spectroscopy has largely been used for quantitation of simple one-component solutions.

Derivative spectroscopy is a technique whereby the information in a UV-VIS spectrum is presented in a new and potentially more useful form. Derivative spectroscopy has the potential of greatly increasing the applications of UV-VIS spectroscopy.

The idea to use derivative spectra is not new, Giese and French [3] developed a first-derivative technique for examining spectra of pigments in plant photochemistry in 1955. Only one year later, Collier and Singleton patented the using of second and higher-order derivative spectra in the field of infrared spectroscopy (see IR) [4].

But, the practical use of derivatives in recent years followed the development of microcomputer technology that allows the almost instant generation of derivative spectra [5].

Derivative spectra are obtained by differentiating the absorbance spectra (A) of a sample with respect to the wavelength. First-, second-, or higher-order derivatives may be generated:

$$A = f(W) ,$$

$$dA/dW = f'(W) ,$$

$$d^2A/dW^2 = f''(W) , \text{ etc.}$$

Taking a GAUSSian curve as a model function, one can see (Fig. 35) that the odd derivatives are disperse functions with little similarity to the original band. The

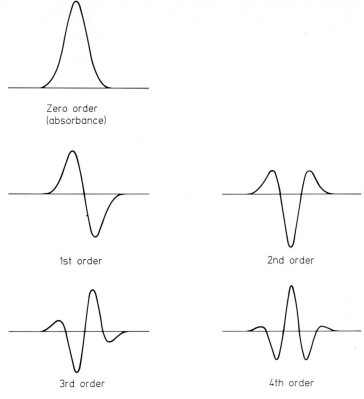

Zero order
(absorbance)

1st order   2nd order

3rd order   4th order

Fig. 35

first-order derivative can be used for a precise determination of W_{max} or detection of so-called "hidden" peaks, since $dA/dW = 0$ at peak maxima.

Higher-order odd derivatives have attracted only slight interest. Even-order derivatives have a central peak of alternating sign that is narrower than, but coincident with, the original band. Because of this similarity, the even-order spectra have generally been found to have more importance in analytical chemistry than odd functions.

Derivative spectra can be generated optically, electronically or mathematically. The most important optical method is the wavelength modulation technique, where the wavelength of incident radiation is modulated very rapidly over a small range by an electromechanical device. Derivatives can now more commonly be measured under using analog resistance-capacitance devices [6]. Higher derivative spectra may be obtained by adding further such devices as required. But these analog derivatives are affected by a lot of instrumental parameters such as scan speed, which can cause unexpected wavelength shifts, slit, width and amplifier gain. In addition, each of the devices used decreases the signal-to-noise ratio (see SNR) by a factor of nearly two, so that they are generally limited to the generation of no higher than fourth-order derivatives.

Mathematical procedures for obtaining UV-VIS derivative spectra give the best results. A variety of methods are available, but the most commonly used utilizes the Savitsky/Golay algoritms [7] to generate derivatives with a variable degree of "smoothing".

Enhancement of spectral differences, resolution enhancement selective substraction, and selective descrimination are typical applications of derivative spectra in UV-VIS spectroscopy [5].

References

1. Jaffé HH, Orchin M (1962) Theory and applications of ultraviolet spectroscopy, Wiley, New York
2. Scott AI (1964) Interpretation of the ultraviolet spectra of natural products, Pergamon, Oxford
3. Giese AT, French CS (1955) Appl. Spectrosc. 9: 78
4. Singleton F, Collier GL (1956) Brit. Pat. 760.729
5. Owen T (1987) Internat. Lab., October, p 58
6. Savitzky A, Mayring L, Kreuzer H (1978) Angew. Chem., Int. Ed. Engl. 17: 785
7. Savitzky A, Golay MJE (1964) Anal. Chem. 36: 1627

U

VASS
Acronym for **V**ariable-**a**ngle **s**ample **s**pinning

When dealing with quadrupolar nuclei, spinning around the magic angle (see MAS) does not necessarily lead to the narrowest lines attainable. The modification VASS has been introduced [1–3], which takes account of the fact that the dominant line-broadening mechanism may well arise from second-order quadrupolar interactions and is determined by an asymmetry factor of the electric field gradient tensor.

For the determination of the NMR parameters of liquid-crystalline samples, a VASS system for the regulation of the liquid crystal orientation with respect to the external magnetic field has been constructed [5]. The construction, the stability of rotation, the achievable resolution, and the application possibilities have been discussed in detail. As an example, the determination of $^1J_{CH}$ and the proton-shielding in CH_3J dissolved in liquid crystals has been demonstrated [5].

Much more recently [4], VASS has been demonstrated to have interesting applications for obtaining chemical shielding anisotropies (see CSA) for systems where severe overlapping makes the analysis of static powder patterns either cumbersome or really impossible. Results obtained for 1,3,5-trimethoxybenzene and *p*-dimethoxybenzene were in excellent agreement with those determined via single-crystal studies [4].

References
1. Oldfield E, Schramm S, Meadows MD, Smith KA, Kinsey RA, Ackerman J (1982) J. Am. Chem. Soc. 104: 919
2. Schramm S, Oldfield E: J. Chem. Soc., Chem. Commun. 1982: 980.
3. Ganapathy S, Schramm S, Oldfield E (1982) J. Chem. Phys. 77: 4360
4. Sethi NK, Grant DM, Pugmire RJ (1987) J. Magn. Reson. 71: 476
5. Väänänen T, Jokisaari J, Seläntaus M (1987) J. Magn. Reson. 72: 414

VCD
Abbreviation of **V**ibrational **C**ircular **D**ichroism

VCD is a typical kind of modulation spectrometry in the infrared region of absorption spectroscopy (see IR) and was first applied successfully for the measurement of those spectra on a grating spectrometer in 1974 [1]. In this and subsequent [2] experiments, radiation emerging from the exit slit of a monochromator was first linearly polarized and then passed into a photoelastic modulator (see PEM) to create circularly polarized radiation following the proposals of Grosjean and Legrand [3,4]. This device consists of an isotropic optical element (ZnSe is commonly used in the infrared region) that is periodically compressed and expanded by one or more piezoelectric transducers (see PZT). The linear polarizer is set at 45° with respect to the orthogonal stress axes of the PEM so that equal intensities of polarized light lie along these axes. Stretching and compressing by PZT cause the indices of refraction of the ZnSe optical element along the stress axes to vary sinusoidally. Thus the components of the polarized light along the axes will be advanced or retarded with respect to one another. When the retardation limits are $\pm 90°$, modulation between left and right circular polarization is achieved. The beam emerging from the PEM is passed

V

through the sample of interest and is then focused on a photodetector. The VCD spectrum is measured as the monochromator is scanned.

The same kind of experiment can be performed on FOURIER transform-infrared (see FT-IR) spectrometers [5]. In this case, FELLGETT's and JACQUINOT's advantages (see pp 101, 152) should improve the signal-to-noise ratio (see SNR) above the best obtainable with a monochromator. Radiation from the source is modulated by a MICHELSON interferometer at frequencies between $2v\bar{v}_{max}$ and $2v\bar{v}_{min}$, the so-called FOURIER frequencies (see FF). As radiation is modulated twice, once by the interferometer and again by the PEM, this is sometimes referred to as double-modulation spectrometry. For further technical details see the literature [5].

Today, it can be seen that the state of art for VCD spectrometry has advanced to the point that spectra of single enantiomers can be measured with little ambiguity [6]. Much of the early published VCD spectral data were obtained from relatively small molecules at high concentrations, but now much more larger biochemically significant molecules are starting to be investigated. 1984 the first spectra-structure correlations for sugars (prepared as 1 molar solutions in deuterated dimethyl sulfoxide) were reported [7]. It is surely only a question of time before VCD spectra of biologically important species in water will be reported. Recently the use of FT-IR spectrometers for VCD measurements on matrix-isolated molecules has been tested [8].

References

1. Holzwarth G, Hsu EC, Mosher HS, Faulkner TR, Moskowitz A (1974) J. Am. Chem. Soc. 96: 251
2. Nafie LA, Keiderling TA, Stephens PJ (1976) J. Am. Chem. Soc. 98: 2715
3. Grosjean M, Legrand M (1960) C.R. Acad. Sci. (Paris) 251: 2150
4. Velluz L, Grosjean M, Legrand M (1965) Optical circular dichroism, Academic, New York
5. Griffiths PR, de Haseth JA (1986) Fourier transform infrared spectrometry, Wiley, New York (Chemical analysis, vol 83)
6. Polavarapu PL (1984) Appl. Spectrosc. 38: 26
7. Black DM, Polavarapu PL (1984) Carbohyd. Res. 133: 163
8. Henderson DO, Polavarapu PL (1986) J. Am. Chem. Soc. 108: 7111

VEST

Acronym for Selected Volume Excitation Using Stimulated Echoes

V

This new localization technique in nuclear magnetic resonance (see NMR) uses the pulse sequence $90° - 90° - 90°$ and acquires the ensuing stimulated echo from the volume of interest (VOI) [1]. This proposal may be seen as a contribution in realization of the full potential of NMR spectroscopy in clinical diagnostic applications or in biological research *in-vivo*. The procedure can be used to produce either volume-restricted images by spatially encoding the signal in two or three dimensions, or spatially localized spectra by sampling the signal in the absence of any gradients.

Reference

1. Granot J (1986) J. Magn. Res. 70: 488

VOSY

Acronym for **V**olume-**S**elective Spectroscop**y**

This method has been introduced especially for *in-vivo* applications of nuclear magnetic resonance (see NMR) with free chosen volume elements within the probe and with the possibility of water line suppression [1]. In addition, VOSY can be combined with spectra editing pulse sequences such as spin-echo double resonance (SEDOR) for better performance in the field of *in-vivo* NMR spectroscopy.

Reference

1. Kimmich R, Höpfel D (1987) J. Magn. Reson. 72: 379

VUV-CD

Abbreviation of **V**acuum **U**ltra **V**iolet-**C**ircular **D**ichroism

Of the various experimental techniques applicable to the study of polysaccharide conformation, optical spectroscopy has been limited by relatively high energy of the electronic conformation, the first excitation in unsubstituted saccharide being in the VUV region. In the past decade, the development of dedicated VUV-CD spectrometers, including the application of synchroton radiation for excitation, has allowed the direct observation of the chiroptical properties of polysaccharides.

Recently [1] in a brief review, the major CD features of polysaccharides have been summarized, and several applications have been described illustrating the usefulness of VUV-CD spectroscopy in conformational studies of polysaccharides. Examples in this review include the glucans, dextran and acetylated cellulose, the glycosaminoglycan heparin, and the microbial polysaccharides gellan and xanthan.

Table 1. Spectral regions of polysaccharide CD

Wavelength range [nm]	Chromophore	Transition	Accessibility
200–240	acetamide and carboxyl substituents	n–π^*	accessible on commercial instruments
180–200	acetamide and carboxyl substituents	π–π^*	only partially accessible on commercial instruments
140–180	acetal and hydroxyl groups of sugar rings	n–σ^*	accessible on vacuum instruments
<140	all groups	σ–σ^*	currently inaccessible

Table 1 shows a simplified partition of the spectral regions of polysaccharide circula dichroism (see CD). A proposed quadrant rule was the first attempt to place the description of polysaccharide VUV-CD on a theoretical background.

Reference

1. Stevens ES (1986) Photochem. and Photobiol. 44: 287

WAHUHA, WAHUHA-4 or WHH-4

Abbreviation from the names of the three authors **Wa**ugh, **Hu**ber, and **Ha**eberlen who introduced the first successful multiple-pulse sequence in nuclear magnetic resonance (see NMR) proposed for the elimination of dipolar interactions in solids, known as WAHUHA-4 or WHH-4 pulse sequence [1]:

In each cycle of the total length $\tau_c = 6\tau$, the four pulse spaced by τ and 2τ lead to rotated coordinate systems, known as so-called toggling frame. The "average HAMILTONian" (see HAMILTONian) is obtained by averaging the transformed HAMILTONians in the toggling frame. The sequence of four 90° pulses with phases x, $-y$, y and $-x$ has unequal intervals $\tau_0 = \tau_1 = \tau_3 = \tau_4 = \tau$ and $\tau_2 = 2\tau$.

In the meantime, a great number of multiple-pulse sequences with improved properties have been proposed for homonuclear dipolar decoupling.

The interested reader is referred for more details to three excellent reviews on this subject [2–4] and to 6 original papers [5–10].

References

1. Waugh JS, Huber LM, Haeberlen U (1968) Phys. Rev. Lett. 20: 180.
2. Haeberlen U (1976) High resolution NMR in solids, Adv. Magn. Reson., Suppl. 1
3. Mehring M (1983) High resolution NMR spectroscopy in solids, 2nd edn, Springer, Berlin Heidelberg New York.
4. Abragam A, Goldmann M (1982) Nuclear magnetism: Order and disorder, Clarendon, Oxford
5. Haeberlen U, Waugh JS (1968) Phys. Rev. 175: 453
6. Mansfield P, Orchard MJ, Stalker DC, Richards KHB (1973) Phys. Rev. B7: 90
7. Rhim WK, Elleman DD, Vaughan RW (1973) J. Chem. Phys. 59: 3740
8. Rhim WK, Elleman DD, Schreiber LB, Vaughan RW (1974) J. Chem. Phys. 60: 4595
9. Burum DP, Rhim WK (1979) J. Chem. Phys. 71: 944
10. Burum DP, Linder M, Ernst RR (1981) J. Magn. Reson. 44: 173

WALTZ

Acronym for **W**ideband **A**lternating-Phase **L**ow-Power **T**echnique for **Z**ero-Residual-Splitting

WALTZ is an improved sequence for broadband decoupling (see BB) used in carbon-13 nuclear magnetic resonance (see NMR). The inventor of the acronym WALTZ wishes to remain anonymous [1]. Other progressive improvements in broadband decoupling have been achieved with pulse sequences known as MLEV (see p 178).

The principal criteria for decoupling performance may be defined by the following statements [1]:

a) Wide effective proton bandwidth for a given power dissipation.
b) Residual splittings of carbon-13 signals small compared with the line width.
c) Insensitivity to pulse length errors or B_2 inhomogeneities.
d) Insensitivity to errors in the radio-frequency phase shifts.
e) Negligible sidebands due to sampling within the decoupling cycle [2].
f) Simplicity and economy in programming.

Answers in this field have been given by different proposals such as WALTZ-4 [1], WALTZ-8 [1], WALTZ-16 [1,2], etc. WALTZ-16 has become the most used sequence on commercially available spectrometers.

References

1. Shaka AJ, Keeler J, Frenkiel T, Freeman R (1983) J. Magn. Reson. 52: 335
2. Shaka AJ, Keeler J, Freeman R (1983) J. Magn. Reson. 53: 313

WATR
Acronym for **W**ater **A**ttenuation by T_2 **R**elaxation

One of the new techniques for solving the water resonance problem in nuclear magnetic resonance (see NMR) has received the name WATR [1]. In this method, the effective spin-spin relaxation time of the water protons is made shorter than those of the solute protons by exchange with the labile protons of an added water proton exchange reagent. The water resonance is selectively eliminated from one-dimensional spectra by having a T_2 relaxation period between the 90° pulse and acquisition of the free induction decay (see FID). If the length of spin-spin relaxation period is long enough, the water resonance can be completely eliminated, and the solute resonances and the frequency of the water resonance can be observed. In order to avoid loss of sample resonances due to magnetic field inhomogeneity effects and to suppress dephasing of the transverse magnetization during the T_2 relaxation period, the spectrum is measured by the Carr-Purcell-Meiboom-Gill pulse sequence (see CPMG). WATR can be used in one-dimensional and COSY spectra (see COSY) of aqueous solutions [1]. Recently [2] a pulse sequence has been described for the measurement of two-dimensional J-resolved ^{1}H NMR spectra of aqueous solutions which incorporates the WATR method. The pulse sequence used in this 2D J-resolved experiment is a combined CPMG/spin-echo sequence:

$$90° \, (\theta) - (\tau - 180° \, (\beta) - \tau)_n - t_1/2 - 180° \, (\phi) - t_1/2\text{-acquisition}$$

where β is phase-shifted by 90° relative to θ, θ and ϕ are phase cycled to EXORCYCLE (see p 98), as proposed [3] to compensate for imperfect 180° pulses, and the receiver follows the usual cycle (0123) to provide quadrature detection (see QPD) in both dimensions. The CPMG part is intended first to create the transverse magnetization and then to eliminate that from the water protons by spin-spin relaxation [1] while the spin-echo part serves to encode the spin-spin coupling information in the second dimension [4]. This kind of experiment has been called WATR-2DJ [2].

Analytical aspects for the application of the WATR method in proton magnetic resonance have been discussed in detail [5].

Much more recently WATR has been evaluated at 80 MHz, using both ammonium and guanidinium chloride, with and without the addition of buffer salts [6]. The pH of maximum effects has been found to depend on magnetic field strength and to be significantly by the buffer salts, especially phosphates. Under carefully controlled pH conditions the method is both simple and effective [6].

References

1. Rabenstein DL, Fan S, Nakashima TT (1985) J. Magn. Reson. 64: 541
2. Rabenstein DL, Srivatsa GS, Lee RWK (1987) J. Magn. Reson. 71: 175
3. Aue WP, Karhan J, Ernst RR (1976) J. Chem. Phys. 64: 4226
4. Bodenhausen G, Freeman R, Turner DL (1977) J. Magn. Reson. 27: 511
5. Rabenstein DL, Fan S (1986) Anal. Chem. 58: 3178
6. Dickinson NA, Lythgoc RE, Waigh RD (1987) Magn. Reson. Chem. 25: 996

WEFT

Acronym for **W**ater **E**limination in **FO**URIER Transform Nuclear Magnetic Resonance Spectroscopy

WEFT is one of the various kinds of solvent-suppression techniques (see SST) based on pulse sequences normally used in inversion-recovery experiments (see IRFT) which rely on a difference in spin-lattice relaxation time T_1 between the solvent (water) and the solute resonances [1–5].

References

1. Patt SL, Sykes BD (1972) J. Chem. Phys. 56: 3182
2. Mooberry ES, Krugh TR (1975) J. Magn. Reson. 17: 128
3. Inubishi T, Becker ED (1983) J. Magn. Reson. 51: 128
4. Haasnoot CAG (1983) J. Magn. Reson. 52: 153
5. Gupta RK (1976) J. Magn. Reson. 24: 461

WNNLSA

Abbreviation of **W**hite **N**oise **N**on-**L**inear **S**ystem **A**nalysis used conceptionally in molecular spectroscopy

Recently a comprehensive review dealing with *white noise non-linear system analysis* in NMR-spectroscopy has been published by Blümich [1]. In the field of non-linear systems VOLTERRA's functional series [2–4], WIENER's series, cross-correlation, symmetry properties, and the frequency domain analysis have been discussed in detail. In the chapter about polarization Blümich [1] describes non-linear susceptibilities, the time evolution of the density matrix, quantum-mechanical susceptibilities, multi-dimensional FOURIER spectroscopy, and the interpretation of kernels and susceptibilities. Then the GAUSSian white noise theory of *stochastic NMR spectroscopy* has been treated followed by a description of developments in stochastic resonance such as imaging, echoes, and two-dimensional interferometry.

W

References

1. Blümich B (1987) Progr. NMR Spectroscopy 19: 331
2. Volterra V (1959) Theory of functionals of integral and differential equations, Dover, New York
3. Volterra V, Rend, R. Academia dei Lincei, 2 Sem: 97, 141, 153 (1887).
4. Fréchet M (1910) Ann. Sci. L'Ecole Norm. Sup., 3rd Ser. 27: 193

WRS
Abbreviation of **W**avequide **RAMAN S**pectroscopy

The capability of investigating submicron thin organic or polymeric films using RAMAN spectroscopy (see RAMAN) did not exist prior to 1980 due to the extremely small scattering volume and internal optical field intensities available in such a thin film. Since then, integrated optical techniques have been combined with RAMAN scattering [1] to allow films as thin as 200 Å to be studied somewhat routinely [2,3]. This new technique has been called *wavequide RAMAN spectroscopy* and abbreviated with WRS [3,4]. Figure 36 illustrates the general principle of WRS [4] in the top panel, whereas the lower panels of Fig. 36 contain some illustrative examples which have been investigated over the last six years [4].

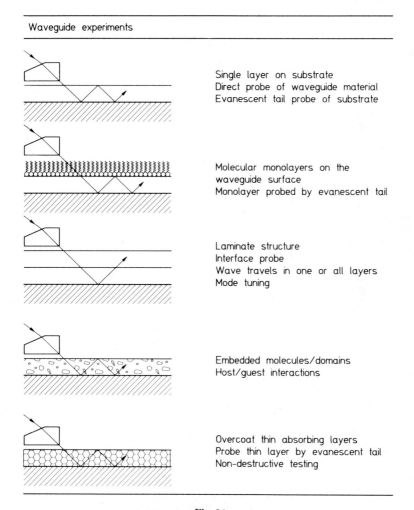

Waveguide experiments

Single layer on substrate
Direct probe of waveguide material
Evanescent tail probe of substrate

Molecular monolayers on the waveguide surface
Monolayer probed by evanescent tail

Laminate structure
Interface probe
Wave travels in one or all layers
Mode tuning

Embedded molecules/domains
Host/guest interactions

Overcoat thin absorbing layers
Probe thin layer by evanescent tail
Non-destructive testing

Fig. 36

If the film whose spectrum is desired has a higher refractive index than both the substrate and the third component (in this case: air) comprising the asymmetric slab wavequide, then laser light (see LASER) can be prism coupled into this film as shown in part (a) of Fig. 35. Under these experimental conditions the beam will be confined to this layer undergoing reflections at both the film/air and the film/substrate interfaces thus resulting in a narrow "streak" within the film which can be imaged on the entrance slit of a double-monochromator. In addition to the obvious increase in sampling volume (nearly up to a factor of 100) there is a sizeable increase in optical field intensity (approximate gain of 100) due to the fact that a 100 micron diameter laser beam has been constrained—at least in one dimension—to propagate in a thin film of submicron dimension. These combined effects therefore allow minute amounts of material to be studied. Today it has become clear that WRS is an extremely flexible technique in RAMAN spectroscopy [2,3] and continues to show hopeful promise as a non-destructive procedure for studying thin films and interfaces [4].

References

1. Rabolt JF, Santo R, Swalen JD (1980) Appl. Spectrosc. 34: 517
2. Rabe JP, Swalen JD, Rabolt JF (1987) J. Chem. Phys. 86: 1601
3. Rabolt JF, Swalen JD (1987) In: Hester R, Clark R (eds) Advances in IR and RAMAN spectroscopy Wiley, London
4. Rabolt JF (1987) paper presented at the Pittsburgh Conference, Atlantic City, NJ, March 9–13

WST

Abbreviation of **W**ater-**S**uppression **T**echniques (see Solvent-Suppression Techniques, too)

In nuclear magnetic resonance (see NMR) spectroscopy of complex molecules in dilute aqueous solutions one- and two-dimensional proton spectra are complicated by the strong water resonance. Not only does the water resonance obscure a large part of the spectrum, it also makes it difficult to detect the much more weaker resonances from the solute molecule of interest because of the limited dynamic range of the analog-to-digital converters (see ADC).

A lot of different methods have been developed to overcome this problem in one-dimensional proton NMR spectra [1–10]. Of these, the most widely used involve water suppression via a selective saturation pulse [1,2] and selective excitation pulse sequences which will excite only the spectra regions of interest [5–8]. While both procedures have been used in several two-dimensional proton experiments [11–15], thus solving the dynamic range problem mentioned above, a disadvantage typical to both is that resonances in the region of the water resonance are also eliminated. This is indeed a serious limitation in the study of peptides and proteins. Recently [16] a new method has been described for this purpose called WATR-2DJ (see WATR). But, this approach has to compete with another one [17].

W

References

1. Jesson JP, Meakin P, Kniessel G, (1973) J. Am. Chem. Soc. 95: 618
2. Campbell ID, Dobson CM, Jeminet G, Williams RJP (1974) FEBS Lett. 49: 115
3. Platt SL, Sykes BD (1972) J. Chem. Phys. 56: 3182
4. Benz FW, Feeney J, Roberts GCK (1972) J. Magn. Reson. 8: 14

5. Redfield AG, Kunz SD, Ralph EK (1975) J. Magn. Reson. 19: 114
6. Plateau P, Guéron M (1982) J. Am. Chem. Soc. 104: 7310
7. Glore GM, Kimber BJ, Gronenborn AM (1983) J. Magn. Res. 54: 170
8. Hore PJ (1983) J. Magn. Reson. 55: 283
9. Bryant RJ (1985) J. Magn. Reson. 64: 312
10. Rabenstein DL, Fan S, Nakashima TT (1985) J. Magn. Reson. 64: 541
11. Wider G, Hosur RV, Wüthrich K (1983) J. Magn. Reson. 52: 130
12. Cutnell JD (1982) J. Am. Chem. Soc. 104: 363
13. Hore PJ (1984) J. Magn. Reson. 56: 535
14. Guittet E, Delsuc MA, Lallemand JY (1984) J. Am. Chem. Soc. 106: 4278
15. Prestgard JH, Scarsdale JN (1985) J. Magn. Reson. 62: 136
16. Rabenstein DL, Srivatsa GS, Lee RWK (1987) J. Magn. Reson. 71: 175
17. Wang C, Pardi A (1987) J. Magn. Reson. 71: 154

WST in 2D-NMR
Acronym for **W**ater **S**uppression **T**echnique in **Two-D**imensional Nuclear **M**agnetic **R**esonance

Recently a new *water suppression technique* using a combination of "*hard*" and "*soft*" pulses has been suggested [1]. But, this and all other approaches mentioned above, as well as a large number of other schemes proposed so far, are only well suited for running conventional one-dimensional spectra (see 1D-NMR).

However, when generating phase-sensitive two-dimensional spectra (see 2D-NMR), baseline distortions remaining after the first FOURIER transformation give rise to serious distortions in the final 2D-NMR spectrum, especially in the direct vicinity of intense resonances. Even relatively small distortions [1,2] can cause strong baseline problems in the 2D spectrum. Today, only a few methods are available that yield NMR spectra without any phase distortion. These include

a) the $90^{\circ}_x - 90^{\circ}_{-x}$ (1-1 or jump-and-return) sequence [3],
b) a "*soft*" GAUSSian-shaped 90°_x pulse followed by a non-selective 90°_{-x} pulse [1], and
c) the so-called 1-1 echo scheme [4].

Whereas both schemes (a) and (b) provide relatively poor water suppression, scheme (c) provides high suppression, but has an undesirable $\sin^3$ offset dependence of the excitation profile.

Much more recently [5], a new procedure, of the "*two-stage*" type, has been published, that offers a very good H_2O suppression and a nearly ideal excitation profile. The proposed sequence starts with a "*soft*" 90°_{-y} pulse that rotates the water magnetization to the x axis of the rotating frame. A subsequent non-selective 90ϕ pulse followed by a short (about 2 ms) spin-lock is used to measure the remaining z-magnetization. Water suppression from a single experiment is limited by the radio-frequency inhomogeneity and the width of the "*hump*" of the H_2O resonance. Further suppression can be achieved by an appropriate phase cycling of ϕ and ψ.

W

References
1. Sklenár V, Tschudin R, Bax A (1987) J. Magn. Reson. 75: 352
2. Levitt MH, Roberts MF (1987) J. Magn. Reson. 71: 576
3. Plateau P, Guéron M (1982) J. Am. Chem. Soc. 104: 7310
4. Sklenár V, Bax A (1987) J. Magn. Reson. 74: 469
5. Sklenár V, Bax A (1987) J. Magn. Reson. 75: 378

X-Filtering
Name for a new concept for simplifying two-dimensional homonuclear proton magnetic resonance spectra (see 2 D-NMR)

The work with synthetic macromolecules and biopolymers benefits in particular from techniques capable of simplifying 2D-NMR spectra and thus enhancing the accessible information content.

Examples of those techniques are multiple quantum spin filters [1–4], X-relayed magnetization transfer [5,6], and editing techniques based on gated heteronuclear decoupling [7].

X-Filtering is a new method for the simplification of 2D-^{1}H NMR spectra, which is based on spin-spin couplings between protons and a NMR-active X-nucleus [8]. The X-filtered 2D-NMR spectra contain exclusively cross peaks and diagonal peaks from protons coupled to X. The procedure has been applied to ^{1}H-2D correlated spectroscopy (see COSY) and ^{1}H-2D nuclear OVERHAUSER and exchange spectroscopy (see NOESY).

Potential applications include studies of small molecules as well as biological macromolecules which contain NMR-active X-nuclei. For example, with the use of X-filtering technique the coordination sites of NMR-active metal ions may be identified, which could otherwise often only be achieved with laborious chemical isotope labeling procedures.

References
1. Piantini U, Sørensen OW, Ernst RR (1982) J. Am. Chem. Soc. 104: 6800
2. Shaka AJ, Freeman R (1983) J. Magn. Reson. 51: 169
3. Rance M, Sørensen OW, Bodenhausen G, Wagner G, Ernst RR, Wüthrich K (1983) Biochem. Biophys. Res. Commun. 117: 479
4. Boyd J, Dobson CM, Redfield C (1985) FEBS Lett. 186: 35–40
5. Delsuc MA, Guittet E, Trotin N, Lallemand JY (1984) J. Magn. Reson. 56: 163
6. Neuhaus D, Wider G, Wagner G, Wüthrich K (1984) J. Magn. Reson. 57: 164
7. Griffey RH, Redfield AG (1985) J. Magn. Reson. 65: 344
8. Wörgötter E, Wagner G, Wüthrich K (1986) J. Am. Chem. Soc. 108: 6162

XCORFE
Acronym standing for X-H Correlation with a Fixed Evolution Time Experiment

The XCORFE procedure has been described by Reynolds and co-workers [1]. The applied pulse sequence removes proton/proton couplings for sensitivity enhancement, retaining only vicinal H–H couplings for the case of two-bond J_{CCH}.

This makes it possible to distinguish between two- and three-bond CH correlations within a long-range heteronuclear correlation experiment (see HETCOR).

The interested reader will find more details in the most recent literature [2].

X

References
1. Reynolds WF, Hughes DW, Perpick-Dumont M, Enriquez RG (1985) J. Magn. Reson. 63: 413
2. Gray GA (1987) Introduction to two-dimensional NMR methods. In: Croasmun WR, Carlson RMK (eds) Two-dimensional NMR spectroscopy: Applications for chemists and biochemists, Verlag Chemie, Weinheim

z-filtered COSY
The name for a new variant in Correlation Spectroscopy in Nuclear Magnetic Resonance (see COSY)

Progress toward an automated assignment of two-dimensional NMR spectra by pattern recognition techniques [1,2] can be greatly facilitated if 2D correlation spectra [3,4] can be recorded where both the cross- and diagonal-peaks have simple multiplet structures. Conventional COSY spectra (see COSY) obtained with a mixing pulse with a small flip angle of nearly 20° indeed yield such simple cross-peaks, since coherence transfer occurs primarily between connected transitions that share a common energy level, but the diagonal multiplets have mixed phase peakshapes when the phase is adjusted so that cross-peaks are in pure two-dimensional absorption [3,5]. This important problem can be circumvented by the elegant E. COSY (see p 84) method [6] or by a new procedure, abbreviated to "z-filtered COSY" [7]. In this new method in the field of two-dimensional correlation of directly and remotely connected transitions [8] one uses a NOESY-like sequence (see NOESY) where the duration of the mixing period is chosen to be too short to allow exchange phenomena to play a role, and merely serves to eliminate coherence-transfer pathways [10] via single-, multiple-, and zero-quantum coherences [11,12], while retaining only transfer processes by way of longitudinal magnetization, in the manner of z-filters [13].

References
1. Meier BU, Bodenhausen G, Ernst RR (1984) J. Magn. Reson. 60: 161
2. Pfändler P, Bodenhausen G, Meier BU, Ernst RR (1985) Anal. Chem. 57: 2510
3. Aue WP, Bartholdi E, Ernst RR (1976) J. Chem. Phys. 64: 2229
4. Bax A, Freeman R (1981) J. Magn. Reson. 44: 542
5. Ernst RR, Bodenhausen G, Wokaun A (1987) Principles of nuclear magnetic resonance in one and two dimensions, Clarendon, Oxford
6. Griesinger C, Sørensen OW, Ernst RR (1985) J. Am. Chem. Soc. 107: 6394
7. Oschkinat H, Pastore A, Pfändler P, Bodenhausen G (1986) J. Magn. Reson. 69: 559
8. For definitions see Anderson et al. [9], and Ernst et al. [5].
9. Anderson WA, Freeman R, Reilly CA (1963) J. Chem. Phys. 39: 1518
10. Bodenhausen G, Kogler H, Ernst RR (1984) J. Magn. Reson. 58: 320
11. Macura S, Huang Y, Suter D, Ernst RR (1981) J. Magn. Reson. 43: 259
12. Rance M, Bodenhausen G, Wagner G, Wüthrich K, Ernst RR (1985) J. Magn. Reson. 62: 497
13. Sørensen OW, Rance M, Ernst RR (1984) J. Magn. Reson. 56: 527

ZECSY
Acronym for Zero-Quantum-Coherence Correlation Spectroscopy

Z

Normally, a chemist has access to a wide array of different methods for the assignment of individual resonances in proton magnetic resonance spectra to specific protons. But, no such methods yet exist for zero-quantum-coherence (see ZQC) spectra [1–3]. Such spectra are extremely difficult to interpret, a problem which is further exacerbated by the unfamiliarity of their parameters [4].

The importance of this problem is readily apparent in the field of *in-vivo* applications of NMR imaging (see NMRI) and spectroscopy, where unavoidably large sample volumes make it nearly impossible to get high-resolution single-quantum spectroscopic information. Under these circumstances high-resolution spectra can still be obtained via ZQC [5–7].

Recently [8] an experiment has been described, called ZECSY, which offers a solution to the problem mentioned above.

A ZECSY spectrum contains additional peaks to those of the conventional ZQ spectrum, at the frequency midpoint between pairs of ZQCs with a common spin. This provides a relatively facile method for tracing spin-spin coupling networks among the ZQCs, and hence aids in their assignment. However, other peaks besides those halfway between ZQCs containing a common spin are produced. These are also a type of correlation peak and have also been interpreted [8].

References

1. Aue WP, Bartholdi E, Ernst RR (1976) J. Chem. Phys. 64: 2229
2. Wokaun A, Ernst RR (1977) Chem. Phys. Lett. 52: 407
3. Bodenhausen G (1981) Prog. Nucl. Magn. Reson. Spectrosc. 14: 137
4. Pouzard G, Sukumar S, Hall LD (1981) J. Am. Chem. Soc. 103: 4209
5. Hall LD, Norwood TJ: J. Chem. Soc. Chem. Commun. 1986: 44
6. Hall LD, Norwood TJ (1986) J. Magn. Reson. 67: 382
7. Hall LD, Norwood TJ (1986) J. Magn. Reson. 69: 18
8. Hall LD, Norwood TJ (1986) J. Magn. Reson. 69: 585

ZF
Abbreviation of **Z**ero-**F**ield Magnetic Resonance

The principal sources of broadening of the resonance lines in solid-state nuclear magnetic resonance (see NMR) are strong dipolar interactions and, in disorderd systems such as powders and amorphous solids, superpositions of different orientations, lead to inhomogeneously broadened resonance signals.

One approach for the elimination of inhomogeneous line broadening in dipolar- or quadrupolar-coupled spin systems is ZF magnetic resonance [1–6].

In the absence of a magnetic field all the spins will experience the same interactions, independent of the orientation of the crystallite under investigation. Most terms of inhomogeneous line broadening therefore vanish to zero while spin-spin couplings and quadrupolar interactions remain.

The HAMILTONian (see p 122) of the studied spin system does not depend on the orientation of the individual crystallite and space becomes essentially isotropic from the point of view of NMR [7].

Recently the possibilities of manipulating the spin degrees of freedom by applying sequences of magnetic field pulses in low or zero static magnetic fields have been analyzed theoretically and compared with other well-known manipulations of the HAMILTONian and higher magnetic fields [7].

Z

References

1. Ramsay NF, Pound RV (1951) Phys. Rev. 81: 278
2. Weitekamp DP, Bielecki A, Zax D, Zilm K, Pines A (1983) Phys. Rev. Lett. 50: 1807
3. Miller JM, Thayer AM, Bielecki A, Zax D, Pines A (1985) J. Chem. Phys. 83: 934
4. Thayer AM, Pines A (1987) Acc. Chem. Res. 20: 47

5. Pines A (1988) In: Maraviglia B (ed) Proceedings of the 100th FERMI School on Physics, Varenna/Italy, 8–18 July 1986, North-Holland, p. 467
6. Kreis R, Suter D, Ernst RR (1985) Chem. Phys. Lett. 118: 120
7. Lee CJ, Suter D, Pines A (1987) J. Magn. Reson. 75: 110

ZF-NQR
Abbreviation of **Z**ero **F**ield-Nuclear **Q**uadrupole **R**esonance

Zero-field deuterium nuclear quadrupole resonance (see NQR) has become a well-established method in the recent years [1–4]. ZF-NQR with pulsed field cycling provides an alternative approach of obtaining high-resolution quadrupolar spectra of polycrystalline solids since it removes the orientational anisotropy. By using the sudden transition [1] or the selective indirect detection [2] zero-field experiments, small quadrupolar coupling constants (see QCC) and asymmetry parameters can be determined directly from the observed frequencies in the spectra. Furthermore, two-dimensional versions of the time-domain zero-field experiment will yield information on the frequency connectivities [5]. Applications of ZF-NQR has been published recently [6].

References
1. Bielecki A, Murdoch JB, Weitekamp DP, Zax DB, Zilm KW, Zimmermann H, Pines A (1984) J. Chem. Phys. 80: 2232
2. Millar JM, Thayer AM, Bielecki A, Zax DB, Pines A (1985) J. Chem. Phys. 83: 934
3. Zax DB, Bielecki A, Zilm KW, Pines A, Weitekamp DP (1985) J. Chem. Phys. 83: 4877
4. Kreis R, Suter D, Ernst RR (1985) Chem. Phys. Lett. 118: 120
5. Thayer AM, Millar JM, Pines A (1986) Chem. Phys. Lett. 129: 55
6. Millar JM, Thayer AM, Zimmermann H, Pines A (1986) J. Magn. Reson. 69: 243

ZQC
Abbreviation of **Z**ero-**Q**uantum **C**oherence

Homonuclear ZQC's [1] have been increasingly studied in recent years, both in conventional isotropic solution nuclear magnetic resonance (see NMR) [2–4] and more recently because of their independence of B_0 field inhomogeneity [5,6] in the context of NMR imaging [7] (see NMRI) as an alternative to chemical shift-resolved imaging experiments [8–12] which are fatally susceptible to B_0 field inhomogeneities.

The use of broadband decoupled ZQC has been demonstrated by Hall and Norwood [13]. This technique reduces, in a zero-quantum proton NMR spectrum, every multiplet to a singlet and uses a pulse sequence which also allows individual coherences to be edited from the spectrum. The method has been extended to produces images resolved as a function of chemical shift in an inhomogeneous magnetic field, which can also be edited.

In a recent communication Hall and Norwood [14] have demonstrated a pulse sequence whereby ZQC's can be made to display single-quantum coherence coupling constants at high resolution. This permits the conventional single-quantum coherence to be reassembled, even when the original data are acquired in an inhomogeneous field.

References

1. Bodenhausen G (1981) Prog. NMR Spectrosc. 14: 137
2. Wokaun A, Ernst RR (1978) Mol. Phys. 36: 317
3. Poudzard G, Sukumar S, Hall LD (1981) J. Am. Chem. Soc. 103: 4209
4. Braunschweiler L, Bodenhausen G, Ernst RR (1983) Mol. Phys. 48: 535
5. Wokaun A, Ernst RR (1977) Chem. Phys. Lett. 52: 407
6. Maudsley AA, Wokaun A, Ernst RR (1978) Chem. Phys. Lett. 55: 9
7. Hall LD, Norwood TJ (1986) J. Magn. Reson. 67: 382
8. Lauterbur PC, Karmer DM, House WV, Chieng-Nien C (1975) J. Am. Chem. Soc. 97: 6866
9. Bendel P, Lai C-M, Lauterbur PC (1980) J. Magn. Reson. 38: 342
10. Cox SJ, Styles P (1980) J. Magn. Reson. 40: 209
11. Brown TR, Kincaid BM, Ugurbil K (1982) Proc. Natl. Acad. Sci. USA 79: 3523
12. Hall LD, Sukumar S (1984) J. Magn. Reson. 56: 314
13. Hall LD, Norwood TJ (1986) J. Magn. Reson. 69: 391
14. Hall LD, Norwood TJ (1986) J. Magn. Reson. 69: 397

ZQS

Abbreviation of **Z**ero-**Q**uantum **S**pectroscopy used in Nuclear Magnetic Resonance (see NMR)

Homonuclear AS in its two-dimensional version offers several advantages over other correlation spectroscopies in the analysis of scalar coupling networks—insensitivity to magnetic field inhomogeneties [1], smaller spectral range in ω_1 and a concomitant increase in ω_1-resolution for and free-induction decay (see FID) data set of fixed size [2], and the interesting ability to discern the relative signs of certain pairs of scalar couplings from the ω_1-multiplet structure [2].

However, ZQS has found limited application due in part to its inability to separate zero-quantum coherence (see ZQC) and longitudinal magnetization during t_1 by phase-cycling [3,4].

The presence of longitudinal magnetization during t_1 gives rise to the so-called axial peaks in the 2D-ZQ spectrum. Therefore, development of an efficient and reliable procedure whereby "axial" peaks may be purged from 2D-ZQ spectra was highly desirable.

The first attempt at this task was published by Bolton [5].

The pulse sequence suggested by Müller [6] to realize an improved excitation of ZQC in homonuclear spin systems also suppresses "axial" peaks to a large extent.

But, application of multi-quantum filters in ZQS proposed recently [7] may be designated as a much more efficient approach in this field (see MQ-NMR and MQF-COSY).

References

1. Aue WP, Bartholdi E, Ernst RR (1976) J. Chem. Phys. 64: 2229
2. Pouzard G, Sukumar S, Hall LD (1981) J. Am. Chem. Soc. 103: 4209
3. Macura S, Wüthrich K, Ernst RR (1982) J. Magn. Reson. 47: 351
4. Macura S, Wüthrich K, Ernst RR (1982) J. Magn. Reson. 46: 269
5. Bolton PH (1984) J. Magn. Reson. 60: 342
6. Müller L (1984) J. Magn. Reson. 59: 326
7. Farmer BT II, Ramachandran R, Brown LR (1987) J. Magn. Reson. 75: 534

Index

E. Pretsch, T. Clerc, J. Seibl, W. Simon

Tables of Spectral Data for Structure Determination of Organic Compounds

Translated from the German by K. Biemann

2nd ed. 1989. XIII, 415 pp. (Chemical Laboratory Practice) ISBN 3-540-51202-0

From the contents: Introduction. – Abbreviations and Symbols. – Summary Tables. – Combination Tables. – ^{13}C-Nuclear Magnetic Resonance Spectroscopy. – Proton Resonance Spectroscopy. – Infrared Spectroscopy. – Mass Spectrometry. – UV/VIS (Spectroscopy in the Ultraviolet or Visible Region of the Spectrum). – Subject Index.

This book represents a compilation of spectroscopic reference data in the format of tables and charts and their correlation to molecular structure. It is intended to aid the interpretation of UV-, IR-, NMR- and mass spectra and complements text- and reference books on these techniques. It is designed to be of use to students as well as to the every day practitioner and expert in the field.

This second edition has been improved by among other things adding data on new compound classes and on C13-NMR-spectra.

Springer-Verlag Berlin Heidelberg New York London Paris Tokyo Hong Kong

Springer

D. J. Gardiner, Newcastle upon Tyne;
P. R. Graves, Oxfordshire, UK (Eds.)

Practical Raman Spectroscopy

With contributions by H. J. Bowley,
D. J. Gardiner, D. L. Gerrard, P. R. Graves,
J. D. Louden, G. Turrell

1989. VIII, 157 pp. 87 figs., 11 tabs.
ISBN 3-540-50254-8

Contents: Introduction to Raman Scattering. – Raman Sampling. – Instrumentation for Raman Spectroscopy. – Calibration and Data Handling. – Non-Standard Physical and Chemical Environments. – Raman Microscopy. – Further Reading. – Subject Index.

N. Chandrakumar, S. Subramanian,
Madras, India

Modern Techniques in High Resolution FT-NMR

1987. IX, 388 pp. 259 figs.
ISBN 3-540-96327-8

Contents: Introduction and General Theory. – One Dimensional Experiments in Liquids. – Coherence Transfer. – Two Dimensional Experiments in Liquids. – Multiple-Quantum Spectroscopy. – High Resolution Pulse NMR in Solids. – Experimental Methods. – Appendices 1 to 6. – Selected Bibliography. – Index.

Springer-Verlag Berlin
Heidelberg New York London
Paris Tokyo Hong Kong

Springer